GAGING
PRACTICAL DESIGN
AND APPLICATION

Edward S. Roth
Editor

Robert E. King
Assistant
Manager

Published by

Society of Manufacturing Engineers
Marketing Services Division
One SME Drive
P.O. Box 930
Dearborn, Michigan 48121

GAGING
PRACTICAL DESIGN AND APPLICATION

SME wishes to express its acknowledgement and appreciation to the following contributors for supplying the various articles reprinted within the contents of this book.
Appreciation is also extended to the authors of papers presented at SME conferences or symposia as well as to the authors who generously allowed publication of their private work.

The American Society of Mechanical Engineers
345 East 47th Street
New York, NY 10017

James B. Bryan
Lawrence Livermore National Laboratory
P.O. Box 808
Livermore, CA 94550

C.I.R.P.
19 rue Blanche
75009 Paris
France

Founding, Welding, Production Engineering
Box 31548
Braamfontein 2017
Republic of South Africa

Lawrence Livermore Laboratory
University of California
Livermore, CA 94550

John M. Leaman
University of Wisconsin-Extension
Dept. of Engineering & Applied Science
929 North Sixth Street
Milwaukee, WI 53203

Machine Design
Penton Plaza
Cleveland, OH 44114

Manufacturing Engineering
Society of Manufacturing Engineers
One SME Drive
P.O. Box 930
Dearborn, MI 48121

Marine Technology
Society of Naval Architects
& Marine Engineers
1 World Trade Center
Suite 1369
New York, NY 10048

Modern Machine Shop
Gardner Publications
6600 Clough Pike
Cincinnati, OH 45244

National Standards Association, Inc.
5161 River Road
Bethesda, MD 20816

Precision Engineering
P.O. Box 63 Westbury House
Bury Street, Guildford
Surrey GU2 5BH
England

Production Engineering
Penton Plaza
Cleveland, OH 44114

George Pruitt
Naval Weapons Center
Geometric Dimensioning
& Tolerancing Center
China Lake, CA 93555

Quality
Hitchcock Publishing Co.
Hitchcock Building
Wheaton, IL 60187

Edward S. Roth
Productivity Services
1413 San Carlos Dr. S.W.
Albuquerque, NM 87104

Tooling & Production
Huebner Publications Inc.
5821 Harper Avenue
Solon, OH 44139

VDI International Magazine
NC & Computerized Automation
Am Wasser 91
8049 Zurich
Switzerland

Cover photo courtesy of *Manufacturing Engineering* magazine.

PREFACE

This, the second edition of Gaging: Practical Design and Application, provides insight into the range of gaging applications currently available. Special emphasis is placed on Coordinate Measuring Machines (CMMs) which are fast becoming the predominant inspection methodology. CMMs also can be utilized as functional gaging systems by replacing the standard probes with collets, or precision chucks, to hold gage pins or bushings, and with noncontact electro-optical sensors and microscopes for unique inspection applications.

Since the CMM's do not achieve their ultimate performance and reliability in the shop environment, they will probably be used for post-process evaluation, which is the least productive. The most productive gaging and inspection is on-line. On-line systems assure that each preceding work station is correct, and process evaluation is continuous. The present alternative is to provide off-line, environmentally protected CMM inspection cells at strategically located positions. Flexible Manufacturing Systems (FMS) should provide adequate environments so that CMMs will fit easily into these systems.

The paper on 'gauging principles' is repeated in this edition because the only U.S. standard, Mil-Std-120, is hopelessly out-of-date.

Most gaging and inspection is critically related to either the fixture design or the setup employed. To be meaningful, these fixtures and setups should position the part as it would be when assembled to the mating part. Most of our current product specifications identify nonfunctional datum features or rely on "implied datums." The major fraction of the parts now rejected, reworked, or deviated are in this catagory. ANSI Y14.5, and all the previous editions of this standard going back to Mil-Std-8, have covered datum referencing in adequate depth. Since engineering personnel are not required to learn or apply this critical standard while in school, we continue to scrap around 10% of the product built in this country.

It is easy to blame the production workers for our productivity problems, yet they did not design the system they operate. Manufacturing systems should be designed with full knowledge of product function, and therefore, require team input in order to be highly productive.

Manufacturing process capability is largely unknown in U.S. industry because inspection and gaging process capability has not been assessed. Existing machine tools and manufacturing systems need to be evaluated for accuracy and repeatability before gaging and inspection systems, with accuracy ratios of 10 to 1, can be designed. In general, the workplace environment has been deteriorating over the past 20 years and many of our new machine tools never reach their performance potential. Since accuracy ratios are generally less than 10 to 1, errors in the acceptance systems are taking much more than 10% of the part toleranced to the detriment of higher productivity.

Designers of inspection systems are urged to work with design and manufacturing engineers primarily because recent mechanical engineering graduates are taught that tight tolerances are equal to quality or increase the state-of-the-art in manufacturing. Since hands-on gage, fixture, and process designers know more about the process capability, they can hopefully convince the design engineer to increase tolerances, and thus reduce the number of engineering change orders. In many companies, that appears to be all that the engineers have time to do. Such work is so disruptive that it smothers both creativity and productivity.

"Do it right the first time," "Prevent mistakes," "We don't want any surprises," do not appear to be in the vocabulary of manufacturing company managers. However, unlimited time and money are expended to fix the mistakes caused by inadequate lead time. Designers of inspection and gaging systems must prepare assembly drawings themselves in order to understand the true function of the parts they are accepting. I urge them to do this well.

When Eli Whitney achieved interchangeability, it was because he designed both the product and the production line as one system. Under these conditions, tolerances were based on fact and function—not fear; and fixturing and gaging was meaningful. A return to this concept, with all the engineering disciplines pooling their wealth of knowledge, will aid us in reaching maximum productivity.

The articles and papers collected in this volume should aid the reader in increasing quality and productivity in gaging application.

I wish to thank all of the companies, organizations, publishers, and authors who gave permission to have their articles reprinted in this volume. Thanks also to the Marketing Services staff at SME for their assistance in the research and development required in making this book possible.

Edward Roth, P.E., CMfgE
Productivity Services

ABOUT THE EDITOR

Edward S. Roth is a Productivity Consultant, providing professional services in product design for manufacturing, manufacturing process evaluations and functional gaging and inspection. His career includes extensive work as a Project Leader of Exploratory systems at  Sandia Laboratories in Albuquerque, New Mexico.

A noted speaker and author, Mr. Roth has written more than 50 technical articles and papers on a wide range of topics. He authored the books *Functional Gaging* and *Functional Inspection Techniques*. Mr. Roth has lead numerous national and international seminars, clinics, workshops and courses for the Society of Manufacturing Engineers and other technical societies, universities and the U.S. Department of Commerce.

Mr. Roth is the 1978 President of the Society of Manufacturing Engineers. He has served SME as International Secretary, Treasurer and Vice President and has been a member of the Board of Directors. Mr. Roth is also a member of the Institute of Production Engineers in Great Britain.

SME

The informative volumes of the Manufacturing Update Series are part of the Society of Manufacturing Engineers' many faceted effort to provide the latest information and developments in engineering.

Technology is constantly evolving. To be successful, today's engineers must keep pace with the torrent of information that appears each day. To meet this need, SME provides, in addition to the Manufacturing Update Series, many opportunities in continuing education for its members.

These opportunities include:

- Monthly meetings through three associations and their more than 230 chapters which provide a forum for member participation and involvement.

- Educational programs including seminars, clinics, programmed learning courses, as well as videotapes and films.

- Conferences and expositions which enable engineers and managers to examine the latest manufacturing concepts and technology.

- Publications including the periodicals *Manufacturing Engineering, Robotics Today,* and *CAD/CAM Technology,* the *SME Newsletter,* the *Technical Digest,* and a wide variety of text and reference books covering everything from the basics to manufacturing trends.

- The SME Manufacturing Engineering Certification Institute formally recognizes manufacturing engineers and technologists for their technical expertise and knowledge acquired through experience and education.

- The Manufacturing Engineering Education Foundation was created by SME to improve productivity through education. The foundation provides financial support for equipment development, laboratory instruction, fellowships, library expansion, and research.

- A database, accessible through SME containing Technical Papers and publication articles in abstracted form. The Information on Technology In Manufacturing Engineering database is only one of several accessible through the Society.

SME is an international organization with more than 70,000 members in 60 countries worldwide. The Society is a forum for engineers and managers to share ideas, information, and accomplishments.

The Society works continuously with organizations such as the American National Standards Institute, the International Organization for Standardization, and others, to establish and maintain the highest professional standards.

As a leader among professional societies, SME assesses industry trends, then interprets and disseminates the information. SME members have discovered that their membership broadens their knowledge and experience throughout their careers. The Society is truly industry's partner in productivity.

MANUFACTURING UPDATE SERIES

Published by the Society of Manufacturing Engineers, the Manufacturing Update Series provides significant, up-to-date information on a variety of topics relating to the manufacturing industry. This series is intended for engineers working in the field, technical and research libraries, and also as reference material for educational institutions.

The information contained in this volume doesn't stop at merely providing the basic data to solve practical shop problems. It also can provide the fundamental concepts for engineers who are reviewing a subject for the first time to discover the state-of-the-art before undertaking new research or application. Each volume of this series is a gathering of journal articles, technical papers and reports that have been reprinted with expressed permission from the various authors, publishers or companies identified within the book. SME technical committees, educators, engineers, and managers working within industry, are responsible for the selection of material in this series.

We sincerely hope that the information collected in this publication will be of value to you and your company. If you feel there is a shortage of technical information on a specific manufacturing area, please let us know. Send your thoughts to the Manager of Educational Resources, Marketing Services Department at SME. Your request will be considered for possible publication by SME—the leader in disseminating and publishing technical information for the engineer.

TABLE OF CONTENTS

CHAPTERS

3 COORDINATE MEASURING MACHINES

4 PRODUCTION GAGING SYSTEMS

APPENDIX

CHAPTER 1

GAGING CRITERIA

Presented at the American Society of Mechanical Engineers
Metals Engineering and Production Engineering Conference, June 1965.
Author's Note for *Gaging: Practical Design and Application*

Thermal Effects in Dimensional Metrology[1]

JAMES B. BRYAN

Chief Metrologist, Lawrence
Radiation Laboratory, University
of California, Livermore, Calif.

ELDON R. McCLURE

Assistant Professor of Mechanical
Engineering, Oregon State University,
Corvallis, Ore.

WILLIAM BREWER

Graduate Student of Mechanical
Engineering, Michigan State
University, East Lansing, Mich.

J. W. PEARSON

Mechanical Engineer, Lawrence
Radiation Laboratory, University
of California, Livermore, Calif.

A Lawrence Radiation Laboratory investigation of thermal effect in dimensional metrology shows that, in the field of close-tolerance work, thermal effect is the largest single source of error, large enough to make corrective action necessary if modern measurement systems and machine tools are to attain their potential accuracies. This paper is an effort to create an awareness of the thermal environment problem and to suggest some solutions. A simple, quantitative, semi-experimental method of thermal-error evaluation is developed. A relatively simple device to monitor the thermal environment and automatically effect error compensation is proposed.

[1]Work performed under the auspices of the U. S. Atomic Energy Commission.

Contributed by the Production Engineering Division for presentation at the Metals Engineering and Production Engineering Conference, Berkeley, Calif., June 9-11, 1965, of The American Society of Mechanical Engineers. Manuscript received at ASME Headquarters, March 8, 1965.

Written discussion on this paper will be accepted up to July 13, 1965.

Copies will be available until April 1, 1966.

Note: The editor has included the following paper, published in 1965, because it is the only English language paper that we could locate that adequately discusses the sensitivity of precision machines to the frequency of temperature variation as well as the amplitude. This paper is also of interest because of its explanation of the consequences of working at average temperatures other than the 68 degrees F International Standard.

Thermal Effects in Dimensional Metrology

JAMES B. BRYAN   **WILLIAM BREWER**   **ELDON R. McCLURE**   **J. W. PEARSON**

During a routine test of a measuring machine at the Lawrence Radiation Laboratory (LRL), it was observed that the measurements varied significantly with time. It was thought, at first, that the electronic gage used to make the measurements was the cause, but a careful check showed that the electronic drift was negligible. The measurement system was then monitored by means of a sensitive temperature pickup mounted in the air near the gage. Temperature and gage output were recorded over long time periods. The results showed a high degree of correlation between temperature variation and measurement variation.

Similar tests were conducted on many different types of measurement systems with similar results. Fig.1 is an example of such results. The significance of this effect is clear when the observed drift is compared with the working tolerance of the gage. The drift is 100 microin. and the working tolerance of the gage is only 100 microin. In the case shown, the drift accounts for 100 percent of the tolerance of the gage.

As a result of this disturbing development, the Metrology Section of LRL began an investigation of thermal effect in dimensional metrology. As the investigation progressed, it became increasingly clear that:

1 In the field of close tolerance work, thermal effect is the greatest single source of error.

2 The usual efforts to correct for thermal error by applying expansion "correction," or by air conditioning the working area do not always solve the problem and are based on an incomplete understanding of the problem.

3 The specified accuracies of modern precision tools and gages are attainable only if the thermal environment matches the requirements of each measurement system.

It has been helpful to think of the temperature problem in terms of (a) the effects of average temperatures other than 68 F, (b) the effects of temperature variation about this average. The paper organization reflects this arbitrary division of the problem. There is also a discussion of ways and means of reducing thermal errors. Appendix 1 is a glossary of terms used to discuss thermal-effects problems. Appendix 2 is a detailed procedure of a "drift" check, and Appendix 3 is an outline of a method of determining the thermal frequency response of a measuring system.

EFFECTS OF AVERAGE TEMPERATURES OTHER THAN 68 F

An inch is the distance between two fixed points in space. It is defined as 41,929.398742 wavelengths of the orange-red radiation of krypton-86 when propagated in vacuum. An inch does not vary with temperature. This fact is obscured because the lengths of the more common representations of the inch such as gage blocks, lead screws, and scales do vary with temperature. The lengths of most of the materials we deal with also change with temperature. In April, 1931, the International Committee of Weights and Measures meeting in Paris agreed that when we describe the length of an object we automatically mean its length when it is at a temperature of 68 F. This agreement was preceded by intensive international debate and negotiation (1,2,3,4,5,11).[2] This agreement means that it is not necessary to specify the measurement temperature on every drawing (no more necessary than it is to define the inch on every drawing).

If dimensions are only correct at 68 F, how have we been getting by all these years by measuring at warmer temperatures? The answer is that if our work is steel and our scale is steel the two expand together and the resultant errors tend to cancel. If, however, the work is another material such as aluminum, the errors are different and they do not cancel. We refer to this error as "differential expansion." We can get into the same trouble if our work is steel, but we are measuring with the "honest" inches that come from an interferometer. As a result of the discovery of the laser and the development of practical laser fringe-counting interferometers, we expect to be using more of these "honest" inches and will have to be very careful of this problem.

Knowledgeable machinists have always made differential expansion corrections. The thing that is sometimes overlooked, however, is that these corrections are not exact. Our knowledge about average coefficients of expansion is meager and we can never know the exact coefficient of each part. This inexactness we call "uncertainty of differential expansion."

This inexactness or uncertainty is zero when the average temperature is 68 F, and increases ac-

---

[2] Underlined numbers in parentheses designate References at the end of the paper.

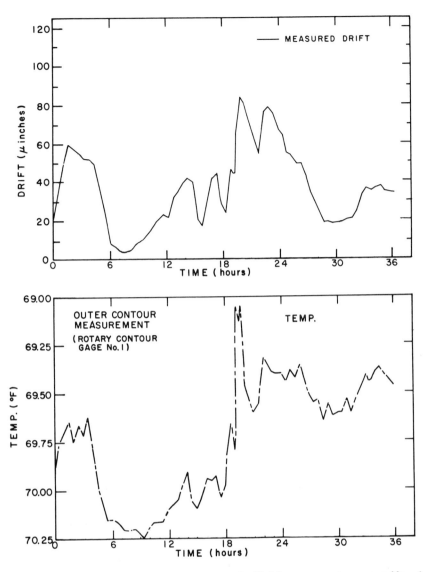

Fig.1 Thermally induced drift of 15-in. Sheffield rotary contour gage No. 1
with steel part

cording to the thermal distance from 68 F. Its
magnitude varies for different materials. We have
reason to think that it is at least 5 percent for
gage steel and on up to 25 percent for other materials. One metallurgist, consulted in the course
of our investigation, stated that the coefficient
of expansion of cast iron may vary as much as 4
percent between thin and thick sections. This uncertainty factor also includes the possibility of
differences in expansion of a material in different directions. Differences between the actual
thermal expansion and the handbook or "nominal"
expansion occur because of experimental errors and
because of dissimilarities between the experimental material and the material of our workpiece.

Complete studies of the errors introduced in
the estimates of thermal expansion are notably absent. The data presented by Goldsmith et al (9)
show the range of disagreement among several in-

vestigators in determining the coefficient of expansion of common materials. This disagreement
might be expected for some of the more exotic materials, but intuition would indicate that the
knowledge of the properties of steel would be more
exact. Not necessarily so, as Richard K. Kirby of
the National Bureau of Standards reports:[3]

"The accuracy of a tabulated value of a coefficient of thermal expansion is about ± 5 percent
if the heat and mechanical treatment of the steel
is indicated. The precision of the coefficient
(a) among many heats of steel of nominally the
same chemical content is about ± 3 percent, (b)
among several heat treatments of the same steel is

_____
[3] A personal communication from Richard K. Kirby,
in charge, Thermal Expansion Laboratory, Length
Section, Metrology Division, U.S. National Bureau
of Standards, Washington, D.C.

about $\pm$ 10 percent, and (c) among samples cut from different locations in a large part of steel that has been fully annealed is about $\pm$ 2 percent (hot or cold rolling will cause a difference of about $\pm$ 5 percent)."

Corrections for uncertainty of differential expansion cannot be made. The error can be reduced by establishing more accurate nominal coefficients of thermal expansion, by improving the uniformity of coefficient of expansion from part to part through better chemical and metallurgical controls, by determining individual part and gage expansions, and by limiting the room-temperature deviation from 68 F.

Control of uncertainty of differential expansion is the primary reason for maintaining a 68 F average temperature. Even if we had an exact knowledge of all coefficients, the confusion and possibility of mistakes in making corrections is a second reason for maintaining 68 F. As our study progressed, it became necessary to establish a more exact definition of terms to facilitate rapid and clear communication. The reader should now refer to Appendix 1: Glossary of Terms, definitions No. 15 through 23, which are pertinent to the discussion in this section.

TYPICAL EXAMPLES

Three examples will be given to illustrate the consequences of average temperatures other than 68 F. Possible errors are shown to be 13, 37, and 20 percent of the working tolerance. These errors do not include the effect of temperature variation which is covered in the next section. They do not include the other errors of measurement such as accuracy of standards and comparison technique. The traditional rule of ten to one allows only 10 percent of the working tolerance for all measurement error.

Example No. 1

A 10-in-long steel part with a tolerance of plus or minus one half-thousandth (500 μ in.) is measured in a C-frame comparator by comparing it to a 10-in. gage block in a room which averages 75 F. A handbook lists the nominal coefficient of expansion (K) for the gage block as 6.5 μ in/in/deg F. The K for the steel part is assumed to have the same value. The uncertainty of nominal coefficient of expansion (UNCE) for the gage block is estimated at plus or minus 5 percent and for the part at 10 percent (its exact composition is unknown). For this case, the nominal differential expansion (NDE) is zero. The uncertainty of nominal differential expansion (UNDE) is, however, significant. It is the sum of the two uncertainty of nominal expansion (UNE) values.

NDE = no correction necessary = 0

UNE gage block = 10 in. x 6.5 μ in/in/ deg x 7 deg F x 5% = 22 μ in.

UNE part = 10 in. x 6.5 μ in/in/ deg x 7 deg F x 10% = 44 μ in.

UNDE = 66 μ in.

$\frac{66}{500}$ x 100 = 13% of working tolerance

Example No. 2

A 10-in-long plastic part with a tolerance of plus or minus 0.002 in. is measured on a surface plate using an indicator stand to compare it to the readings of a Cadillac gage. The room temperature averages 75 F (7 deg temperature offset). A handbook lists the nominal coefficient of expansion (K) for the gage steel assumed to be used in the Cadillac gage as 6.5 μ in/in/deg F. The K-value for the plastic is listed by the manufacturer as 40 μ in/in/deg F. The uncertainty of nominal coefficient of expansion (UNCE) for the gage steel is estimated at 10 percent since we do not know the exact composition or heat-treatment. Because of past experience with plastics and a lack of any information to the contrary the UNCE for the plastic is estimated at 25 percent. The inspector making the measurement is thoroughly familiar with differential expansion. He computes the NDE correctly and applies it in the proper direction to the dial indicator reading which is used to transfer the Cadillac gage reading. A correction for UNDE cannot be made. Its possible value is computed as follows:

NDE = corrections are made = 0

UNE$_{gage}$ = 10 in. x 6.5 μ in/in/deg F x 7 deg x 10% = 46 μ in.

UNE$_{part}$ = 10 in. x 40 μ in/in/deg F x 7 deg x 25% = 700 μ in.

UNDE = 746 μ in.

$\frac{746}{2000}$ x 100 = 37% of working tolerance

Example No. 3

A 10-in-long aluminum part with a tolerance of plus or minus 0.001 in. is measured on a surface plate using an indicator stand to compare it to the readings of a Cadillac gage. The room-temperature averages 70 F. (2 deg F temperature offset.) As in the preceding example, the NCE for the gage is assumed to be 6.5 μ in/in/deg F. The NCE for the aluminum part is assumed to be 13.5 μ in/in/deg F. The UNCE for the gage is estimated at 10 percent and for the aluminum part at 20 percent. The inspector in this case does not appreciate the magnitude of NDE, and arbitrarily de-

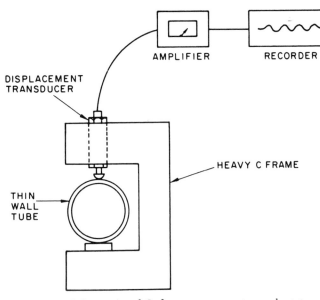

Fig.2 Schematic of C-frame comparator and part

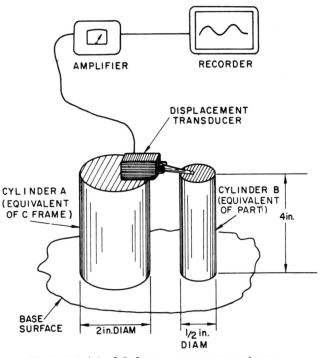

Fig.3 Model of C-frame comparator and part

cides that 70 F is "close enough," he does "not bother" with an NDE correction to his readings. The possible error is computed as follows:

$$NDE = 10 \text{ in. } (13.5 - 6.5)$$
$$\times 2 \text{ deg F} = 140 \mu \text{ in.}$$

UNDE:
$$UNE_{gage} = 10 \text{ in. } \times 6.5 \times 2 \text{ deg F} \times 10\% = 13 \mu \text{ in.}$$
$$UNE_{part} = 10 \text{ in. } \times 13.5 \times 2 \text{ deg F} \times 20\% = \underline{52} \mu \text{ in.}$$

$$UNDE = \underline{65} \mu \text{ in.}$$

$$NDE \text{ plus } UNDE = 205 \mu \text{ in.}$$

$$\frac{205}{1000} \times 100 = 20\% \text{ of working tolerance}$$

In the foregoing examples we assumed that the average temperature of the gage and part were the same as the average temperature of the room. If adequate time has been allowed for the gage and part to "soak out" and reach thermal equilibrium this is a reasonable assumption. Unfortunately, this assumption does not apply to the instantaneous temperature of these components. Instead, the environment is continually varying around some mean value. The result is that differences in temperature in the various components are dynamically induced in the system. The next section discusses the errors caused by variation in thermal environment.

EFFECTS OF VARIATION IN THERMAL ENVIRONMENT

The Two-Element System

All length-measuring apparatus can be viewed as consisting of a number of individual elements arranged to form a "C." Fig.2 shows a schematic of a C-frame comparator measuring the diameter of a short section of hollow tubing. The comparator frame and the part form two elements. If the co-efficient of expansion of the comparator is exactly the same as the part, the gage head will read zero after soak-out at any uniform temperature that we might select. If we induce a change in temperature, however, the relatively thin section of the tubing will react sooner than the thick section of the comparator frame and the gage head will show a temporary deviation. The amount of the deviation will depend on the rate of change of temperature. If the rate is slow enough to allow both parts to keep up with the temperature changes, there will be a small change in gage-head reading. If the rate is so fast that even the thin tubing cannot respond, there will again be a small change in reading. Somewhere in between these extremes there will be a frequency of temperature change that results in a maximum change in reading. This is somewhat similar to resonance in vibration work.

To confirm our intuition on the nature of these effects the foregoing model was further simplified to that shown in Fig.3. Sample heat-transfer calculations were made for this model and programmed on an analog computer. The cylinder with the displacement pickup can be considered the comparator. Both cylinders are made of steel and are 4 in. long. Cylinder A is 2 in. dia and cylinder B is 1/2 in. dia. Fig.4 shows the computer-predicted changes in length of the two cylinders as a result of a plus and minus 1 deg F sinusoidal change in air temperature having a frequency of 1 cph. The thick cylinder shows less than one third of the temperature change of the thin cylinder and its temperature lags the thin cylinder by 3 or 4 min. The dotted line in Fig.4 shows the predicted gage - head reading which is the same as the instantaneous difference in the lengths of the two cylinders. We call this the "thermal drift" (definition 24) of the system. The effect of varying the "thermal vibration" frequency is plotted in Fig.5. As our intuition predicted, the drift is small for very high or low frequencies and reaches a maximum amplitude at a point in between which we call resonance. Fig.5 is called the "frequency response" (definition 14) of the system. In this case resonance occurs at 1/2 cph and has a value of 15 μ in. This error would occur even if the time average temperature of the environment and all mechanical elements was 68 F exactly.

Fifteen μ in. may appear to be a negligible magnitude, but real measuring machines and machine tools, of course, do not have uniform coefficients throughout, and real workpieces can have quite different coefficients. This makes the effects of temperature variation much worse. If the part were lucite, for example, these responses would be much more severe. Real systems generally have much longer overall lengths and more severe differences in mass between elements. Magnitudes of 150 μ in. per deg F at resonant frequency are not unusual. Rather than mass alone, the more significant factor is the ratio of cubic inches of volume to the square inches of surface exposed to the air. This ratio is proportional to the "time constant" (definition 13) of the element. The time constant is discussed in the following heat-transfer calculations, which support the foregoing results. A complete understanding of these equations is not necessary, because the important conclusions have been presented already.

We have used a gage frame to illustrate the effect of temperature variation, but it should be emphasized that the same thing happens to machine-tool frames. Deflection due to temperature variations is common to all machine structures whether they be measuring machines or machine tools.

Calculations for Frequency Response of
Two-Element System Model

To simplify the calculations, the following assumptions have been made:

(a) The bodies always have uniform temperatures; i.e., there is no resistance to heat transmission between the parts of the body and any heat

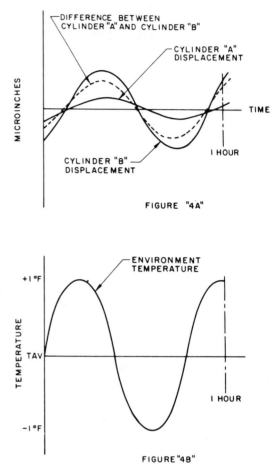

FIGURE "4A"

FIGURE "4B"

Fig.4 Individual and net drift thermal response of two - element model shown in Fig.3

added simply raises the temperature at all points uniformly and instantaneously.

(b) The temperature of the air surrounding the bodies is uniform at T_e.[4]

(c) All heat transmission to and from the body is governed by Newton's law of cooling:

$$q = hA(T_e - T)$$

where A is the surface area of the body in ft^2, h is a film coefficient defining the ability of heat to pass from air to the body, in Btu/hr-ft^2 deg F, and q is the rate of heat in-flux, Btu/hr.

(d) The heat stored in the body is proportional to the thermal capacitance of the body or that

$$q_s = c_p \rho V \, dT/dt$$

where q_s is the rate of heat storage, c_p is the

[4] Keeping in mind that radiative and conductive environments can exist, we shall limit the following discussion to the effect of a convective environment on the measurement process and the resulting error.

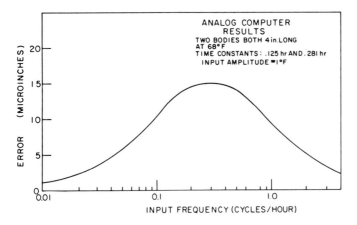

Fig. 5 Computed frequency response of C-frame comparator and part

Fig. 6 Partial view of 15-in. Sheffield rotary gage with thin-walled steel part

specific heat in Btu/lb deg F, V is the volume of the body in ft^3, ρ is the density, lb_m per ft^3, and dT/dt is the rate of change of body temperature with time. Since the heat influx must equal the heat stored in any interval of time

$$q = hA(T_e - T) = q_s = c_p V\rho\ dT/dt$$

Solving for T_e yields the differential equation describing the system:

$$T + \tau\ dT/dt = T_e \qquad (1)$$

where

$$\tau = \frac{c_p V\rho}{hA} = \frac{\text{thermal capacitance}}{\text{thermal resistance}}$$

Because of assumption (a), equation (1) is only approximately correct. However, for metallic objects the thermal conductivity is high enough to make the approximation reasonable.

Equation (1) is well known in the literature on the analysis of linear systems ($\underline{10}$) in which all elements it represents are called "time-constant elements" and τ is called "time constant" of the element.

Given that T_e varies sinusoidally around some mean T_{e_0}, i.e.

$$T_e = \left(T_{e_{max}} - T_{e_0}\right) \sin \omega t \qquad (2)$$

where ω is the frequency of oscillation in radians per unit time, the solution to equation (1) gives:

$$T = \left(T_{e_{max}} - T_{e_0}\right) \sin (\omega t - \phi)(1 + \omega^2 \tau^2)^{-1/2} \qquad (3)$$

where the phase lag angle ϕ is given by

$$\phi = \tan^{-1} \omega\tau$$

When this solution is applied to the two-element system we obtain the relationships between the temperatures of the two elements and the environment temperature as discussed above.

Drift Check

Measurement of the drift in a measurement system is called a "drift check" (definition 26). To make a drift check it is merely necessary to indicate from the comparator to the master, or part as the case may be, and record the relative motion between the elements under the normal conditions of the measurement process. This procedure has been made possible by the development of high-sensitivity, drift-free, displacement transducers and recorders. The electronics drift check (definition 25) provides a simple means of proving the stability of these devices. Our experience indicates that "electronics drifts" of less than 3 μ in. per day for ± 3 deg F environments can be expected. Details of drift-check procedures are given in Appendix 2.

Predicting the Effects of Temperature Variation

The mathematical approach just given allows us to make quantitative observations about the effects of thermal vibration in systems for which we know all the time constants. The drift check provides us with a practical means of error evaluation on real systems in a given environment regardless of their complexity. Neither of these approaches can provide us with a means of determining how large the errors will be in real systems before they are installed in a given environment. Such information is necessary if rational design decisions are to be made. Therefore, our investigation included a study of the means to experimentally determine the dynamic response of

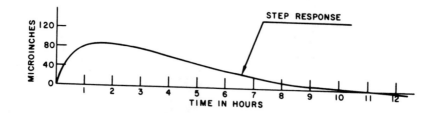

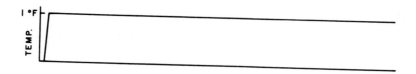

Fig. 7   Step input response on 15-in. Sheffield rotary contour gage No. 2

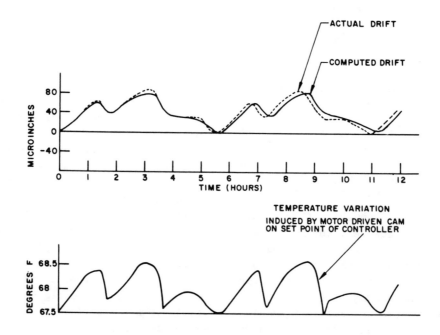

Fig. 8   Forced drift check on 15-in. Sheffield rotary gage No. 2

measurement systems and to find ways to predict, from this information, what the drift will be for any system in any environment.

A study of the literature (10) on the analysis of linear systems shows that it is possible to conduct "step input" tests, the results of which provide a means of approximating the effects of any kind of change. To determine the feasibility of applying this procedure to a real measurement process, a series of experiments was conducted in an LRL inspection shop. In these experiments the apparatus consisted of a 15-in. Sheffield rotary

contour gage measuring a hollow steel hemispherical part, as shown in Fig. 6.

The Sheffield gage chosen was particularly suited to these experiments because it was located in a room that had a particularly good air-conditioning system. Room-air temperatures in the vicinity of the gage responded to a 1 deg change in the set point of the air-conditioning controller within several minutes.

Linearity of the system was established in three experiments which consisted of suddenly raising the set point of the controller 1 deg in

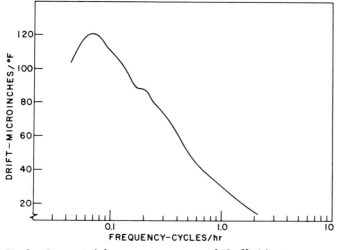

Fig. 9 Computed frequency response of Sheffield rotary contour gage No. 2

the first, lowering it 1 deg in the second, and raising it 2 deg in the third. Air temperature at a point just above the part was recorded using a thermister magnetically held in contact with a 10-in. piece of 0.010-in. shim stock. Resulting drift was recorded by the equipment on the gage.

The results from this series of experiments were compared. All three drift curves showed the effects of a high degree of linearity. They differed only in magnitude and this disagreement was less than 10 percent.

Subsequently, an arbitrary temperature fluctuation was imposed on the room by driving the controller set point with a motor-driven cam mechanism. Temperature and drift were recorded as before.

The recorded drift and the corresponding recorded temperature changes for the 1-deg step-input change experiment were used as shown in Appendix 3 to compute a theoretical drift from the forced-drift temperature data. Fig. 7 shows the step-input temperature change and drift profiles, and Fig. 8 the recorded and computed drift. Considering the fact that the experiment continued over a period of about 6 weeks, we believe the results fully justify the applicability of this type of system testing.

These results encouraged us to use the computation method to calculate the frequency response of the system with the results shown in Fig. 9. Comparing these data with those in Fig. 5, we see the typical pattern as well as the distorting effect of additional time-constant elements.

The next question to be answered is: "Can data obtained on a system in one environment be applied to a similar system in another environment?" If the answer is yes, this means that a

gage manufacturer, by conducting these simple step-input tests in his laboratory, can provide information that will allow the customer to decide whether or not his environment is suitable for the gage.

To answer the question, we conducted a normal drift check, shown in Fig. 1, on a second 15-in. Sheffield rotary contour gage located in a different room with a different environmental control system. We used the recorded temperature variation from this system and the frequency-response information obtained from the first system to compute a predicted drift. The results of this experiment are shown in Fig. 10. The correspondence between the computed and actual drift is impressive, and though the method used must be tried for a large variety of cases before we know how general it is, we feel confident that the affirmative answer has been obtained.

Application of this method requires a high-quality, temperature-controlled room large enough to house the completed machine. The room must have the ability to hold a given temperature to a tolerance that makes the step-input change significant. The time required for the room to stabilize at the new temperature should be a small fraction of the soak-out time of the machine.

It would be more convenient if we could arrive at a method for making these predictions by analyzing the results of an ordinary drift check taken in an ordinary environment. At the present time, achievement of this goal appears to be difficult, but possible. The difficulty is the dependence of the microinch drift on the frequency of temperature variation as well as its amplitude. Work on the solution of this problem is now under way.

The Three-Element System

All length-measurement systems can be discussed in terms of three elements, a part, a master, and a comparator used to compare the part with the master. The master is sometimes obscured because it is combined with the comparator, as in a micrometer. In a micrometer, the screw is the master and the rest of the device is the comparator. If the master and the comparator are not combined, a time element is introduced into the measurement process because the comparator, such as a pair of calipers, cannot be mastered, or set, at the same time it is used to indicate on a part. This time lapse is the difference between a two-element system and a three-element system. We have already seen that the different response of part and comparator in a two-element system causes a drift error. A similar error will occur between master and comparator. This means that, in a three-element system, there is a master-comparator

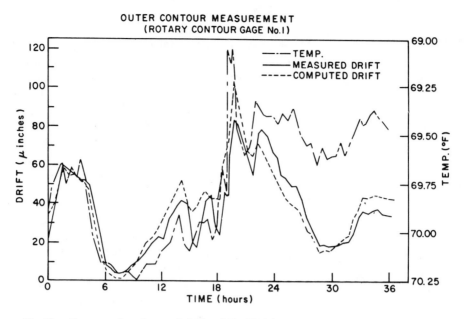

OUTER CONTOUR MEASUREMENT
(ROTARY CONTOUR GAGE No.1)

Legend:
—·—·— TEMP.
——— MEASURED DRIFT
– – – – COMPUTED DRIFT

Fig.10   Computed and actual drift of Sheffield rotary contour gage No. 1 with steel part

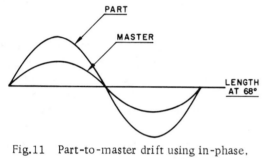

Fig.11   Part-to-master drift using in-phase, sinusoidal curves

drift error that must also be considered to get the maximum temperature-variation error. The two drift curves, between part and comparator and between master and comparator, can be used to approximate this temperature-variation error in a three-element system for any mastering time cycle.

If the mastering time is zero or insignificantly small, then the comparator is slaved to the master and the temperature-variation error is the drift between part and master over a representative time period. This is equivalent to the two-element system discussed previously. The representative time period is usually a working day, but may be shorter or longer depending on environment control and work habits. It should be long enough to cover the entire temperature cycle of each measurement situation. The drift of part and master cannot usually be compared directly but can be compared indirectly by comparing the part-comparator drift curve with the master-comparator drift curve. The maximum excursion of the two

curves for the same temperature phase and amplitude over the representative time period will provide the maximum part-to-master drift error. This error is an approximation because the temperature conditions of the two drift checks will never be identical.

Fig.11 shows part and master-drift curves. For simplicity they are made sinusoidal and in phase. The curves show absolute drift in length from an average temperature of 68 F at which point they are equal length. It can be seen that measuring the part at any time other than when the part and master curves are at the same point will result in an error reading. The maximum error will occur when the part is measured at the point of maximum difference. The part is, of course, measured with the comparator, but with zero mastering time, the comparator length is held to the master length at measuring time.

If the comparator cannot be used to indicate on a part at the same time it is mastered, then the drift of the comparator with respect to part and master becomes an additional source of error. It can be shown that the maximum possible temperature variation error for a finite mastering cycle time will not be greater than the already determined maximum error from part-to-master drift unless either the total part-to-comparator drift or the total master-to-comparator drift during the mastering cycle time is more than twice the maximum part-to-master drift error.

Fig.12 shows the absolute drifts for part, master, and comparator. The phase lag shown is typical for the varying responses to temperature

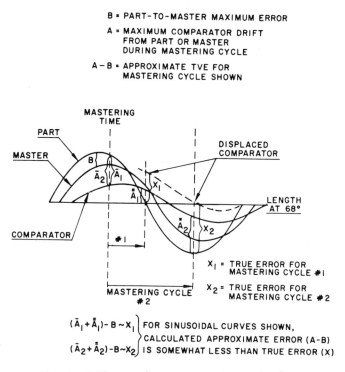

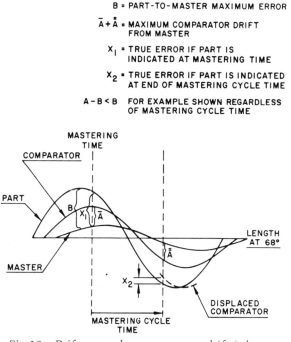

B = PART-TO-MASTER MAXIMUM ERROR

A = MAXIMUM COMPARATOR DRIFT
FROM PART OR MASTER
DURING MASTERING CYCLE

A-B = APPROXIMATE TVE FOR
MASTERING CYCLE SHOWN

X_1 = TRUE ERROR FOR
MASTERING CYCLE #1

X_2 = TRUE ERROR FOR
MASTERING CYCLE #2

$(\bar{A}_1 + \bar{\bar{A}}_1) - B \sim X_1$ FOR SINUSOIDAL CURVES SHOWN,
CALCULATED APPROXIMATE ERROR (A-B)
$(\bar{A}_2 + \bar{\bar{A}}_2) - B \sim X_2$ IS SOMEWHAT LESS THAN TRUE ERROR (X)

Fig. 12   Drift error for two mastering cycle times

B = PART-TO-MASTER MAXIMUM ERROR

$\bar{A} + \bar{\bar{A}}$ = MAXIMUM COMPARATOR DRIFT
FROM MASTER

X_1 = TRUE ERROR IF PART IS
INDICATED AT MASTERING TIME

X_2 = TRUE ERROR IF PART IS INDICATED
AT END OF MASTERING CYCLE TIME

A-B < B   FOR EXAMPLE SHOWN REGARDLESS
OF MASTERING CYCLE TIME

Fig. 13   Drift error when comparator drift is between
master and part

variation. In the example shown, the maximum part-to-master drift (B) is the maximum temperature variation error for zero mastering cycle time. Mastering the comparator at the time shown will displace the comparator drift curve. It can be seen that the subsequent maximum temperature variation error will now vary with the mastering cycle time between mastering and indicating. The true error for mastering cycle time number 1 is X_1. This can be approximated by measuring the peak-to-valley drift of part-to-comparator or master-to-comparator, whichever is greater, and subtracting the part-to-master drift. In the example shown, the part-to-comparator drift, A, is greater so that A - B conservatively approximates the true error X_1. The same is true for mastering cycle time number 2. In each case, A - B is greater than B so A - B is the maximum temperature variation error for those conditions. If B were greater than A - B, B would remain the maximum temperature variation error because indicating could be done at any time during the mastering cycle including shortly after mastering.

Fig. 13 shows what happens when the drift rate of the comparator is between that of the master and of the part. In this example the part-to-master drift error is B and the master-comparator drift is A. Fig. 13 shows that A - B can never be greater than B, so part-to-master drift remains the maximum temperature-variation error regardless of mastering cycle time. This condition of a comparator drift rate between that of master and part

drift rates becomes apparent when the two drift curves, part comparator and master comparator, are compared for part-to-master drift. If the two drift curves when aligned for temperature phase are out of phase, the comparator drift must be between master and part and the maximum temperature-variation error becomes that of maximum excursion between the two curves over the representative time period.

The general case for maximum temperature variation error approximation can be stated as:

For zero or insignificantly small mastering cycle time

$$TVE = \left[\begin{array}{l} \text{Maximum excursion of part-comparator} \\ \text{drift from master-comparator drift over} \\ \text{representative time period.} \\ \quad \text{(part-to-master drift error)} \end{array} \right]$$

For a significant mastering cycle time, take either the maximum part-comparator drift or the maximum master-comparator drift, whichever is greater, for the mastering cycle time period chosen and subtract the part-to-master drift error previously chosen.

$$TVE = \left[\begin{array}{l} \text{Master-comparator drift} \\ \text{or part-comparator drift,} \\ \text{whichever is greater, for} \\ \text{mastering cycle time period} \\ \text{chosen} \\ \quad\quad A \end{array} \right] - \left[\begin{array}{l} \text{Part-to-} \\ \text{master} \\ \text{drift} \\ \\ \quad B \end{array} \right]$$

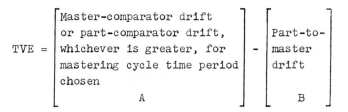

13

If the result of A - B is greater than B then that result is the maximum temperature-variation error for the chosen mastering cycle time. If B is greater, then the part-to-master drift error remains the maximum temperature-variation error. A will not be more than 2B, and will therefore not add to the part-to-master drift error if the comparator is made to have a drift rate between that of the part and the master regardless of mastering cycle time. Also, A will not be more than 2B if the mastering cycle time is kept short enough to prevent the peak-to-valley drift of the comparator from either part or master from exceeding 2B. Thus the proper mastering cycle time for preventing the temperature-variation error from exceeding the part-to-master maximum drift error will depend on the characteristics of the measurement system. Also, efforts to reduce master and comparator drift may add to the maximum temperature-variation error if the part cannot be treated similarly. The ideal condition is to have identical drift rates of all three elements. This calculated maximum temperature-variation error is an approximation because the varying drift-curve shapes and phase relations under real conditions will not allow more precise handling of the drift curves without adding greatly to complexity.

If the temperature conditions of frequency and amplitude are not repetitive enough to allow a good temperature phase and amplitude match when comparing the two drift curves, the maximum TVE can be determined by adding one half the part-comparator drift to one half the master-comparator drift over the representative time period. This will give the maximum TVE under the worst conditions of the representative time period, though it may, for many special conditions, overstate the error.

EVALUATION OF THERMAL ERRORS IN EXISTING MEASURING PROCESSES

If the possible existence of significant errors of measurement due to thermal-environment conditions is recognized, then it follows that a definite need exists for a means of evaluating their magnitude. It is, of course, desirable to determine the total thermal error (definition 28) but this is usually not possible. An evaluation plan has therefore been developed which provides an estimate of the total thermal error. It is an index of the true error and has been named thermal error index (TEI). In this procedure, a drift check is performed for both the master-comparator combination and the part-comparator combination while simulating the actual conditions of the measurement process. Both the master-comparator

and the part-comparator drift checks are analyzed to determine the value of the maximum drift of each occurring within a time period equal to the mastering cycle. Both the master-comparator and part-comparator drift checks are then analyzed to determine the maximum excursion of the drift curves that occurs within a "representative" time period. This representative time period is somewhat difficult to define. It should be long enough to reveal the full pattern of temperature variation. In most cases, a period of 24 hr is sufficient. The results of the drift-check analyses are then substituted into the following expression for temperature variation error (TVE):

For zero or small mastering cycle time.

$$
TVE = \left[\begin{array}{l} \text{Maximum excursion of part-comparator and} \\ \text{master-comparator drift curves when curves} \\ \text{are aligned for in-phase temperature con-} \\ \text{ditions over representative time period.} \\ \\ \text{(Part-to-master drift)} \end{array} \right]
$$

For significant mastering cycle times.

$$
TVE = \left[\begin{array}{l} \text{Master-comparator drift or} \\ \text{part-comparator drift,} \\ \text{whichever is greater, for} \\ \text{chosen mastering cycle time} \end{array} \right] - \left[\begin{array}{l} \text{Part-to-} \\ \text{master} \\ \text{drift er-} \\ \text{ror as} \\ \text{above} \end{array} \right]
$$

Use whichever of the foregoing TVE is greater.

The temperature-variation error is combined with the nominal differential expansion (NDE) and the uncertainty of nominal differential expansion (UNDE) to obtain the thermal error index (TEI). The plan consists of:

1 Computing the nominal differential expansion (NDE).

2 Computing the uncertainty of nominal differential expansion (UNDE).

3 Determining the thermal variation error (TVE) by evaluation of drift-check data.

4 Summing the absolute values obtained in 1, 2 and 3.

5 If NDE corrections are made, NDE is not included in the foregoing sum.

The time-honored rule of 10/1 suggests that the total measuring error be limited to 10 percent of the working tolerance. We have found, however, that the error due to temperature is, in most cases, so large that in order to stay within economic reality we must plan on giving up the full 10 percent and more for temperature alone.

The following example shows how the evaluation plan is used in practice. This example is the same as the one used in the second section but now includes the effects of temperature variation:

A 10-in-long steel part with a tolerance of

plus or minus a half thousandth (500 u in.) is measured in a C-frame comparator by comparing it to a 10-in. gage block in a room which averages 75 F. A handbook lists the nominal coefficient of expansion (K) for the gage block as 6.5 μ in/in/deg F. The K for the steel part is assumed to have the same value. The uncertainty of nominal coefficient of expansion (UNCE) for the gage block is estimated at plus or minus 5 percent and for the part at 10 percent (its exact composition is unknown). For this case, the nominal differential expansion (NDE) is zero. The uncertainty of nominal differential expansion (UNDE) is, however, significant. It is the sum of the two uncertainty of nominal expansion (UNE) values.

A 24-hr drift check between the comparator and master gage block shows a 300-u in. peak-to-valley drift. The comparator is normally remastered every 2 hr. Interpreting the drift checks for maximum drift in 2 hr gives a maximum value of 30 μ in. Because the part has fewer cubic inches of volume per square inch of surface than the gage block (its time constant is smaller) the time-constant mismatch to the relatively heavy comparator frame is worsened. The part-comparator drift is found to be 350 μ in. in 24 hr. Substituting into the evaluation plan for the foregoing conditions yields the following:

NDE   = no correction necessary   = 0

UNE gage block = 10 in. x 6.5 μ in/in/deg F x 7 deg F x 5% = 22 μ in.

UNE part   = 10 in. x 6.5 μ in/in/deg F x 7 deg F x 10% = 44 μ in.

              UNDE = 66 μ in.

$$TVE = \begin{bmatrix} \text{Maximum comparator drift} \\ \text{from part or master over} \\ \text{24 hr = 30 } \mu \text{ in.} \end{bmatrix} - \begin{bmatrix} \text{Part-to-mas-} \\ \text{ter drift =} \\ \text{50 } \mu \text{ in.} \end{bmatrix}$$

Use A - B or B, whichever is greater

(30 - 50) < 50

Therefore

         TVE = 50

         TEI = 116

$\dfrac{116}{500}$ x 100 = 23% of the working tolerance

This example shows a thermal-error index of more than 10 percent and corrective action is indicated. If, however, the tolerances increased or we decided to accept a higher percentage thermal index the situation would return to normal. A "bad" environment would suddenly become a "good"

environment which does not justify the cost of any improvements. The evaluation plan is a way of estimating the temperature problem for each shop, each machine, and each job. It can tell us whether or not we need to improve our temperature control and by how much. The plan provides concrete economic justification for investment of the large amount of money that may be necessary to control the temperature problem. It can also prevent overdesign in the situations where it has become stylish to have special temperature-controlled areas. It substitutes an orderly thinking process for emotion or arbitrarily set rules. Natural priorities are established to indicate where our improvement efforts should be made. Should we try to move closer to 68 F or should we try to reduce our temperature variation? The plan not only answers these questions, it gives a positive response to any improvements that may be made.

In spite of the advantages of the plan, some objections have been raised. One objection is that the plan pretends to be an exact procedure when obviously we are still estimating. Our answer to this objection is to agree that the plan is not perfect and not exact. It may be in error by 25 percent or more and still be a significant advancement over no plan at all. No plan at all means that we must depend on the opinion of experts who arbitrarily decide that this or that environment is, or is not, acceptable.

METHODS FOR DECREASING THERMAL ERROR INDEX

Average Temperature Other Than 68 F

The possibilities for controlling the error resulting from average temperatures other than 68 F are limited. They can be summarized in one sentence: The error can be reduced by making nominal differential expansion corrections, by establishing more accurate nominal coefficients of expansion, by improving the uniformity of coefficient of expansion from part to part through better chemical and metallurgical controls, by determining individual part expansions, and by limiting the room temperature deviation from 68 degrees.

Temperature-Variation Error

What are some of the things we can do to improve the ability of a gaging system to withstand temperature variation? Our first reaction is to make the thermal response of the master and comparator equal. This will result in zero drift between the master and comparator. Shortening the mastering cycle has the same effect. This is a false goal, however, because we may create an increased mismatch to the part. A worthwhile goal is to make the thermal response of all three elements the same. This completely eliminates the problem but is not a practical approach because most gages are used for more than one part. The best compromise is to design the comparator drift to be about halfway between the part drift and the master drift (this is discussed in more detail in the preceding section). Adjustment of thermal response can be accomplished in several ways. The use of Invar is quite practical. Invar is readily obtainable at a reasonable cost and has a coefficient of only 1 μ in/in. Time constants can be controlled by use of insulation and by proper design of wall thickness.

Unfortunately, none of these solutions can be applied to the part itself. We cannot insulate it; we cannot change its coefficient; we can't change its wall thickness. The only thing we can do is improve the environment.

What are some of the things that can be done to improve the environment? Our first reaction is simply to reduce the temperature excursion of the whole room. This is effective, but also expensive. It may be cheaper to control the temperature excursion in a small area around the machine. The Moore Special Tool Company of Bridgeport, Conn., uses this approach in comparing and calibrating its ultraprecise step gages to an accuracy of one part in ten million.

Another approach is the possibility of increasing the rate of cycling of the room. The frequency-response diagram of the rotary contour gage, Fig.9, shows the advantage of mismatching the environmental frequency and the resonant frequency of the gage. Because the resonant frequencies of real gaging systems are so slow (in this case 14 hr per cycle), this mismatching is best accomplished by increasing the environmental frequency. Interpreting Fig.9 we see that a plus or minus 1 deg F temperature control at 0.07 cph gives the same drift as a plus or minus 4 deg F control would give at 1 cph. In some cases it is possible to increase the rate of room cycling by a simple readjustment of the thermostat. The results can be quite dramatic.

High cycling rates are generally achieved by circulating large volumes of air. High-volume air circulation is not too expensive and offers several advantages. A greater volume of air requires a smaller temperature difference between the inlet and outlet to maintain the same room average. This is simply a matter of removing the same number of Btu with more pounds of air at a smaller temperature difference. Another advantage of air volume is that the increased velocity tends to scrub the whole gaging system and remove the heat that may be coming from external point sources of heat such as motors, lights, people, and radiation from the sun. Stated more exactly, the increased air velocity increases the convective heat-transfer coefficient and decreases the thermal resistance between the gage and the room air, which is the thing that is being controlled. Still another advantage of high air flow is increased operator comfort. The decreased difference between inlet and outlet air temperatures means fewer cold drafts which are the real source of discomfort.

The benefits of high air flow, high cycling rates, and close containment of sensitive equipment have recently been demonstrated at LRL. A new rotary contour gage has just gone into service which is completely enclosed in a plexiglass box. Air is admitted through a plenum chamber at the top and leaves through a plenum chamber at the bottom. The circulation rate is one complete change of air every 3 sec. The cycling rate is 25 cph. The room-temperature variation is 0.7 deg F, but a 24-hr drift check shows less than 3 μ in. of drift!

The problem of standardization of room air temperature measurement is illustrated by the different values obtained on this system with three different ways of measuring. High-sensitivity mercury thermometers show less than 0.05 deg F variation. The thermister recorder-controller for the enclosure shows a 0.4-deg F variation and a high-frequency response thermograph shows 0.7 deg F.

An Automatic Error-Correcting Device

In reviewing our experiences with computing thermal drift from knowledge of system-frequency response and measurement of temperature variation, J.W. Routh, of LRL, suggested that we consider the possibility of automatic error correction. Preliminary investigation of this idea has convinced us that it should be possible to design a thermal model of the system that can sense the room temperature and provide an electrical output equal to the drift. This output can be used to zero shift the coordinate system of the gage and provide direct, on-line compensation for thermal error. As it is now visualized, this device would be completely automatic once set for the specified part to be measured. The operational settings required would be nominal coefficient of expansion, time constant, and size of the part. The response of the gage would be built into the device. Some adjustment might be required for different setups that might be encountered. If the time constant of the master could be tailored to match the gage, the bulk of the thermal error could be eliminated. Error due to uncertainty of nominal differential expansion would still remain.

While this manuscript was being prepared, a

63.2 percent of its total change after a sudden step change in the environment.

14 _Frequency Response_. The frequency response of a measurement system is defined as the ratio of the amplitude of the drift in microinches to the amplitude of a sinusoidal environment temperature oscillation in degrees Fahrenheit for all frequencies of temperature oscillation.

15 _Thermal Expansion_. The difference between the length of a body at one temperature and its length at another temperature is called the thermal expansion of the body.

16 _Coefficient of Expansion_:

(a) The true coefficient of expansion, α, at a temperature, t, of a body is the rate of change of length of the body with respect to temperature at the given temperature divided by the length at the given temperature.

$$\alpha = \frac{1}{L} \frac{dL}{dt}$$

(b) The average true coefficient of expansion of a body over the range of temperatures from 68 F to t is defined as the ratio of the fractional change of length of the body to the change in temperature.

Fractional change of length is based on the length of the body at 68 F:

$$\alpha_{68,t} = \frac{L - L_{68}}{L_{68}(t - 68)}$$

Hereinafter the term "coefficient of expansion" shall refer only to the average value over the range from 68 F to another temperature, t.

17 _Nominal Coefficient of Expansion_. The estimate of the coefficient of expansion of a body shall be called the nominal coefficient of expansion. To distinguish this value from the average true coefficient of expansion ($K_{68,t}$) it shall be denoted by the symbol K.

18 _Uncertainty of Nominal Coefficient of Expansion_. The maximum possible percentage difference between the actual coefficient of expansion, α, and the nominal coefficient of expansion shall be denoted by the symbol δ, and expressed as a percentage of the true coefficient of expansion:

$$\delta = \frac{\alpha - K}{\alpha}(100)\%$$

Variations in material composition, forming processes, and heat-treatment as well as inherent anisotropic properties and effects of preferred orientation cause objects of supposedly identical composition to exhibit different thermal-expansion characteristics. Also, differences in experimental technique cause disagreement among thermal expansion measurements. As a result, it is difficult, solely from published information, to obtain an exact coefficient of expansion for any given object.

This value like that of K itself must be an estimate. Various methods can be used to make this estimate. For example:

(a) The estimate may be based on the dispersion found among published data.

(b) The estimate may be based on the dispersion found among results of actual experiments conducted on a number of like objects.

Of the two possibilities given above, (b) is the recommended procedure.

Because the effects of inaccuracy of the estimate of the uncertainty are of second order, it is considered sufficient that good judgment be used.

19 _Nominal Expansion_. The estimate of the expansion of an object from 68 F to its time-mean temperature at the time of the measurement shall be called the nominal expansion and it shall be determined from the following relationship:

$$NE = L(t - 68)(K)$$

20 _Uncertainty of Nominal Expansion_. The maximum difference between the true thermal expansion and the nominal expansion is called the uncertainty of nominal expansion. It is determined from

$$UNE = L(t - 68)\left(\frac{\delta}{100}\right).$$

21 _Differential Expansion_. Differential expansion is defined as the difference between the expansion of the part from 68 F to its time-mean temperature at the time of the measurement and the expansion of the master from 68 F to its time-mean temperature at the time of the measurement.

22 _Nominal Differential Expansion_. The difference between the nominal expansion of the part and of the master is called the nominal differential expansion:

$$NDE = (NE)_{part} - (NE)_{master}$$

23 _Uncertainty of Nominal Differential Expansion_. The sum of the uncertainties of nominal expansion of the part and master is called the uncertainty of nominal differential expansion:

$$UNDE = (UNE)_{part} + (UNE)_{master}$$

24 _Thermal Drift_. Drift is defined as the differential movement of the part or the master and the comparator in microinches caused by time-variations in the thermal environment.

25 _Electronics Drift Check_. An experiment conducted to determine the drift in a displacement transducer and its associated amplifiers and re-

corders when it is subjected to a thermal environment similar to that being evaluated by the drift check itself. The electronics drift is the sum of the "pure" electronics drift and the effect of the environment on the sensing head, amplifier, and so on. The electronics drift check is performed by blocking the transducer and observing the output over a period of time at least as long as the duration of the drift test to be performed. Blocking a transducer involves making a transducer effectively indicate on its own frame, base, or cartridge. In the case of a cartridge-type gage head, this is accomplished by mounting a small cap over the end of the cartridge so the plunger registers against the inside of the cap. Finger-type gage heads can be blocked with similar devices. Care must be exercised to see that the blocking is done in a direct manner so that the influence of temperature on the blocking device is negligible.

26 _Drift Check._ An experiment conducted to determine the actual drift inherent in a measurement system under normal operating conditions is called a _drift check_. Since the usual method of monitoring the environment (see definition 28) involves the correlation of one or more temperature recordings with drift, a drift check will usually consist of simultaneous recordings of drift and environmental temperatures. The recommended procedure for the conduct of a drift check is given in Appendix 2.

27 _Temperature Variation Error, TVE._ An estimate of the maximum possible measurement error induced solely by deviation of the environment from average conditions is called the _temperature variation error_. TVE is determined from the results of two drift checks; one of the master and comparator, and the other of the part and the comparator.

For zero or small mastering cycle time:

$$TVE = \begin{bmatrix} \text{Maximum excursion of part-comparator and} \\ \text{master-comparator drift curves when curves} \\ \text{are aligned for in-phase temperature con-} \\ \text{ditions over representative time period.} \\ \text{(Part-to-master drift)} \end{bmatrix}$$

For significant mastering cycle times:

$$TVE = \begin{bmatrix} \text{Master-comparator drift} \\ \text{or part-comparator drift,} \\ \text{whichever is greater, for} \\ \text{chosen mastering cycle} \\ \text{time} \end{bmatrix} - \begin{bmatrix} \text{Part-to-mas-} \\ \text{ter drift} \\ \text{error as} \\ \text{above} \end{bmatrix}$$

Use whichever of above TVE is greater.

28 _Total Thermal Error._ Total thermal error is defined as the maximum possible measurement error resulting from temperatures other than a uniform, constant temperature of exactly 68 F. It is, of course, desirable to determine the total thermal error induced in any measurement. However, this is usually not practical to do, and in many cases, not even possible. Therefore, an alternative procedure is outlined below.

29 _Thermal Error Index._ The evaluation technique proposed in this section does nothing more than estimate the maximum possible error caused by thermal environment conditions affecting a particular measurement process. It does _not_ establish the actual magnitude of any error. It serves to remove doubt about the existence of the errors and to establish a system of rewards and penalties to processes which are combinations of techniques, some of which may be "good" and some "bad."

The thermal error index shall apply only so long as conditions do not change.

The proposed plan consists of:

(a) Computing the nominal differential expansion, NDE. In this computation (and in the next), the temperature offset is assumed to be the average difference between 68 F and the air temperature in the vicinity of the process over the mastering cycle of the process.

(b) Computing the uncertainty of NDE, UNDE.

(c) Determining the thermal variation error, TVE, by means of a drift check.

(d) Summing the absolute values obtained in (a), (b), and (c) to obtain an index related to the quality of the process, yields the temperature error index.

$$TEI = NDE + UNDE + TVE$$

(f) If an effort is made to correct the measurement by computing the NDE, part (a) is to be deleted.

The plan penalizes a measurement process on two counts:

(i) Existence of environment temperature offset, resulting in differential expansion.

(ii) Existence of environment variations.

The plan rewards good technique by reducing the thermal error index for:

(i) Attempting a correction for differential expansion.

(ii) Keeping environmental variations to a minimum.

Thermal-error index can be used as an administrative tool for certification of measurement processes as is discussed in the next section. It can also be used as an absolute index of acceptability of the process. For example, a good rule of thumb for establishing the acceptability of a measurement process with respect to thermal errors is to limit the acceptable thermal error index to 10 percent of the working tolerance of the part.

30 _Monitoring_. To perpetuate the thermal-error index it will be necessary to monitor the process in such a way that significant changes in operating conditions are recognizable.

The recommended procedure is to establish a particular temperature recording station which has a demonstrable correlation with the magnitude of the drift. In a "convective environment" this could simply be the "environment temperature."

The temperature of the selected station should be recorded continuously during any measurement process to which the index is to be applied. If the temperature shows a significant change of conditions, the index is null and void for that process, and a reevaluation should be accomplished, or the conditions corrected to those for which the index applies.

In addition to continuous monitoring of environmental conditions, it is recommended that efforts be made to establish that the process is properly soaked out. This may be done by checking the temperature of all elements before and after the execution of the measurements.

APPENDIX 2

DRIFT-CHECK PROCEDURE

The following is the recommended procedure for the conduct of a drift check for a process in which the proposed monitoring method is based on the measurement of environment temperatures:

(a) _Equipment_. The major equipment necessary includes very sensitive displacement transducers and sensitive, drift-free temperature sensors with associated amplifiers and recorders. A linear variable differential transformer with provision for recorder output has proven quite successful. Also, various resistance-bulb thermometers with recording provision have proven successful as temperature monitoring devices.

The required sensitivity of the displacement transducers used may be adjusted according to the rated accuracy of the measurement system.

(b) _Equipment Testing_. The temperature-measuring and recording apparatus should be thoroughly tested for accuracy of calibration, response and drift. The availability of sensitivities of at least 0.1 deg F is desirable. Time constants of sensing elements of about 30 sec are recommended.

Before the displacement transducers and associated apparatus are used they should be calibrated and checked for drift _in the environment_. An "electronics drift check" should be performed by blocking the transducer and observing the output over a period of time at least as long as the duration of the drift test to be performed.

"Blocking" a transducer involves making a transducer effectively indicate on its own frame, base, or cartridge.

(c) _Preparation of System for Test_. An essential feature of the drift check is that conditions during the check must duplicate the "normal" conditions for the process as closely as possible. Therefore, before the check is started, "normal" conditions must be determined. The actual step-by-step procedure followed in the subject process must be followed in the same sequence and with the same timing in the drift check. This is especially important in terms of the actions of any human operators in mastering and all preliminary setup steps. With as little deviation from normal procedure as possible, the displacement transducers should be introduced between the part (or master, depending on the type of drift check) and the rest of the C-frame such that it measures relatively displacement _along the line of action_ of the subject measurement process.

The temperature-sensing pickup must be placed to measure a temperature which is correlatable with the drift. Some trial and error may be necessary. In the extreme case, temperature pickups may have to be placed to measure the temperature of all of the active elements of the measurement loop.

(d) _Representative Time Period For a Drift Check_. Once set up the drift check should be allowed to continue as long as possible, with a minimum of deviation from "normal" operating conditions. In situations where a set pattern of activity is observed its duration should be over some period of time during which most events are repeated. When a 7-day work week is observed in the area, and each day is much like any other, a 24-hr duration is recommended. If a 5-day work week is observed, then either a full-week cycle should be used or checks performed during the first and last days of the week.

(e) _Postcheck Procedure_. After the drift test, the displacement transducers and the temperature recording apparatus should be recalibrated.

(f) _Evaluation of the Drift Check (Drift-Check Report)_. Following the drift check, the data should be assessed for the following values:

(1) _Nonperiodic effects_. The effects of the operator tend to disappear with elapsed time. These and similar effects should be described and the portion of this error not compensated by soak out should be included in the TVE.

(11) _Temperature Variation Error (TVE)_. For zero or small mastering cycle time:

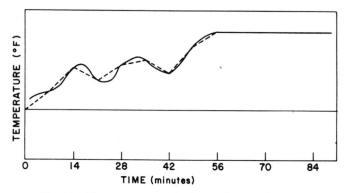

Fig.14 Temperature variation and approximation

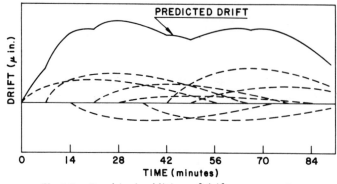

Fig.15 Graphical addition of drift components

$$TVE = \begin{bmatrix} \text{Maximum excursion of part-comparator and} \\ \text{master-comparator drift curves when curves} \\ \text{are aligned for in-phase temperature con-} \\ \text{ditions over representative time period.} \\ \\ \text{(Part-to-master drift)} \end{bmatrix}$$

For significant mastering cycle times:

$$TVE = \begin{bmatrix} \text{Master-comparator drift} \\ \text{or part-comparator drift,} \\ \text{whichever is greater, for} \\ \text{chosen mastering cycle} \\ \text{time} \end{bmatrix} - \begin{bmatrix} \text{Part-to-mas-} \\ \text{ter drift} \\ \text{error as} \\ \text{above} \end{bmatrix}$$

Use whichever of above TVE is greater.

A complete report of the drift check findings should include the following:

Thermal Drift-Check Reports Outline. Items in parentheses are suggested as a guide to what might be pertinent under a heading.

1 Description of System
 (a) Identification. Manufacturer, model, pertinent specifications, and dimensions.)
 (b) Component motions. (Active elements, lines of action.)
 (c) Operations:
  (1) Type of operation
  (ii) Typical workpiece: Sizes; materials; minimum tolerances.
 (iii) Method of mastering.
  (iv) Cycle times: Operating, mastering.
2 Environment description
 (a) Room features: Size; solar exposure; exits; wall, floor, ceiling, and other heat sources.
 (b) System features: (i) Location with respect to "room features"; (ii) internal heat sources: motors, lamps, electronics.
 (c) Air circulation: Inlet-outlet locations, sizes, numbers, drafts, air

volume circulated.
 (d) Temperature monitoring and control.
3 Test apparatus description:
 (a) Temperature monitoring: Identification, response, sensitivity, location.
 (b) Displacement monitoring: Identification, response, sensitivity, location.
4 Procedure
 (a) Stepwise description of testing.
5 Results. Displacement-temperature versus time graphs; maximum displacements and temperature variations; cycle times; causes if known.
6 TVE.
7 Recommendations.

APPENDIX 3

METHOD FOR DETERMINING FREQUENCY RESPONSE
OF A MEASUREMENT SYSTEM

Data obtained from step-change experiments performed on the 15-in. Sheffield rotary contour gage gave (a) an indication that the system was linear with respect to thermal variations; and (b) the set of data correlating temperature variation with drift.

The basic characteristic of a linear system is that the output (in this case, drift) corresponding to any input (temperature variation) is the sum of outputs corresponding to the components of the input.

This characteristic permits the use of the data in Fig.7 in computing a predicted drift for a variation of temperature in the environment of the Sheffield gage as follows:

Suppose that the temperature variation is recorded as shown in Fig.14. This record can be approximated by straight lines over 7-min increments. This procedure decomposes the temperature variation into a series of components similar to that in Fig.7.

The drift corresponding to any one of these

components can be determined by scaling the data in Fig.7.

Fig.15 shows the resulting set of drift components and their sums which is an approximation to the drift caused by the temperature variation of Fig.14.

This computation procedure is cumbersome when done by hand but is easily and quickly done by a digital computer. The results shown in Figs.8 and 10 of the paper were computed using an IBM 7094 digital computer at the Lawrence Radiation Laboratory.

There are two practical considerations to be observed in making computations of this kind:

1 It must be possible to make an accurate approximation to the temperature data. For example, in the case described above, a temperature variation consisting of a sine wave with a period of less than 7 min cannot be approximated by the data available.

2 Because of the "memory" of the system, the computed drift is in error until a period of time equal to the soak-out time has elapsed. For example, in computing the data for Fig.8, the drift computed for the first 12 hr of temperature variation record was inaccurate and was omitted.

ACKNOWLEDGMENTS

The authors wish to express special gratitude to Mr. I.H. Fullmer of NBS for his aid in assembling bibliographical material, and to Mr. T.R. Young of NBS for suggestions concerning the uncertainties of thermal expansion properties.

REFERENCES

1. Ch. Cochet, "On the Choice of a Uniform Temperature for the Calibration of Measuring Instruments (Sur le choix d'un degre uniforme de temperature pour l'etalonnage des instruments de mesure)," Rev. Gen. de l'Electricite, Vol. 4, Nov. 16, 1918, pp. 740-2. Report of Commission de Normalisation des Ingenieurs des Arts et Metiers de Boulogne-sur-Seine recommending adoption of 0 degree centigrade as standard.

2. A.Perard, " Temperature of Adjustment of Industrial Gages (La temperature d'adjustage des calibres industriels)," Genie Civil, Vol. 90, June 25, 1927, pp. 621-4. Plea to German Industrial Standards Committee to readopt adjustment of 0 degree instead of 20 degree centigrade; discusses requisite precision for industrial gages and necessity for standardization of temperature of adjustment.

3. L. Graux, " Temperature Adjustment of Length Standards (La temperature d'adjustage des etalons industriels, des verificateurs et des pieces mecaniques)," Genie Civil, Vol. 91, Sept. 10, 1927, pp. 250-2. Discusses adjustment on basis of standard meter at 0 degree or at +20 degree centigrade.

4. G. E. Guillaume, "Temperature of Standardization of Industrial Gages (La temperature de definition des calibres industriels)," Rev. Gen. de l'Electricite, Vol. 23, Jan. 14, 1928, pp. 73-6. Object of article is to emphasize importance of decision made by German Commission fixing 20 degree centigrade as temperature of standardization; after stating necessity of unification of temperature in all countries and difficulties

which have arisen, author gives some reasons which have moved him in stating troublesome effect German decision can have on desired unification.

5. A. Perard. "Concerning Standardization Temperature of Industrial Gages (A propos de la temperature de definition des calibres industriels)," Rev. Gen. de l'Electricite, Vol. 23, Jan. 28, 1928, pp. 169-70. Discussion of reasons for not considering temperature of 20 degree centigrade as temperature for standardization rather than 0 degree centigrade; takes into account coefficient of expansion.

6. P. Grand, "Contribution a la metropole industrielle des longueurs," Rev. Gen. Mecanique, Vol. 36, Nos. 41, 42, 45, May 1952, pp. 135-40; June, pp. 167-70; Sept., pp. 293-6. Contribution to industrial measurements of length, with reference to parts of large dimension; influence of ambient temperature in workshops; permissible deviation between temperature of gage and object being measured; measurement of steel parts with steel and wooden gages, and light alloy parts with steel, light alloy, and wooden gages; diagrams.

7. P. Grand, "Influence de la temperature sur l'etalonnage des jauges," Rev. Gen. de Mecanique, Vol. 39, Nos. 74, 79, 83, 84, Feb. 1955, pp. 39-45; July, pp. 247-54; Nov., pp. 395-400; Dec., pp. 445-50; Vol. 40, Nos. 86, 88, Feb., 1956, pp. 81-7; Apr., pp. 159-64. Influence of temperature on calibration of gages; environmental temperature in measuring shop: temperature of materials subjected to different environments; calibration on measuring machine; comparator calibration; errors of calibration due to temperature variations.[*] (UCRL Translation-1145(L).)

8. R. Noch and H. Huhn, "Investigations of the influence of temperature variations on the readings of a comparator with large mechanical magnification," Z. Instrumkde, Vol. 67, No. 11, Nov., 1959, pp. 285-8. in German. Mechanical comparators with high magnification for linear measurements can alter their readings due to the influence of small changes of temperature. The paper deals with the influence of a slight heat source, for example, an operator or a lamp, on comparators with a graduation of 0.0001 mm.

9. A. Goldsmith, T. E. Waterman, and H. J. Hirschhorn, Handbook of

---

[*]This reference was not received until after the completion of this paper. Mr. Grand has covered much of the same material presented here. His approach to the problem is different, however, and the duplications do not justify withdrawal of this paper.

Thermophysical Properties of Solid Materials, Pergamon. N.Y., 1961.

10. D. K. Cheng, Analysis of Linear Systems, Addison-Wesley, Redding, 1961.

11. I. H. Fullmer, "Comments on Various Temperature Combinations," National Bureau of Standards, 1960, 4 pages. Gives historical background of three standard temperatures; discusses importance of knowing coefficient of thermal expansion accurately, and discusses proposed change of standard temperature to 23° C. (Not published.)

12. Evan L. S. Lum, "Dynamic Corrections of Thermally Induced Measuring Machine Errors," (M. E. 299 report to Prof. H. Thal-Larsen) Department of Mechanical Engineering, University of California, Berkeley, January 1965.

Presented at the SME Functional Gaging and Inspection Techniques Program
June 1983

Gages Reject Too High A Percentage Of Good Parts

By Edward S. Roth
Productivity Services Inc.

Final acceptance gages have always rejected a percentage of parts that function perfectly at assembly. This reject percentage is increasing with the advent of more precise machine tools because part accuracy is approaching gage accuracy.

Gages are usually designed to reject all out-of-tolerance parts. Unfortunately many in-tolerance parts are also rejected, since the gage designer can take up to 15% of the part tolerance for the gage tolerance and wear allowance (10% for gage tolerance, 5% for wear allowance). Since the part must usually fit freely into a receiver type gage, this fit requirement, when added to the 15% tolerance, can cause the rejection of a very high percentage of good products.

Parts I and II (Figure I) illustrate this problem. These two parts are fastened at assembly with four 0.750 diameter bolts. When the male and female pilot diameters are at their most critical interchangeable size or at maximum material condition (MMC)[1], there is an intentional fit allowance of 0.002 (10,000 female - 9.998 male diameter).

NOTE: The 0.750 diameter gage pins and bushings will not be toleranced to keep this article short but the same tolerancing and wear allowance techniques discussed will apply.

Gages I and II (for Parts I and II respectively) reflect current receiver type gage design practice and take up to 10% (consider that this 10% includes the wear allowance to simplify this presentation) of the part tolerance (.0005) as their tolerance. (See Figure II.)

Tables I and II show the assembly relationship of Part I and Gage I, and Part II and Gage II, respectively. Since acceptable parts must fit freely in the gage, some of the parts will be rejected. It is certainly possible to fit a rather soft part into a rather hard gage, even when the part and gage are identical in size if sufficient force is applied. (Even more force may be required to remove the part.) It is quite safe to say that an inspector would reject any part that had to be forced into or out of a gage.

When Part I is 9.998 and Part II 10.000, they should fit together--providing the fit allowance is appropriate--yet both parts will certainly be rejected by the inspector as there could be interference between part and gage. Even when Part I is 9.997 and Part II is 10.001 (there is actually twice the fit allowance initially specified) the inspector could again reject each part because they might have to be forced into their respective gages. Tables I and II indicate that even nominal size parts (9.9955 and 10.0025) have a rejection probability.

---

[1]External features are at MMC when finished to their largest specified size; internal features are at MMC when finished to their smallest specified size.

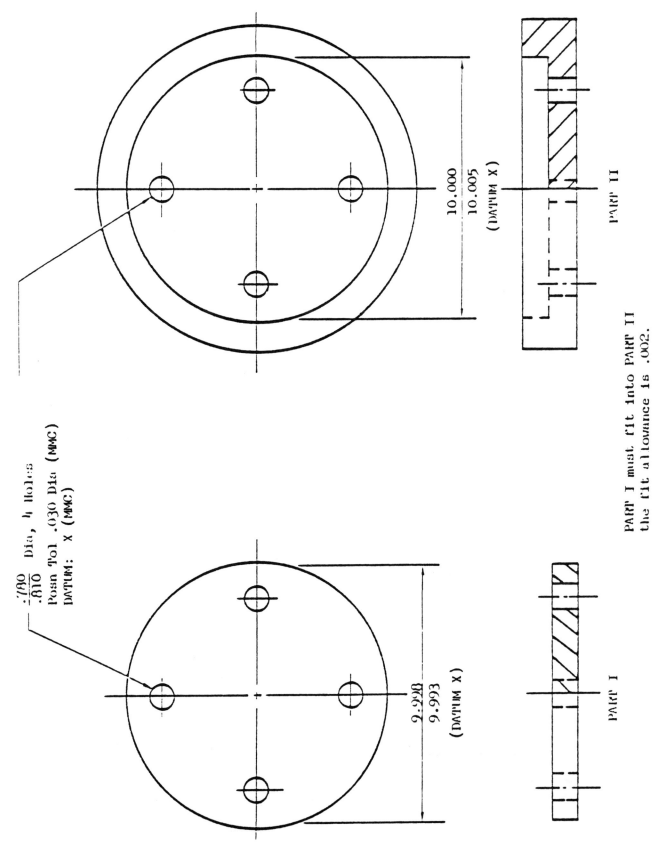

.790
.810 Dia, 4 Holes
Posn Tol .030 Dia (MMC)
DATUM: X (MMC)

10.000
10.005
(DATUM X)

9.990
9.993
(DATUM X)

PART II

PART I

PART I must fit into PART II the fit allowance is .002.

Figure I

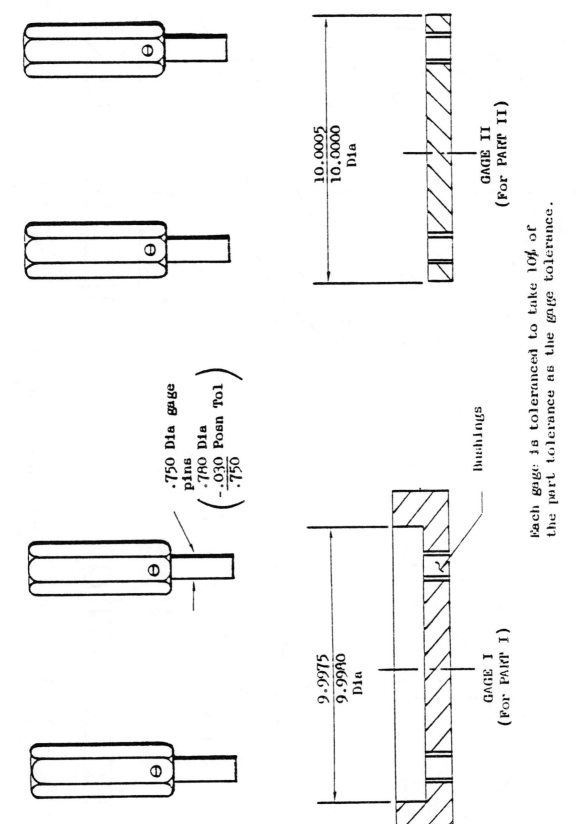

.750 Dia gage pins
$$\left(\begin{array}{c}.780 \text{ Dia} \\ -.030 \text{ Posn Tol} \\ \hline .750\end{array}\right)$$

10.0005
10.0000
Dia

GAGE II
(For PART II)

9.9975
9.9960
Dia

bushings

GAGE I
(For PART I)

Each gage is toleranced to take 10% of the part tolerance as the gage tolerance.

Figure II

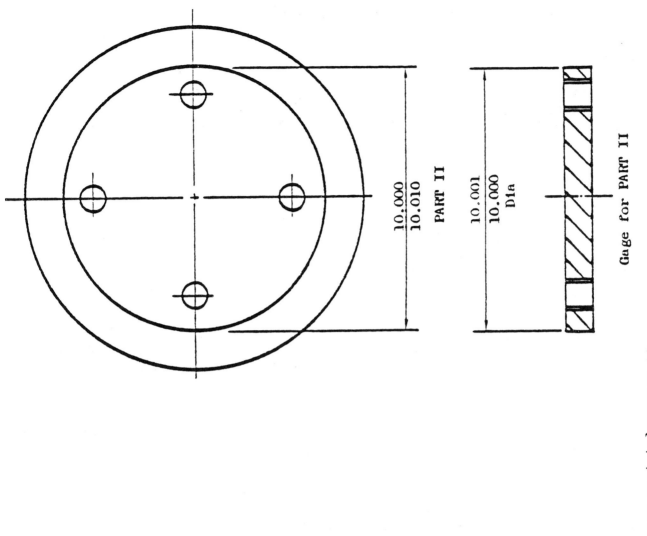

PART II

10.000
10.010

10.001
10.000
Dia

Gage for PART II

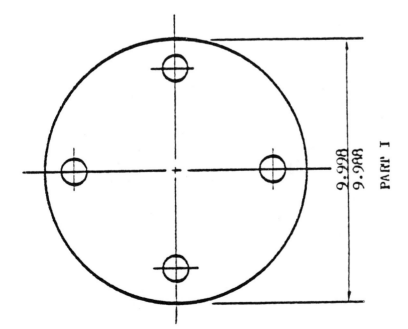

PART I

9.998
9.988

9.997
9.998
Dia

Gage for PART I

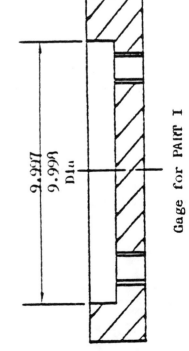

Increase part tolerances so
gages will accept more parts.

Solution I

The following solutions should enable product designers to choose an approach that is compatible with either personal or company standards:

SOLUTION I

Increase the size tolerance so that more parts will fit in the gage (even though the gage tolerance will increase). If Part I had a size tolerance of 9.998/9.988, nominal parts (9.993 diameter) would certainly fit a 9.998/9.997 gage.

SOLUTION II

Reduce the fit allowance to zero or specify a slight interference fit. Since each gage requires a fit allowance, any two parts that passed their respective gages should fit together with _twice_ the fit allowance required. This solution automatically increases the size tolerance if the least critical size limits (9.993 and 10.005) are not changed.

SOLUTION III

The product designer may specify the gage allowance on the product drawing. The gage designer will make the gage the size specified and then take up to 15% of the part tolerance as the gage tolerance in the usual way.

SOLUTION IV

Use the bilateral (split-tolerance) system and apply the gage tolerance equally from nominal.

SOLUTION V

Standardize a table of gage tolerances and fit allowances to be used by both product and gage designers. See Table III.

SUMMARY

Basic inspection policy should be to acept--not to reject--all possible in-tolerance products, and the author hopes that the solutions offered will be helpful to companies who are plagued with the problem presented in this article. Since a final acceptance gage should usually look like the mating part at its most critical interchangeability (MMC) and configuration, the designer must consider gage design concurrently with product design. The author feels that Solution III enables the designer to directly control end-product quality because the sole acceptance criterion is specified directly on the product drawing and the designer is forced to consider the gage.

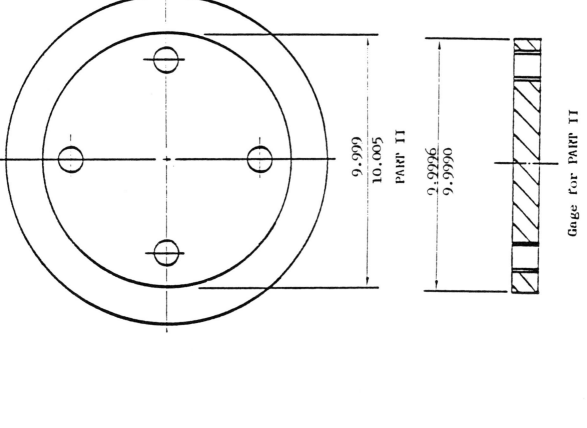

9.999
10.005

PART II

9.9996
9.9990

Gage for PART II

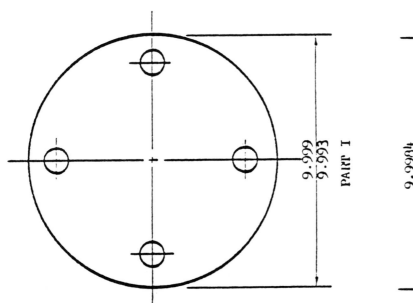

9.999
9.993

PART I

9.9994
9.9990

Gage for PART I

Increase part tolerances by
reducing the fit allowance.

Solution II

31

TABLE I

Gage diameter and tolerance (for Part I)	Actual part diameter	Allowance (gage to part)	Part accepted by gage operator
	9.9980	0.0000 to 0.0005 interference	No
	9.9970	0.0005 to 0.001	* No - This part should always be accepted if gage takes only 10% of part tolerance
9.9975 Diameter 9.9980	9.9960	0.0015 to 0.002	* Possibly
	9.9950	0.0025 to 0.003	* Probably
	9.9940	0.0035 to 0.004	* Yes
	9.9930	0.0045 to 0.005	Yes

* Assembly of gage and part depend on:
(1) Part finish
(2) Part thickness
(3) Material
(4) Temperature
(5) Chamfer on gage
(6) Gage depth
(7) Patience of gage operator

Quite often, part acceptance or rejection is a personal determination.

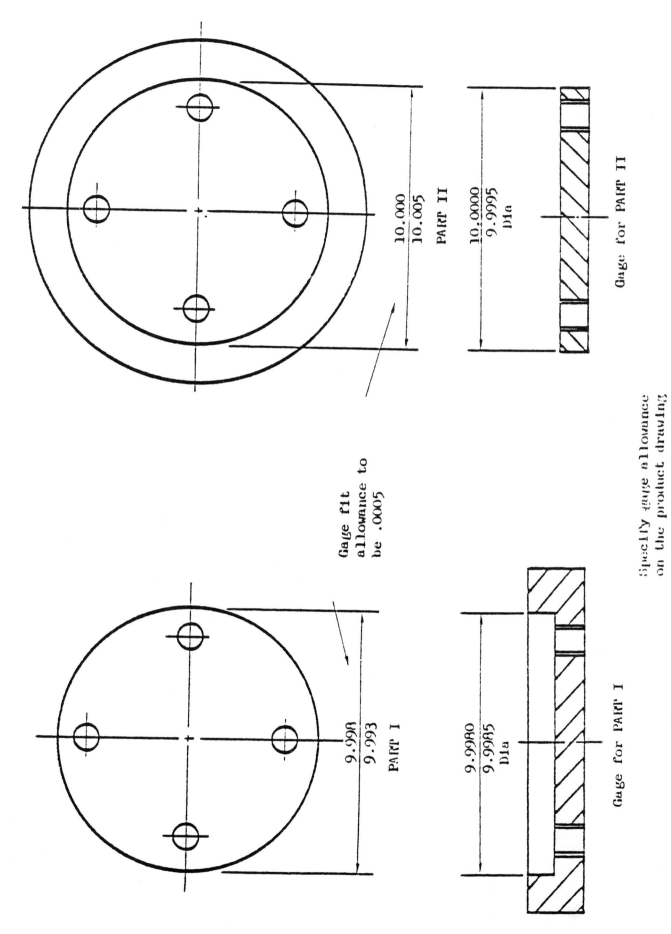

Gage fit
allowance to
be .0005

9.998
9.993
PART I

9.9980
9.9985
Dia

Gage for PART I

10.000
10.005
PART II

10.0000
9.9995
Dia

Gage for PART II

Specify gage allowance
on the product drawing

Solution III

TABLE II

Gage diameter and tolerance (for Part II)		Actual part diameter	Allowance (gage to part)	Part accepted by gage operator
		10.0000	0.0000 to 0.0005 interference	No
		10.0010	0.0005 to 0.001	* No - This part should <u>always</u> be accepted if gage takes only 10% of part tolerance
10.0005 / 10.0000	Diameter	10.0020	0.0015 to 0.002	* Possibly
		10.0030	0.0025 to 0.003	* Probably
		10.0040	0.0035 to 0.004	* Yes
		10.0050	0.0045 to 0.005	Yes

* Assembly of gage and part depend on:
   (1) Part finish
   (2) Part thickness
   (3) Material
   (4) Temperature
   (5) Chamfer on gage
   (6) Gage depth
   (7) Patience of gage operator

Quite often, part acceptance or rejection is a <u>personal</u> determination.

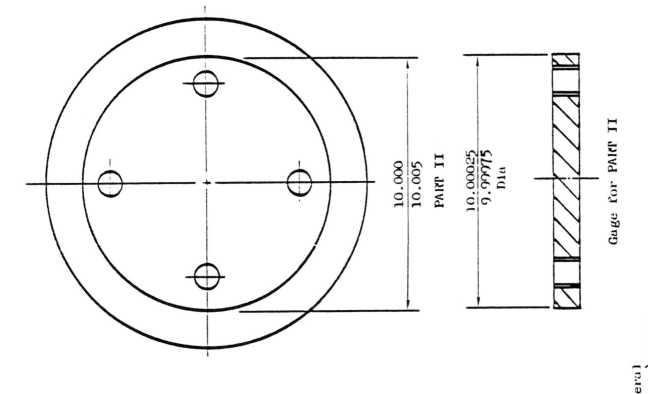

10.000
10.005

PART II

10.00025
9.99975
Dia

Gage for PART II

Specify bilateral
(split tolerance) gages

Solution IV

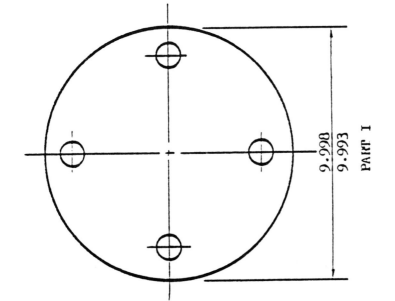

9.998
9.993

PART I

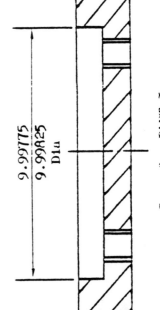

9.99775
9.99825
Dia

Gage for PART I

TABLE III (Tentative)

Gage Fit Allowance and Tolerance Table

Part diameter		Part tolerance				
		0.001-0.005	0.006-0.010	0.011-0.015	0.016-0.020	Over 0.020
Over 0 to 1	Gage allowance	0.0000	0.0000	0.0002	0.0002	0.0005
	Gage tolerance	0.0002	0.0002	0.0003	0.0003	0.0005
Over 1 to 2	Gage allowance	0.0000	0.0002	0.0002	0.0005	0.0005
	Gage tolerance	0.0002	0.0003	0.0003	0.0005	0.0005
Over 2 to 4	Gage allowance	0.0002	0.0002	0.0005	0.0005	0.001
	Gage tolerance	0.0003	0.0003	0.0005	0.0005	0.001
Over 4 to 8	Gage allowance	0.0002	0.0005	0.001	0.001	0.001
	Gage tolerance	0.0003	0.0005	0.001	0.001	0.002
Over 8 to 16	Gage allowance	0.0005	0.001	0.001	0.001	0.002
	Gage tolerance	0.0005	0.001	0.002	0.002	0.002
Over 16 to 32	Gage allowance	0.001	0.001	0.002	0.002	0.002
	Gage tolerance	0.001	0.002	0.002	0.002	0.003
Over 32 to 64	Gage allowance	0.002	0.002	0.002	0.003	0.003
	Gage tolerance	0.002	0.002	0.003	0.003	0.004

Compiled by H. L. Durflinger, Tool Design Engineer, Sandia Corporation.

The tabular values listed should facilitate mating of product and gage and are presented as a basis for decision making. It may be necessary to depart from the listed values to further minimize the probability of accepting bad product and any such departure should be specified per Solution III.

Presented at the SME Qualtest I Conference, October 1982

Design For Certainty

By Clifton D. Merkley
IBM

Design certainty during product development ensures a quality functional unit which has fitness for use at minimum cost of design and manufacturing.

A set of structured steps starting with the product description will define the activities to be completed by all functions. Such steps when defined, will reduce the cost and length of the schedule in the development cycle. This effort will accomplish the goal of a product that is sound in design and manufacturing. The result will be a profitable program by reducing engineering redesign, manufacturing errors, costly field changes, and customer dissatisfaction.

The following sequential, and sometimes parallel, steps are critical and require detailed attention.

1. Writing the product description.
2. Developing the engineering schematic with engineering specifications for review and concurrence by:

    A. Marketing

    B. Service

    C. Manufacturing Engineering

    D. Quality Assurance

    E. Cost Services

3. Developing the complete design layout.

4. Building prototypes to check for extreme fits and clearances.

5. Performing designed experimental testing to identify design controlling parameters.

6. Updating design layout for changes required as a result of designed experiments.

7. Preparing detail part prints from the design layout.

8. Building tools and ordering parts for production.

9. Certifying the process with limited production build.

10. Testing the product to specification.

11. Distributing the product.

12. Selling the Customer.

The method to accomplish each of these steps will depend on the company organization. Regardless of method, however, ignoring these steps exposes the project to problems of cost, schedule, and quality. A careful detailed plan for the tasks in each of these steps should be outlined. Following this plan will prevent re-do, missed events, and duplicated work. The following detail should become part of the planning for the product's development.

1. Writing the Product Description

Prepared by marketing, this statement should completely describe:

° Function

° Competition

° Expected price

° Estimated quantity of sales

° Service objectives

° Time Period to be marketed

° Life Expectancy

° Distribution method

The size and organization of the company will dictate the method by which the product is to be developed. This product may enhance an existing product or may come from market research indicating the need for a new venture. However, to use development time without this product description is to spend money uselessly. The product description becomes not only the source material for marketing sales literature but also the requirement for engineering development.

This effort is critical and cannot be overemphasized. Many products are developed without considering the customer. What is developed and produced must be consistent with demand. Demand is oriented to quantity, quality, and function. Figure 1 describes the full cycle of a product.

2. Developing the Engineering Schematic and Specifications

The engineering schematic and specifications of the product should be prepared by development engineering and should address the requirements as stated by Marketing. See Fig. 2

The engineering schematic allocates space for modules and subassemblies, defines module interfaces, manufacturing planning, product function, customer interface, and identifies hardware presently being used in production.

The schematic must consider the dimensioning system to be used. Items such as datum identification, module interface, and the function of the product must be stated. Subassemblies and replaceable units are identified along with possible methods of manufacture. Repair strategy (by the customer or by service personnel) is identified.

Accompanying the schematic should be an engineering specification detailing the product's internal operating conditions, (horse power requirement, RPM, Input per Unit/Time, Output per Unit/Time, Operating interfaces, "what does the customer do") and the functional operations that the customer will see and what installation activity must be completed.

The review and concurrence of the schematic layout by those named in Step 2 of the outline are imperative to proceed with development.

The review by the participating parties should be completed in detail. Each reviewer should address items particular to his functional activity and the interface with other functions.

Marketing will review the product that engineering has configured to meet the market requirements stated in the product description. Manufacturing engineering will review for abilities and technologies to produce the product within the desired cost and schedule. Quality assurance will determine the testing needed for overall reliability.

The concurrence of each function is a commitment to develop a product for the market at a minimum manufacturing cost, high in quantity and quality, meeting the program schedule. All concerns and objections must be addressed and resolved.

3. Developing the Complete Design Layout

This step completes the design for function, operating conditions and customer interface. A formal review may not be necessary if all concerned functions have input to the design. If not a formal review by the same group who reviewed the product at the engineering schematic level is required. The review should be in detail.

The design layout by development engineering is the first identification of fit and function of parts, subassemblies, and hardware presently being used in production. This design must meet the product objectives agreed to in the engineering schematic and specified by the product description. The master design may contain module outlines with space for how modules fit into the major assembly. The design layout of the subassemblies can be left to the subassembly designer. If this is done, the subassembly design layout must be completed before the final design layout approval. Limits of fit, dimension system, datum dimensions, and final adjustment at test or assembly are identified. From this layout, it should be possible to fabricate some parts for the prototype build. For an illustration, see Figs. 3, 4.

Figure 5, an Owens Corning advertisement, taken from the publication "Appliance", illustrates what may happen when the design effort is not coordinated. Space constraints are violated and function lost. The result is a nonfunctional unit which must be redesigned. Obviously Owens Corning was advertising insulation, not illustrating a design layout.

4. Building Prototypes

Using sketches, the design layout, and prints where applicable, development engineering can build a set of prototypes. An information file of the fits and clearances as built should be generated for reference during the designed experiment phase. Exercising fits to extreme limits (of plus and minus combinations) should be used to determine the units function and the ability to assemble the unit. (The extreme limits of function for controlling variables are covered further in Step 5 "Performing Designed Experimental Testing.") The information file for the fits and clearances can be accumulated from data gathered from receiving inspection or requested from the model maker or vendors. Without this information, failure analysis at testing intervals will be limited.

5. Performing Designed Experimental Testing

Testing prototypes does not necessarily mean running the unit to failure. Product indicators (these variables that affect function, sometimes referred to as the controlling variables) which can be used for process control during production, need to be identified. To produce without using these process control variables (that is, an early warning system) is to expose the manufacturing plant to potentially high reject rates, scrap and rework, and a marginal product.

Designing these experiments calls for detailed attention in the use of statistical designed experiments to determine the ability of the product to meet the market needs.

The completion of the prototype testing identifies the product variables to be used for controlling the process during production. The planning for control of these items goes into initial production processing planning. Care must be given to the number of variables identified. Too much data and too many variables to monitor will only confuse manufacturing and may result in insufficient attention to control of the manufacturing process.

The subject of designed experiments can be further pursued by consulting a statistical experimental design textbook such as PRACTICAL EXPERIMENT DESIGN FOR ENGINEERS AND SCIENTISTS, by William J. Diamond, Belmond, California, Lifetime Publication; or STATISTICAL DESIGN AND ANALYSIS OF EXPERIMENT by Peter W. M. John, New York, Macmillan Company.

6. Updating Design Layout

This effort occurs at all stages of the design cycle. Each engineer must consult the design layout when a change is made which affects space and interfaces between modules. During the designed experiment phase and at its completion, the update should bring everything up to the level ready to prepare the final part prints for building production hardware.

7. Preparing Detail Part Prints

Prints are prepared using the design layout and applying the corporation design standards. Care must be taken to generate prints and documents to the company specification or a recognized national standard (such as the ANSI Series). This minimizes the interpretation conflicts between engineering, manufacturing and vendors. Prints prepared to a recognized set of standards make use of presently available knowledge which comes from experience.

8. Building Tools and Ordering Parts

This step begins the final build of the manufacturing process. During steps two through eight, representatives of manufacturing should have been involved throughout the product design cycle. The considerations for methods of fabrication, assembly, and test are thus defined.

To assume any tooling will not be built prior to this time would be to believe there is no pressing time interval to arrive at the marketplace. Long-lead items may be tooled knowing that the results of testing may call for an engineering change. Dividing the product into major functional modules and subassemblies will allow some areas to proceed faster than others. However, each area must comply with the design layout for an integrated system. Fig. 2 shows the inter-relationship of different modules.

9. Certifying the Process with Limited Production Build

To complete this step requires that the designed experiments find which variables control the process as designed. The result should be a statement describing the process with confidence levels of the repetitive ability of the process.

The certification check list should answer:

1. Is the process operator-sensitive?

2. Can process changes be readily detected to prevent rejects?

3. Are safety items in place?

4. Are applicable government regulations in place?

5. Do we have manufacturing capability to produce the required tolerances?

At the ASQC Midwest '81 Conference, H. J. Harrington outlined two steps to certify a process. These steps have been summarized as follows:

1. Certification—QA certifies an operation or equipment when they have a level of confidence that the operation or equipment will produce products to specification when the documentation is followed. Certification applies to individual operations.

2. Qualification—Qualification is the QA support for a process consisting of many operations. For a process to be qualified by QA, each operation and all test equipment must be certified. Also, the process must have demonstrated that it can produce high-quality products to specification.

Quality Assurance has the responsibility to qualify all manufacturing processes. Often, this responsibility extends to the development pilot lines because the development pilot line provides a base from which the process and the product specifications are derived. It is also the source for all the data that is required to assess manufacturability. Two process lines need to be qualified: the development pilot line and the manufacturing process line.

Qualification becomes even more complex because the manufacturing process goes through at least two, and in most cases, three different phases. These include:

1. Establishing the initial manufacturing process line.

2. Expanding and automating the manufacturing process facilities.

3. Reducing the manufacturing process cost.

To control the process as it moves from line to line and from phase to phase requires developing a qualification program.

10. Testing the Product to Specification

The product description (as stated by marketing) serves as the guide for the test to be performed. The product must be exposed (tested) to the demands that will be placed on it from the customer. In addition, the engineering specifications of internal unit functions and build and test parameters, need to be evaluated by the test. In general, the length of test will depend on product complexity. Important to the test structure is the independence of the test organization and the engineering function. This test evaluates the product as manufactured. With this charter, the quality organization should monitor the activities of the test when such test is conducted by an independent group. When Quality conducts the test, the interface should be Manufacturing Engineering. Completion of the test with problem resolution qualifies the product for first customer ship.

Increase in production quantities may require limited or repeated testing at later intervals to evaluate the process stability.

11. Distributing the Product

Planning for distribution begins early in the product design. Distribution is the physical movement to the point of sale.

Concerns for the type of packaging, customer setup, mail-in returns and damage claims should have already been in place.

12. Selling the Customer

A functional product, fit for use, at a reasonable cost enables the salesman to return the anticipated revenue to the business.

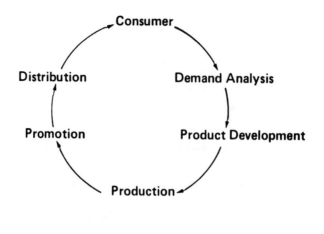

FIG. 1. PRODUCT CYCLE

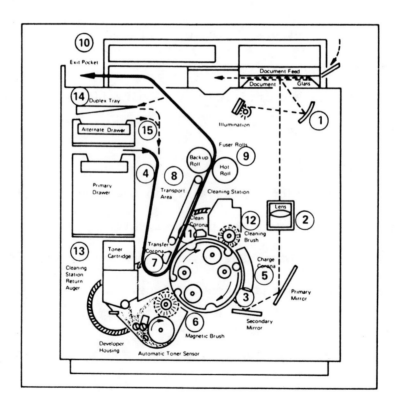

FIG. 2. TYPICAL ENGINEERING SCHEMATIC

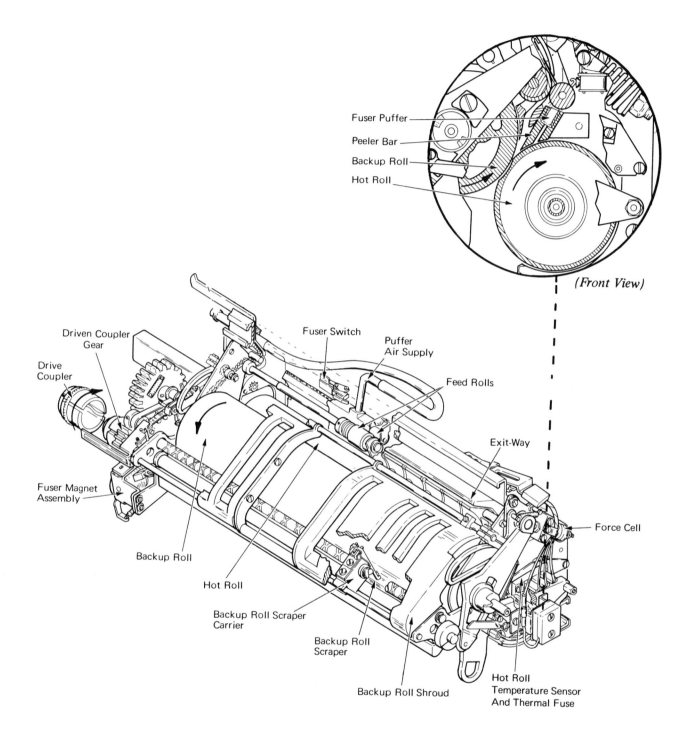

Fuser Puffer

Peeler Bar

Backup Roll

Hot Roll

(Front View)

Driven Coupler
Gear

Fuser Switch

Puffer
Air Supply

Drive
Coupler

Feed Rolls

Exit-Way

Fuser Magnet
Assembly

Force Cell

Backup Roll

Hot Roll

Backup Roll Scraper
Carrier

Backup Roll
Scraper

Hot Roll
Temperature Sensor
And Thermal Fuse

Backup Roll Shroud

FIG. 3. TYPICAL DESIGN LAYOUT (3-Dimensional)

47

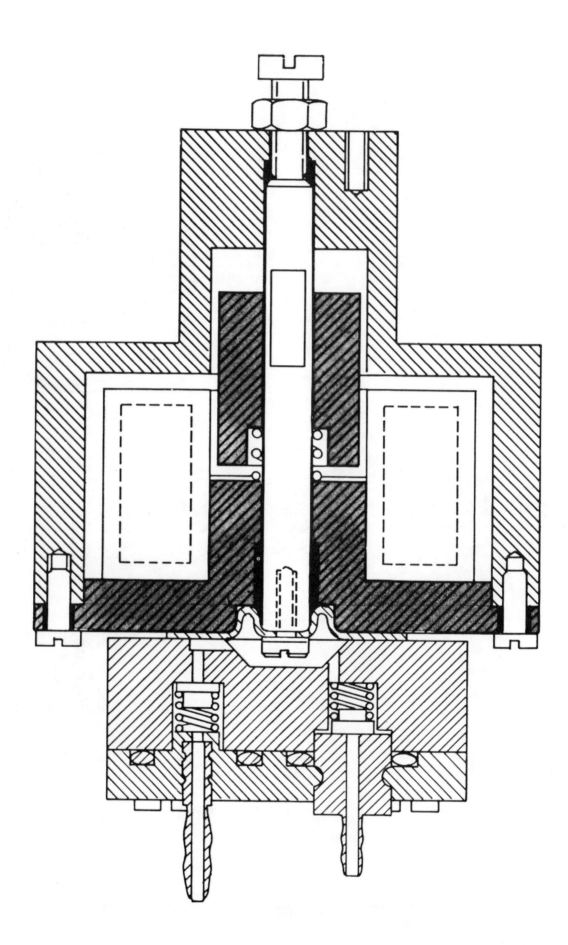

FIG. 4. ENGINEERING LAYOUT (Dimensional and Tolerance Definition)

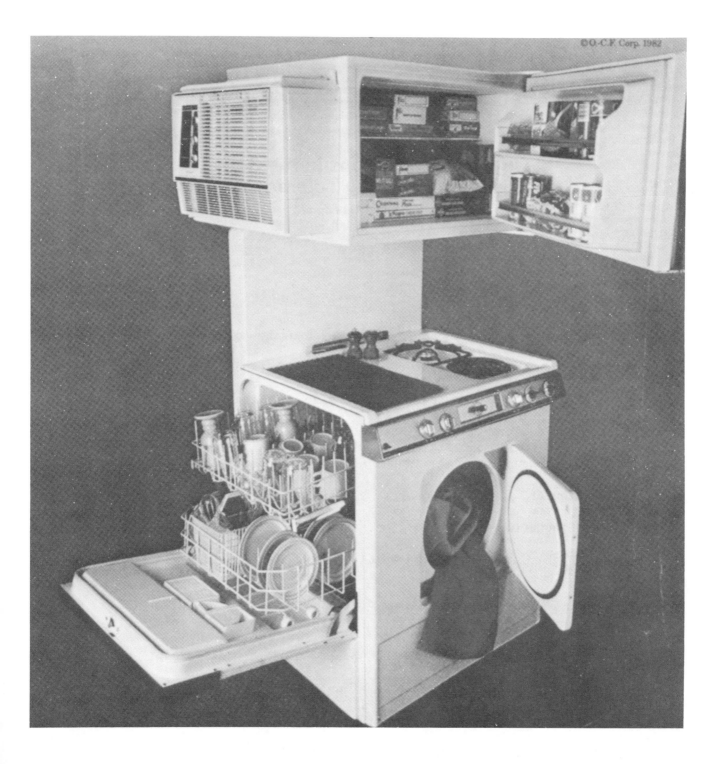

FIG. 5. MISGUIDED DESIGN EFFORT (reprinted with permission)

Reprinted courtesy of John M. Leaman
Milwaukee, Wisconsin

Design Considerations in the Selection of Engineering Tolerances

By John M. Leaman
University of Wisconsin—Extension

One of the most critical responsibilities of a product or machine designer is the establishment of engineering or manufacturing tolerances. It is critical because of the magnitude of cost resulting from the specification of incorrect tolerances. It should be noted that excessive cost can result from tolerances which are both <u>too</u> <u>tight</u>, or <u>too</u> <u>loose</u>. Tight tolerances result in unnecessary machining and inspection time; in the selection of more costly than necessary process equipment, in rejected parts, and in an adversary relationship between design, manufacturing, and inspection. How can excessively loose tolerances also cause high cost? They often result in assembly problems and poor product performance.

The responsibility for selection of tolerances producing the most economical, thus optimum part, lies on the shoulders of the design engineer. He must, therefore, have broad knowledge of engineering metrology or how the part is to be measured. He needs to know the capability of a wide range of manufacturing processes, including metal casting, stamping and forging, plastic molding, as well as the many machining processes available for production of the part. Finally, he needs a clear understanding of the functional requirements of the piece part in assembly and operation with the other components of the system. These areas of knowledge require a great deal of education and experience. Unfortunately, the inexperienced designer will usually request a significantly tighter tolerance than necessary in order to protect himself. This invariably results in a never-ending conflict between manufacturing, inspection, and design.

What are the basic reasons for the tendency, on the part of designers, to select overly tight tolerances?

1. Training in the engineering curriculum promotes the creed of precision.

2. Tight tolerances are specified due to the fear of interference or excessive clearance between assembled parts. It seems safer to err on the side of tight tolerances.

3. Design tolerances are taken from a similar previous design which may have been established on unrealistic tolerances.

4. Tolerances are selected from company, vendor, or industry design standards which tend to favor tight tolerances.

5. Tight tolerances are considered synonomous with good quality.

The purpose of this paper will be to explore information and techniques available as guidance in the establishment of realistic and, where possible, optimum engineering tolerances. Three general topic areas will be discussed:

** Tables, Charts and Standards

** Statistical Tools

** Computer Simulation Methods

I. Tables, Charts and Standards

There are many sources of helpful information relating to tolerancing. Figures I and II[1] are an example of charts developed by L.J. Bayer to reflect the cost of several machining operations vs. tolerance and surface finish.

---

[1]References are listed at the end of this paper.

Figures III through VI[2] are examples from many charts published by Wade in his text on tolerances. Tolerance ranges for commercially available sheet, rod, wire, etc. are shown as well as for forgings, extrusions, castings, etc. are provided in this text. These sources are just one of many which provide guidance for the designer. They are somewhat general, however, and their use requires some experience and are not a complete answer.

Figure VII[3] shows another type of charting cost relationships for various machining processes.

Standards are available, as you are aware, from various industry groups, to specify tolerance ranges. Figures VIII and IX[4] are taken from the American Gear Manufacturing Associations' gear standards. As shown in Figure IX, extremely tight tolerances are specified for high quality gears.

The use of data such as illustrated in these figures, does not assure good design. The tendency for tight tolerances is not avoided as the opportunity exists for the selection of classes of fit which are more precise than the design calls for. They are merely guidelines to provide better understanding of the cost of tolerances and guidance to an understanding of tolerances resulting from various production operations.

II. Statistical Tools

Most application of statistics in design and manufacturing are based on the assumption that the process produces parts which are normally distributed. If the sampling is random and the dimension under consideration is independent, one piece to another, the assumption of normality is a good one.

The distribution of sample data can be evaluated with a frequency plot. Figure X shows plotted data from an example in the text "Statistical Quality Control" by E.L. Grant, McGraw-Hill, pp. 43-4. The measurements are weights of cans of tomatoes, but could be a plot of a critical dimension on a machined part. This plot shows a good approximation to a normal distribution.

Two important statistics can be calculated from sample measurements--the mean value and the sample variance. They provide good estimates of the average and standard deviation for the universe or population, which in our case is the entire production lot.

Figure XI shows the curve for a normal population. While the ordinate of the curve would usually indicate measurement, as on the frequency plot, this curve has been standardized to show the area under the curve at various standard deviations away from the mean value. We thus show that 95% of all production is within ±2 standard deviations and 99.7% within ±3 standard deviations. It is this extremely useful information which we will apply in our examples relating to tolerances.

A. Tolerances of Mated Parts

Figure XII shows the analysis for a typical four-part stacked assembly. From a normal engineering approach, the tolerance for the housing would be calculated by adding the tolerance range of the individual parts as shown. Note that it allows the full range of ±.008 or .016 inches. Is this realistic?

In order to make a statistical analysis, we need to obtain an estimate of the standard deviation, σ, for this production operation. Knowing that for the piece parts to meet an acceptable quality level, such as 99% within tolerance, we can say that our tolerance range will encompass a spread of approximately ±3σ. We therefore obtain our estimate of σ by dividing our tolerance range by six. Note that in the engineering approach we have added these individual estimates to obtain $\sigma_{AE} = .0027$.

Using an important equation from statistical theory, we know that

$$\sigma_{AE} = \sqrt{\sigma_{AB}^2 + \sigma_{BC}^2 + \sigma_{CD}^2 + \sigma_{DE}^2}$$

for the values above:

$$\sigma_{AE} = \sqrt{(.00067)^2 + (.00033)^2 + (.001)^2 + (.00067)^2}$$

or

$$\sigma_{AE} = .0014$$

Our engineering approach has been too conservative.

Note that this value of σ is only one half that obtained in the engineering approach. Knowing that in a random assembly, the probability of all large parts being assembled with large parts, or of all small parts likewise being assembled together, is extremely small, we can, thus, utilize the more realistic statistical approach and choose a tolerance for AE of 1.645 ± .004 inches rather than our original ±.008 inches. Conversely, we could retain the larger housing tolerance, but expend the tolerance on the individual piece parts.

B. Quality Control Charts

A powerful tool for the definition of the potential dimensional accuracy of manufacturing processes is the quality control chart. Unfortunately, few process engineers and essentially no design engineers understand or are even aware of this tool. The quality control chart is based on the assumption of normal distribution, random sampling, and independence between samples. We will first define the quality control chart and then discuss its application in design.

The quality control charting technique was developed by Walter Shewhart, a statistician from Bell Laboratories. In his infinite wisdom, Shewhart realized that with proper sampling and appropriate statistical analysis, a process could literally show in a sequential plot, the limits within which it could perform.

Figure XIII shows a typical control chart for a critical dimension on a rheostat knob. This example was also taken from Grant's text, "Statistical Quality Control".

The piece part in this example is a plastic molded rheostat knob, containing a metal insert. The insert has a drilled hole utilized to pin the knob to its shaft. The dimension plotted locates the pin hole from the end of the molded part. The total variation in the critical dimension is due to variation in the metal insert as well as in the molding operation.

The control charting principle is basically simple. Periodic samples of five parts were randomly selected from the production run. Smaller lots could have been used. The critical dimension was measured and charted as shown. The average; $\overline{X}$, and the range; R, were calculated for each sample lot and plotted. After 25 sample lots had been charted, a simple calculation for $\overline{\overline{X}}$; the average of the total lots, was calculated as well as $\overline{R}$; the average of the ranges. $\overline{\overline{X}}$ shows the central tendency or the value about which the measurements vary, and $\overline{R}$ is representative of the process spread. From $\overline{R}$; an estimate of the standard deviation σ was calculated. Using the tabulated constants based on the size of the sample lot, upper and lower control limits for $\overline{X}$ and R were calculated based on our previously discussed ±3 σ range. These calculations are shown in Figure XIV.

The control chart defines if the process is in control. In this example, the points all fall within the limits, as one would expect from the ±3 σ limits for a process under control. Variations in the readings do exist, and are from normal process variation due to machine clearances, variations in machine settings, changes in material and even measurement error. These are all inherent variations present in all processes. If a measurement falls outside of the limits, we can be 99.7% certain that this error was caused by an assignable cause which is correctable.

In the sample shown it can be seen that while the process is in control, many individual readings are outside of the given process tolerance of .140 inches ± .003 inches. From this, we can say that the process is not capable to meet the specified tolerance. Two choices exist--change the tolerance or utilize another tool or machine.

Many companies maintain control charts in production. Charts are, therefore, available to guide the design office in the establishment of realistic specifications. Where they do not exist, it is strongly recommended that the manager of the design function have the quality control department or industrial engineering obtain charts on various machine tools and other process machines for various operations and materials. A catalog of these charts will guide design decisions and also provide a common ground of understanding with manufacturing and inspection. The money spent to develop this data will pay off in better design and a more cooperative company effort.

Control chart data can also be charted for more convenient use. Figure XI[5] shows a chart of screw machine capabilities for various materials obtained from process control charts.

Finally, the author has found control charting of tremendous value in research work dealing with new product development. The control chart will show when process changes result in product improvement. Quality control charts were used with great benefit in the development of a critical induction welding process.

III. Computer Simulation

We have discussed the importance of the designer understanding the capability of manufacturing processes in producing parts to a given tolerance. We have provided some guidance in his developing this understanding. The need for his understanding the tolerance requirement of his product or design has also been stressed. Computer simulation provides a powerful tool in defining this requirement. Many companies depend on pilot runs to provide this information. This is an expensive and unreliable method to predict the way in which parts from production tooling will perform in assembly. Computer simulation will provide data on the probability of parts of a certain tolerance range interfering at assembly. The designer can then exercise his judgment on the acceptability of the specified tolerance.

To illustrate how computer simulation works, let us look again at a stack-up assembly problem. The design dimensions and tolerances are shown in Figure XVI[6]. We know from experience for this product, that a minimum of .020 inches clearance is desirable for ease in installation at assembly. Figure XVI also shows a pictoral representation of the problem.

For computer simulation it is necessary that a mathematical model of the problem be defined as well as the distribution of the dimensions. This model assumes truncated normal distributions. The truncation occurs in production due to parts being inspected out of the lot. If the distribution is not known, especially for new parts, it can be assumed to be normal. Quality control is often able to improve on this approximation from a study of similar parts.

The computer model for our example will be simple:

$$RND\ (Y) = RND\ (X_4) - RND\ (X_1) - RND\ (X_2) - RND\ (X_3)$$

These values are random variates of our dimensions. If we now ask the computer to run through several thousand runs using random variables of our nominal dimensions, plus or minus random tolerances from the specified ranges and distributions, we will be literally reaching into tote boxes and producing typical assemblies.

Figure XVII shows the results of a simulation of 20,000 iterations conducted by Ford Motor Company on this example problem. The frequency distribution for Y shows that the probability of obtaining clearance below .020 inches is about .00035% or about one assembly in 2,800. Since .020 inches is really not critical, this probability is acceptable and may permit loosening of tolerances if necessary.

Several simulation runs may be required to optimize tolerances. Attention should be focused on cost savings. It may be advantageous to tighten one tolerance to permit increasing tolerance on a part which imposes a larger cost penalty. If a computer program is written in the early stages of design, studies relating to manufacturing or vendor requests for tolerance relief can be conducted quickly, since only the input values need be changed.

All of the automotive companies are using simulation methods with claims for very significant cost savings. The most impressive is General Motors current program using computer simulation in conjunction with computer graphics. Figure XVIII[7] shows schematically, the ultimate in the use of this design tool. Notice the possibility to interactively adjust--both tighten and relax--critical dimensions to produce an optimum design. Their program permits analysis of component sensitivity interactions and resultant correlations, all in real time.

At Fisher Body, the caster-camber variation on the Vega front suspension involving 26 dimensions for critical parts, has been analyzed with this technique. The analysis representing a full week of work with non-graphic methods was completed in one hour at the console.

These advanced techniques are available to the smaller manufacturers, and design managers should be aware of their potential for the production of better designs at reduced cost.

REFERENCES

1. Bayer, L.J. (August 1956), "Manufacturing Cost as Related to Product Design", Tooling and Production, Vol. 22, No. 5, pp. 73-6.

2. Wade, O.R. (1967), Tolerance Control In Design and Manufacturing, Industrial Press, New York.

3. Trucks, H.E. (1974), Designing for Economical Production, Society of Manufacturing Engineers, Dearborn, MI.

4. AGMA GEAR HANDBOOK, "Gear Classification, Material and Measuring Methods for Unassembled Gears," 390.03, American Gear Manufacturers Association, Rosslyn, VA.

5. Juran, J.M. (1962), Quality Control Handbook, McGraw-Hill, New York.

6. Choksi, Suresh C., "Computers Can Optimize Manufacturing Tolerances for an Economical Design", Ford Motor Company report.

7. "A New Dimension in Dimensional Tolerancing", Search, Vol. 9, No. 6, General Motors Research Laboratories.

8. Buckingham, E. (1954), Dimensions and Tolerances for Mass Production, Industrial Press, New York.

9. Conway, H.G. (1966), Engineering Tolerances, Pitman, London.

10. Wakefield, L.P. (1964), Dimensioning for Interchangeability, McMillan, N.Y.

11. Kirkpatrick, E.G., Quality Control for Managers and Engineers, John Wiley and Sons, Inc., New York.

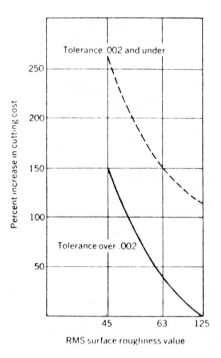

Figure 4.4. Comparing surface quality and dimensional tolerance in face milling. From L.J. Bayer (August, 1956), "Manufacturing Cost as Related to Product Design," Tooling and Production, Vol. 22, No. 5, pp. 73-76.

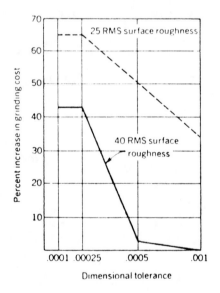

Figure 4.5. Comparing surface quality and tolerance in cylindrical grinding. From L.J. Bayer (August, 1956), "Manufacturing Cost as Related to Product Design," Tooling and Production, Vol. 22, No. 5, pp. 73-76.

Figure I

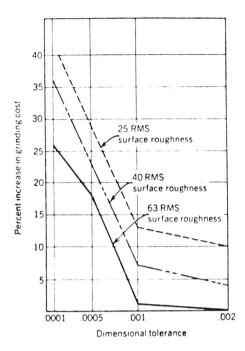

Figure 4.7. Comparing surface quality and tolerance in reciprocating surface grinding. From L.J. Bayer (August, 1956), "Manufacturing Cost as Related to Product Design," Tooling and Production, Vol. 22, No. 5, pp. 73-76.

Figure II

TABLE 72

Cold and Hot Rolled Steel Sheet: B.S. 1449:1956
(Tolerance plus and minus, unit 0.001 in.)

| Sheet Gauge, in. | Width of Sheet, in. | | | | | | | |
| | 18-48 | | 48.1-60 | | 60.1-70 | | 70.1-80 | |
	Cold	Hot	Cold	Hot	Cold	Hot	Cold	Hot
-0.018	2	2	--	--	--	--	--	--
0.019-0.024	3	3	3	--	--	--	--	--
0.025-0.027	3	3	3-5	--	4	--	4	--
0.028-0.030	3	3	3-5	--	4	--	4	--
0.031-0.034	3-5	4	3-5	4	4	5	4	--
0.035-0.039	4	4	4	4	5	5	5	--
0.040-0.044	4	5	4	5	5	6	5	--
0.045-0.049	4	5	4	5	5	6	5	--
0.050-0.055	5	6	5	6	6	7	6	--
0.056-0.062	5	6	6	7	6	7	6	--
0.063-0.069	6	6	6	7	6	7	6	--
0.070-0.078	6	7	7	7	7	8	7	8
0.079-0.088	7	8	8	8	9	9	9	9
0.089-0.099	8	8	9	9	10	10	11	11
0.100-0.117	9	9	10	10	10	11	11	12

Cold and Hot Rolled Steel Strip: B.S. 1449:1956
(8 to 12 in. maximum width approximately)

| Strip Gauge, in. | Tolerance Plus and Minus, unit 0.001 in. | |
	Cold	Hot
0.005-0.0124	0.5	--
0.013-0.028	1.0	--
0.028-0.048*	1.5	3
0.049-0.128	2.0	±½
0.129-0.250	3.0	gauge
0.251-0.500	5.0	spacing

* 0.045 for hot rolled.

Figure III

TABLE 79
American Drop Forging Association Limits
for Steel Stampings
(Up to 100 lb. in Weight)
Tolerance unit = 0.001 in.

Stamping Weight, lb.	Thickness Normal to Die Lane	
	Commercial	Close
0.2	+ 24, − 8	+12, − 4
0.4	+ 27, − 9	+15, − 5
0.6	+ 30, −10	+15, − 5
0.8	+ 33, −11	+18, − 6
1	+ 36, −12	+18, − 6
2	+ 45, −15	+24, − 8
3	+ 51, −17	+27, − 9
4	+ 54, −18	+27, − 9
5	+ 57, −19	+30, −10
10	+ 66, −22	+33, −11
20	+ 78, −26	+39, −13
30	+ 90, −30	+45, −15
40	+102, −34	+51, −17
50	+114, −38	+57, −19
60	+126, −42	+63, −21
70	+138, −46	+69, −23
80	+150, −50	+75, −25
90	+162, −54	+81, −27
100	+174, −58	+87, −29

TABLE 80
American Drop Forging Association Limits for Steel Stampings
(Up to 100 lb. in Weight)
Tolerance unit = 0.001 in.

Limits on Length Parallel to Die Line					
Total = Shrinkage + Die Wear Allowance					
Length, in.	Commercial	Close	Weight, lb	Commercial	Close
1	± 3	± 2	1	±32	±16
2	± 6	± 3	3	±35	±18
3	± 9	± 5	5	±38	±19
4	±12	± 6	7	±41	±21
5	±15	± 8	9	±44	±22
6	±18	± 9	11	±47	±24
For each 1 in.	add ± 3	add ±1½	For each 2 lb.	add ± 3	add ±1½

Figure IV

62

TABLE 86
Limits on Brass and Copper Tubes--Wall Thickness--Copper

Wall Thickness t, in.	Limits		
	Mean t, in.	Max. t, in.	Min. t, in.
0.022	±0.002	+0.003	-0.003
0.028	±0.002	+0.004	-0.004
0.036	±0.003	+0.005	-0.005
0.048	±0.003	+0.006	-0.006
0.064	±0.004	+0.007	-0.007

TABLE 87
Limits on Brass and Copper Tubes--Wall Thickness--Brass

Wall Thickness t, in.	Limits		
	Mean t, in.	Max. t, in.	Min. t, in.
0.018-0.022	±0.002	+0.003	-0.003
0.024-0.036	±0.003	+0.004	-0.004
0.040-0.072	±0.004	+0.006	-0.006
0.080-0.116	±0.006	+0.010	-0.010
0.128-0.192	±0.008	+0.013	-0.013

Figure V

TABLE 104
American Die Casting Institute Tolerances

1. Basic Tolerances in One Die Part

Tolerance ±, unit 0.001 in.		Material or Alloy							
		Zinc		Aluminum		Magnesium		Copper	
Dimension, in.		C	N	C	N	C	N	C	N
-1		3	10	4	10	4	10	7	14
1.01-12	Add per inch	1	1.5	1.5	2	1.5	2	2	3
12.01-	Add per inch	1	1	1	1	1	1	--	--

N = normal dimension; C = central dimension

2. Additional Tolerances to be Added to Above Simply or Jointly

Tolerance ±, unit 0.001 in.	Material or Alloy			
	Zinc	Aluminum	Magnesium	Copper
Projected Area of Die Casting Across Parting Plane, sq in.	Thickness Dimensions, Across Parting Plane			
- 50	4	5	5	5
51-100	6	8	8	--
101-200	8	12	12	--
201-300	12	15	15	--
Projected Area of Portion of Die Casting Affected by Moving Die Part, sq in.	Local Depth Dimensions Affected by Moving Die			
- 10	4	5	5	10
10.1- 20	6	8	8	--
20.1- 50	8	12	12	--
51 -100	12	15	15	--

3. Flatness

Tolerance (unit 0.001 in.) as measured by feeder gauge from flat reference surface, all materials.

Maximum Dimension of Surface	Tolerance
Up to 3 in.	8
Above 3 in., add per inch	3

Figure VI

RANGE OF SIZES		TOLERANCES ±								
FROM	THROUGH									
0.000	0.599	0.00015	0.0002	0.0003	0.0005	0.0008	0.0012	0.002	0.003	0.005
0.600	0.999	0.00015	0.00025	0.0004	0.0006	0.001	0.0015	0.0025	0.004	0.006
1.000	1.499	0.0002	0.0003	0.0005	0.0008	0.0012	0.002	0.003	0.005	0.008
1.500	2.799	0.00025	0.0004	0.0006	0.001	0.0015	0.0025	0.004	0.006	0.010
2.800	4.499	0.0003	0.0005	0.0008	0.0012	0.002	0.003	0.005	0.008	0.012
4.500	7.799	0.0004	0.0006	0.001	0.0015	0.0025	0.004	0.006	0.010	0.015
7.800	13.599	0.0005	0.0008	0.0012	0.002	0.003	0.005	0.008	0.012	0.020
13.600	20.999	0.0006	0.001	0.0015	0.0025	0.004	0.006	0.010	0.015	0.025

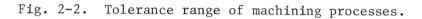

LAPPING & HONING
GRINDING, DIAMOND TURNING, BORING
BROACHING
REAMING
TURNING, BORING, SLOTTING, PLANING, & SHAPING
MILLING
DRILLING

Fig. 2-2.   Tolerance range of machining processes.

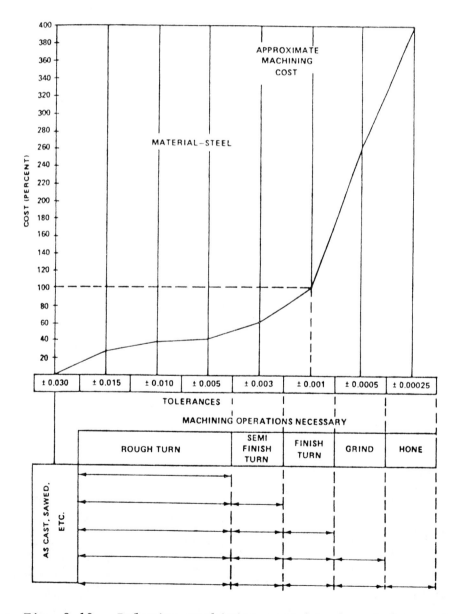

Fig. 2-10.   Relative machining cost based on tolerances.

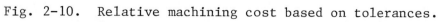

Figure VII

TABLE III COARSE–PITCH GEAR TOLERANCES (Continued)
Tolerances in Ten-Thousandths of an Inch

AGMA QUALITY NUMBER	NORMAL DIAMETRAL PITCH	RUNOUT TOLERANCE — PITCH DIAMETER (INCHES)										PITCH TOLERANCE — PITCH DIAMETER (INCHES)										PROFILE TOLERANCE — PITCH DIAMETER (INCHES)										LEAD TOLERANCE — FACE WIDTH (INCHES)				
		3/4	1½	3	6	12	25	50	100	200	400	3/4	1½	3	6	12	25	50	100	200	400	3/4	1½	3	6	12	25	50	100	200	400	1 and Less	2	3	4	5
8	1/2					146.5	174.5	205.8	242.7	286.3	337.6					19.0	21.7	24.5	27.7	31.3	35.4					42.6	47.7	53.1	59.1	65.7	73.1	5	8	11	13	16
	1				88.8	104.8	124.8	147.2	173.6	204.7	241.4				14.4	16.3	18.6	21.0	23.7	26.8	30.3				28.3	31.5	35.3	39.3	43.7	48.6	54.1					
	2			53.9	63.5	74.9	89.2	105.2	124.1	146.3	172.6			10.9	12.3	14.0	15.9	18.0	20.3	23.0	26.0			18.8	21.0	23.3	26.1	29.0	32.3	36.0	40.0					
	4		32.7	38.5	45.4	53.6	63.8	75.2	88.7	104.6	123.4		8.3	9.3	10.6	11.9	13.6	15.4	17.4	19.7	22.2		12.5	13.9	15.5	17.2	19.3	21.5	23.9	26.6	29.6					
	8	19.8	23.3	27.5	32.5	38.3	45.6	53.8	63.4	74.8	88.2	6.3	7.1	8.0	9.0	10.2	11.7	13.2	14.9	16.8	19.0	8.3	9.3	10.3	11.5	12.8	14.3	15.9	17.7	19.7	21.9					
	12	16.3	19.2	22.6	26.7	31.5	37.5	44.2	52.1	61.5	72.5	5.8	6.5	7.4	8.3	9.4	10.6	12.0	13.6	15.4	17.4	7.0	7.8	8.6	9.6	10.7	12.0	13.3	14.8	16.5	18.4					
	20	12.7	15.0	17.7	20.8	24.6	29.3	34.5	40.7	48.0	56.6	5.2	5.8	6.6	7.4	8.4	9.5	10.7	12.1	13.7	15.5	5.6	6.2	6.9	7.7	8.6	9.6	10.7	11.9	13.2	14.7					
9	1/2					104.7	124.7	147.0	173.4	204.5	241.2					13.4	15.3	17.3	19.5	22.1	24.9					30.4	34.1	37.9	42.2	46.9	52.2	4	7	9	11	13
	1				63.5	74.8	89.1	105.1	124.0	146.2	172.4				10.2	11.5	13.1	14.8	16.7	18.9	21.4				20.2	22.5	25.2	28.1	31.2	34.7	38.6					
	2			38.3	45.4	53.7	63.7	75.2	88.5	104.5	123.3			7.7	8.7	9.8	11.2	12.7	14.3	16.2	18.3			13.5	15.0	16.7	18.6	20.7	23.1	25.7	28.6					
	4		23.3	27.5	32.4	38.3	45.3	53.4	63.0	74.7	88.1		5.8	6.6	7.4	8.4	9.6	10.8	12.3	13.8	15.7		8.9	10.0	11.1	12.3	13.8	15.3	17.1	19.0	21.1					
	8	14.1	16.7	19.7	23.2	27.4	32.6	38.4	45.3	53.4	63.4	4.5	5.1	5.8	6.6	7.4	8.4	9.5	10.7	12.1	13.7	5.9	6.6	7.4	8.2	9.1	10.2	11.4	12.6	14.1	15.6					
	12	11.6	13.7	16.2	19.1	22.5	26.8	31.6	37.2	43.9	51.8	4.1	4.6	5.2	5.9	6.6	7.5	8.5	9.6	10.8	12.2	5.0	5.5	6.2	6.9	7.6	8.6	9.5	10.6	11.8	13.1					
	20	9.1	10.7	12.6	14.9	17.6	20.9	24.7	29.1	34.3	40.4	3.6	4.1	4.6	5.2	5.9	6.7	7.6	8.5	9.6	10.9	4.0	4.4	4.9	5.5	6.1	6.8	7.6	8.5	9.4	10.5					
10	1/2					74.8	89.0	105.0	123.8	146.1	172.3					9.4	10.8	12.2	13.7	15.5	17.6					21.7	24.3	27.1	30.1	33.5	37.3	3	5	7	9	10
	1				45.3	53.5	63.7	75.1	88.5	104.4	123.2				7.2	8.1	9.2	10.4	11.8	13.3	15.0				14.5	16.1	18.0	20.0	22.3	24.8	27.6					
	2			27.5	32.5	38.4	45.3	53.4	63.0	74.7	88.1			5.5	6.2	7.0	7.9	8.9	10.1	11.4	12.9			9.6	10.7	11.9	13.3	14.8	16.5	18.3	20.4					
	4		16.7	19.6	23.2	27.3	32.5	38.4	45.3	53.4	63.0		4.1	4.7	5.3	6.0	6.7	7.6	8.6	9.7	11.0		6.4	7.1	7.9	8.8	9.9	11.0	12.2	13.6	15.1					
	8	10.1	11.9	14.0	16.6	19.5	23.3	27.4	32.4	38.2	45.0	3.1	3.5	4.0	4.5	5.1	5.8	6.5	7.4	8.3	9.4	4.2	4.7	5.3	5.9	6.5	7.3	8.1	9.0	10.0	11.2					
	12	8.3	9.8	11.5	13.6	16.1	19.1	22.5	26.6	31.4	37.0	2.9	3.2	3.7	4.1	4.7	5.3	5.9	6.7	7.6	8.6	3.6	4.0	4.4	4.9	5.5	6.1	6.8	7.6	8.4	9.4					
	20	6.5	7.6	9.0	10.6	12.5	14.9	17.6	20.8	24.5	28.9	2.6	2.9	3.3	3.7	4.2	4.7	5.3	6.0	6.8	7.7	2.9	3.2	3.5	3.9	4.4	4.9	5.4	6.1	6.7	7.5					
11	1/2					53.4	63.6	75.0	88.5	104.3	123.0					6.7	7.6	8.6	9.7	11.0	12.4					15.5	17.3	19.3	21.5	24.0	26.7	3	4	6	7	8
	1				32.4	38.2	45.5	53.6	63.2	74.6	88.0				5.0	5.7	6.5	7.3	8.3	9.4	10.6				10.3	11.5	12.9	14.3	15.9	17.7	19.7					
	2			19.6	23.1	27.3	32.5	38.3	45.2	53.3	62.9			3.8	4.3	4.9	5.6	6.3	7.1	8.1	9.1			6.8	7.6	8.5	9.5	10.6	11.8	13.1	14.6					
	4		11.9	14.0	16.6	19.5	23.2	27.3	32.3	38.1	45.0		2.9	3.3	3.7	4.2	4.8	5.4	6.1	6.9	7.8		4.5	5.1	5.6	6.3	7.0	7.8	8.7	9.7	10.8					
	8	7.2	8.5	10.0	11.8	14.0	16.5	19.6	23.1	27.3	32.2	2.2	2.5	2.8	3.2	3.6	4.1	4.6	5.2	5.8	6.6	3.0	3.4	3.8	4.2	4.6	5.2	5.8	6.4	7.2	8.0					
	12	5.9	7.0	8.2	9.7	11.5	13.7	16.1	19.0	22.4	26.4	2.0	2.3	2.6	2.9	3.3	3.7	4.2	4.8	5.4	6.1	2.5	2.8	3.1	3.5	3.9	4.4	4.9	5.4	6.0	6.7					
	20	4.6	5.5	6.4	7.6	9.0	10.7	12.6	14.8	17.5	20.6	1.8	2.0	2.3	2.6	2.9	3.3	3.7	4.2	4.8	5.4	2.0	2.3	2.5	2.8	3.1	3.5	3.9	4.3	4.8	5.4					
12	1/2					38.1	45.4	53.6	63.2	74.5	87.9					4.7	5.3	6.0	6.8	7.7	8.7					11.1	12.4	13.8	15.4	17.1	19.0	2	3	5	6	7
	1				23.1	27.3	32.5	38.3	45.2	53.3	62.8				3.5	4.0	4.6	5.2	5.8	6.6	7.5				7.4	8.2	9.2	10.2	11.4	12.7	14.1					
	2			14.0	16.5	19.5	23.2	27.4	32.3	38.1	44.9			2.7	3.1	3.5	3.9	4.4	5.0	5.7	6.4			4.9	5.5	6.1	6.8	7.6	8.4	9.4	10.4					
	4		8.5	10.0	11.8	13.9	16.5	19.5	23.1	27.2	32.1		2.0	2.3	2.6	3.0	3.4	3.8	4.3	4.9	5.5		3.3	3.6	4.0	4.5	5.0	5.6	6.2	6.9	7.7					
	8	5.2	6.1	7.2	8.5	10.0	11.9	14.0	16.5	19.5	23.0	1.6	1.8	2.0	2.3	2.6	2.9	3.3	3.7	4.2	4.7	2.2	2.4	2.7	3.0	3.3	3.7	4.1	4.6	5.1	5.7					
	12	4.2	5.0	5.9	6.9	8.2	9.8	11.5	13.6	16.0	18.9	1.4	1.6	1.8	2.0	2.3	2.6	3.0	3.4	3.8	4.3	1.8	2.0	2.2	2.5	2.8	3.1	3.5	3.9	4.3	4.8					
	20	3.3	3.9	4.6	5.4	6.4	7.5	8.9	10.6	12.5	14.7	1.3	1.4	1.6	1.8	2.0	2.3	2.6	3.0	3.4	3.8	1.5	1.6	1.8	2.0	2.2	2.5	2.8	3.1	3.4	3.8					

Figure VIII

TABLE IVA – TOTAL COMPOSITE TOLERANCE (Continued)

(Tolerances in Ten-Thousandths of an Inch)

AGMA Quality Number: 12

Normal Diametral Pitch	Pitch Diameter (Inches)																	
	0.040	0.063	0.100	0.160	0.250	0.400	0.630	1.0	1.6	2.5	4.0	6.3	10.	16.	25.	50.	100.	200.
0.5														27.2	40.4	63.8	76.7	92.3
1.0												18.7	24.0	32.6	38.9	46.0	54.3	64.1
2.0											16.4	19.5	24.5	26.1	28.8	33.5	38.9	45.2
4.0									12.2	13.3	15.1	16.4	18.0	19.7	21.5	24.6	28.1	32.2
8.0								9.8	10.3	11.1	11.7	12.6	13.7	14.9	16.1	18.2	20.6	
12.0							8.4	8.6	9.0	9.3	10.1	10.9	11.7	12.7	13.7	15.3	17.2	
20.0						7.0	7.0	7.0	7.3	7.8	8.4	9.0	9.7	10.4	11.1	12.4		
24.0					6.6	6.6	6.5	6.4	6.9	7.3	7.9	8.4	9.0	9.7	10.4	11.5		
30.0					6.1	6.0	5.9	6.0	6.4	6.8	7.3	7.8	8.3	8.9	9.5			
32.0					6.0	5.9	5.7	5.8	6.2	6.7	7.1	7.6	8.1	8.7	9.2			
40.0				5.6	5.5	5.3	5.1	5.4	5.8	6.2	6.6	7.0	7.5	8.0	8.5			
48.0				5.3	5.1	4.7	4.8	5.1	5.5	5.8	6.2	6.6	7.0	7.4	7.9			
50.0				5.2	5.0	4.6	4.7	5.1	5.4	5.7	6.1	6.5	6.9	7.3				
60.0			5.0	4.8	4.5	4.2	4.5	4.8	5.1	5.4	5.7	6.0	6.4	6.8				
64.0			4.9	4.7	4.4	4.1	4.4	4.7	5.0	5.2	5.6	5.9	6.3	6.7				
80.0			4.5	4.2	3.8	3.9	4.1	4.3	4.6	4.9	5.2	5.5	5.8					
100.0		4.3	4.0	3.7	3.4	3.6	3.8	4.0	4.3	4.5	4.8	5.0	5.3					
200.0	3.4	3.1	2.6	2.6	2.8	2.9	3.1	3.2	3.4	3.6	3.7							

Figure IX

67

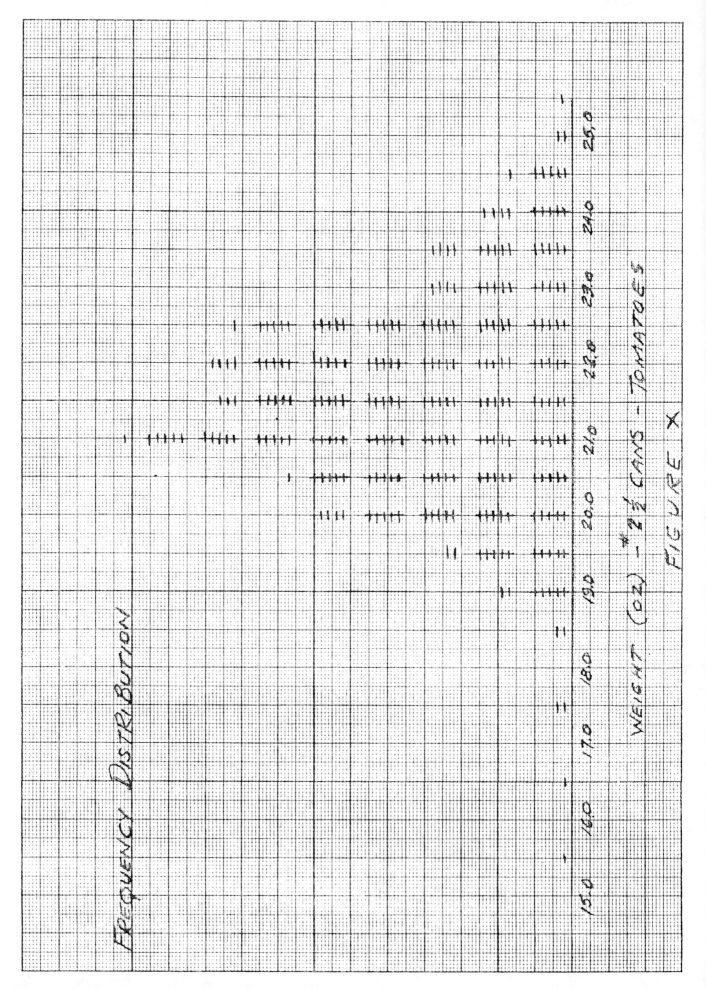

FREQUENCY DISTRIBUTION

WEIGHT (oz) - #2½ CANS - TOMATOES

FIGURE X

15.0 16.0 17.0 18.0 19.0 20.0 21.0 22.0 23.0 24.0 25.0

APPROXIMATE AREAS
FOR A
TYPICAL NORMAL CURVE
(STANDARDIZED)

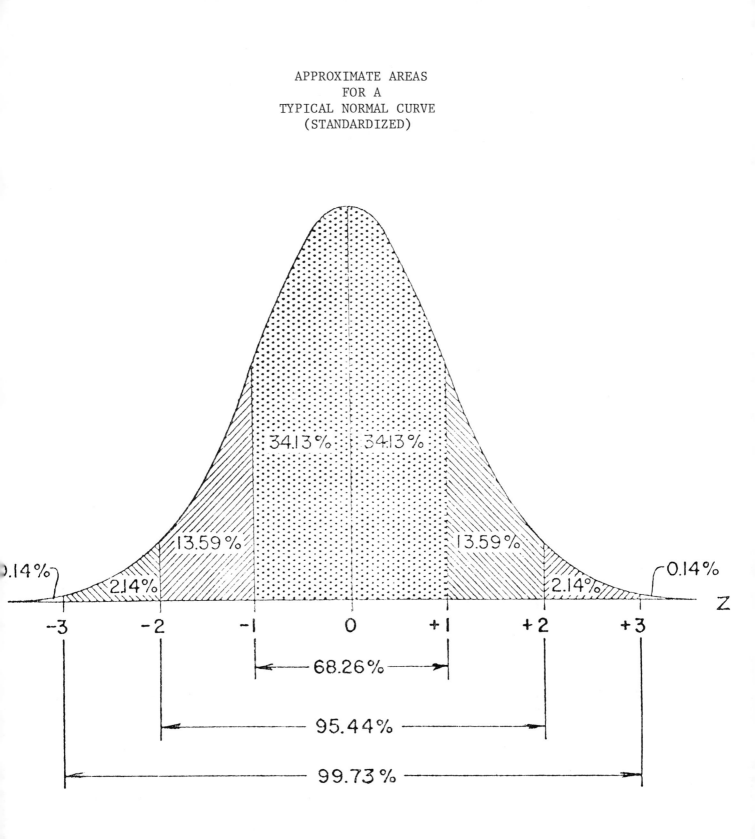

Figure XI

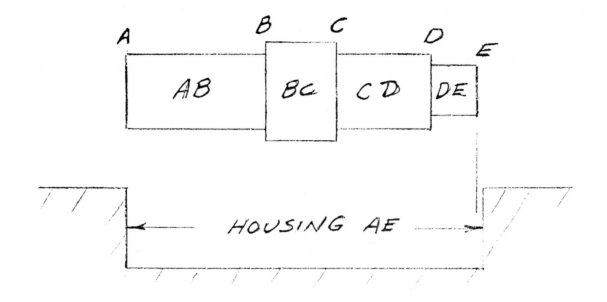

SPECIFICATIONS FROM ENGINEERING
ANALYSIS

PART	SPECS (INCHES)	TOLERANCE RANGE (INCHES)	ESTIMATE FOR σ TOL/6
AB	.750 ± .002	.004	.00067
BC	.320 ± .001	.002	.00033
CD	.475 ± .003	.006	.00100
DE	.100 ± .002	.004	.00067
AE	1.645 ± .008	.016	.0027

Figure XII

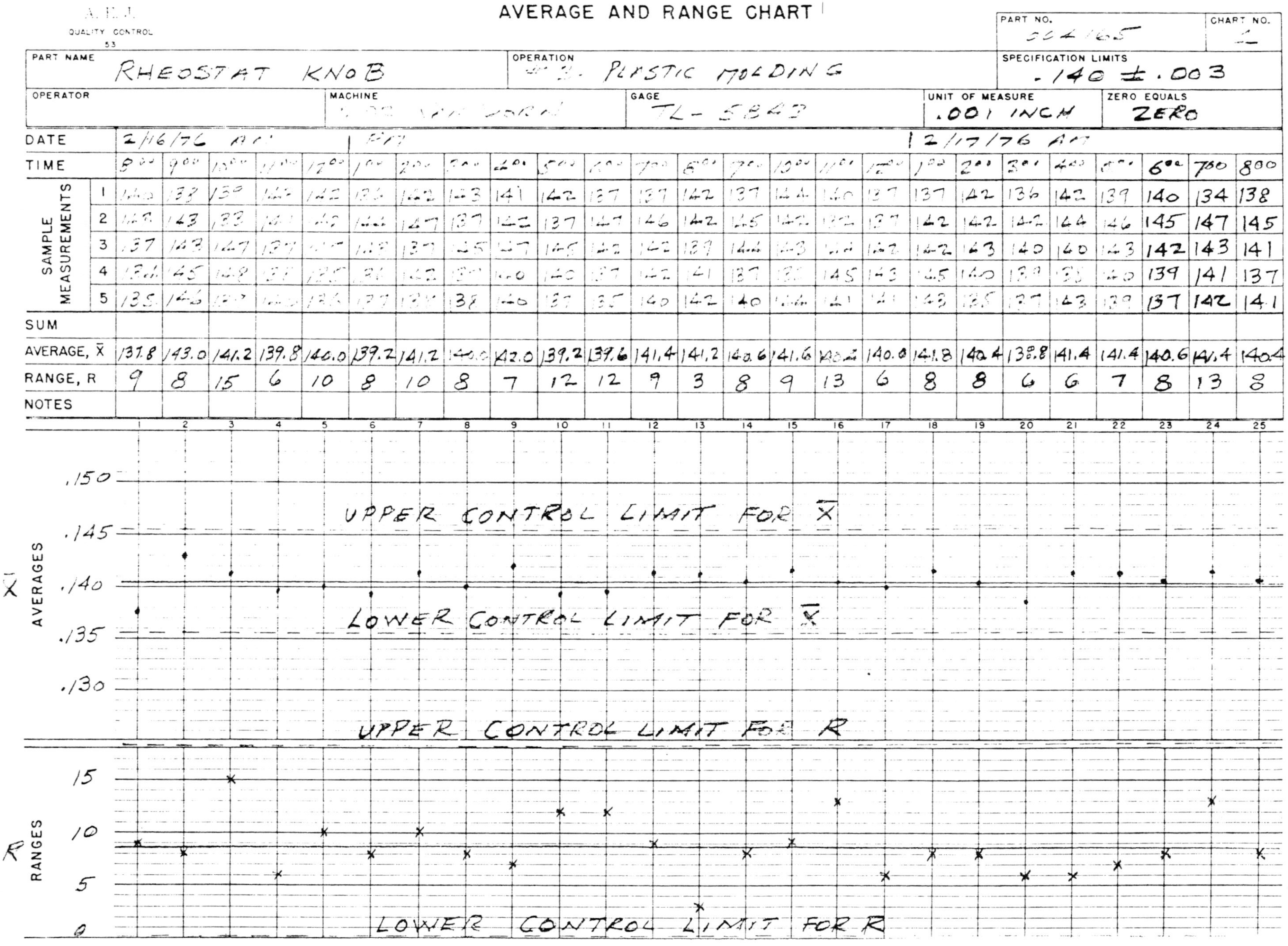

AVERAGE AND RANGE CHART

PART NO.	CHART NO.
PART NAME	OPERATION	SPECIFICATION LIMITS

PART NAME: RHEOSTAT KNOB
OPERATION: PLASTIC MOLDING
SPECIFICATION LIMITS: .140 ± .003

OPERATOR
MACHINE
GAGE: TL-5842
UNIT OF MEASURE: .001 INCH
ZERO EQUALS: ZERO

DATE: 12/17/76

TIME	8⁰⁰	9⁰⁰	10⁰⁰	11⁰⁰	12⁰⁰	1⁰⁰	2⁰⁰	3⁰⁰	4⁰⁰	5⁰⁰	6⁰⁰	7⁰⁰	8⁰⁰
1		137	135									140	134	138
2		143										145	147	145
3		137										142	143	141
4		145										139	141	137
5		135										137	142	141

| SUM | 131.8 | 143.0 | | | | | | | | | | 140.6 | 141.4 | |

| AVERAGE, X̄ | 131.8 | 143.0 | 141.2 | 139.8 | 140.0 | | 142.0 | 139.2 | 141.4 | 140.6 | 141.4 | 140.6 | 141.4 | 140.6 |
| RANGE, R | 9 | 8 | 15 | 6 | 10 | 8 | 7 | 12 | 9 | 3 | 8 | 7 | 8 | 8 |

NOTES

Chart (Averages)

.150
.145
.140
.135
.130

UPPER CONTROL LIMIT FOR X̄

LOWER CONTROL LIMIT FOR X̄

X̄ — AVERAGES

Chart (Ranges)

15
10
5
0

UPPER CONTROL LIMIT FOR R

LOWER CONTROL LIMIT FOR R

R — RANGES

Figure XIII

71

CALCULATION OF LIMITS

59

CONTROL LIMITS

SUBGROUPS INCLUDED 25 SAMPLE LOTS 2/16/76 - 2/17/76

$\bar{R} = \frac{\Sigma R}{k} = \frac{218}{25}$ = 8.72 ____ =

$\bar{\bar{X}} = \frac{\Sigma \bar{X}}{k} = \frac{3511.4}{25}$ = 140.5 ____ =

OR

$\bar{X}'$ (MIDSPEC. OR STD.) = 140.0 =

$A_2\bar{R}$ = .577 x 8.72 = 5.03 x = ____

$UCL_{\bar{X}} = \bar{\bar{X}} + A_2\bar{R}$ = 140.5 + 5.03 = 145.5

$LCL_{\bar{X}} = \bar{\bar{X}} - A_2\bar{R}$ = 140.5 − 5.03 = 135.5

$UCL_R = D_4\bar{R}$ = 2.114 x 8.72 = x = 18.43

LIMITS FOR INDIVIDUALS

COMPARE WITH SPECIFICATION OR TOLERANCE LIMITS

$\bar{\bar{X}}$ = 140.5

$\frac{3}{d_2}\bar{R} = \frac{3}{2.326}$ x 8.72 = 11.25

$UL_X = \bar{\bar{X}} + \frac{3}{d_2}\bar{R}$ = 151.8

$LL_X = \bar{\bar{X}} - \frac{3}{d_2}\bar{R}$ = 129.2

US = 143.0

LS = 137.0

US − LS = 6.0

$6\sigma = \frac{6}{d_2}\bar{R}$ = 22.5

MODIFIED CONTROL LIMITS FOR AVERAGES

BASED ON SPECIFICATION LIMITS AND PROCESS CAPABILITY. APPLICABLE ONLY IF: US−LS > 6σ.

US = LS =

$A_M\bar{R}$ = x = ____ $A_M\bar{R}$ = ____

$URL_{\bar{X}} = US - A_M\bar{R}$ = $LRL_{\bar{X}} = LS + A_M\bar{R}$ =

FACTORS FOR CONTROL LIMITS

n	A_2	D_4	d_2	$\frac{3}{d_2}$	A_M
2	1.880	3.268	1.128	2.659	0.779
3	1.023	2.574	1.693	1.772	0.749
4	0.729	2.282	2.059	1.457	0.728
5	0.577	2.114	2.326	1.290	0.713
6	0.483	2.004	2.534	1.184	0.701

NOTES

Figure XIV

SCREW-MACHINE CAPABILITIES

(In thousandths of an inch)

Machine and operation	Material			
	Soft brass	Steel, free machining, and alloy	Stainless steel	Phosphor bronze
#00G —Brown & Sharpe:				
D—Turn	0.4–0.7	0.5–1.5	1.0*	1.0*
D—Form	0.5–1.0	0.7–1.7	0.8–1.5	0.6–1.5*
D—Burnish	......	......	0.3*	
D—Thread roll	2.0*	3.0*		
L—Form	0.5–1.5*	2.0*		
L—Cut off	0.5–1.5	1.0–2.0	1.0–2.5	2.0*
#0G —Brown & Sharpe:				
D—Turn	1.0*	1.0*	1.0*	1.2*
D—Form	0.8–1.5	0.8–1.5	1.0–1.5*	1.3–2.0*
L—Form	0.5*		0.7*	
L—Cut off	0.8–1.0		1.5*	
L—Counterbore	1.0–2.0			
#2G —Brown & Sharpe:				
D—Turn	1.0–1.2*	1.5*	1.5–2.0*	1.8*
D—Form	0.8–1.5	2.2*	2.5*	2.5*
L—Form	0.5*		0.7*	
L—Cut off	0.5–1.0		1.5*	
9/16-in. Gridley:				
D—Turn	......	1.0–4.0*		
D—Form	......	1.0–4.0*		
1-in. Gridley:				
D—Turn	......	1.5–5.0*		
D—Form	......	1.0–5.0*		
1¼-in. Gridley:				
D—Turn	......	2.0–6.0*		
D—Form				
1 5/8-in. Gridley:				
D—Turn	......	2.0–6.0*		
D—Form	......	2.0–7.0*		
2 5/8-in. Gridley:				
D—Turn	......	5.0–8.0*		
D—Form	......	2.5–7.0*		
2-in. Greenlee:				
D—Form	......	1.5*		
#F & G—National Acme:				
D—Form	......	5.3*		
D—Form (internal)	......	3.2*		
L—Cut off	......	6.1*		
#RB–6—National Acme:				
D—Form	......	3.5*		
D—Form (internal)	......	5.2*		
L—Cut off	......	2.6*		

* Based on data from a single contributor. May not be representative.

Figure XV

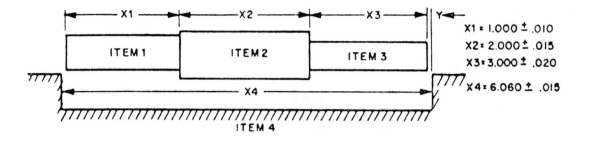

Schematic for Tolerance Stack-Up Calculations

1.000 ± 0.10 X1, item 1   6.060 ± 0.15 X4, item 4
2.000 ± 0.15 X2, item 2   6.000 ± 0.45 X1 + X2 + X3, Assembly
3.000 ± 0.20 X3, item 3   .060 ± .060 Y, clearance
6.000 ± 0.45 X1 + X2 + X3,

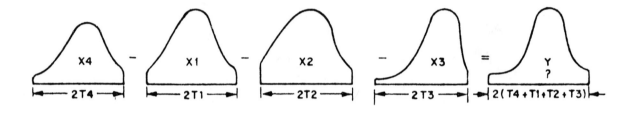

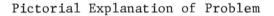

Pictorial Explanation of Problem

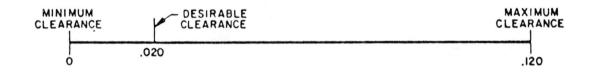

Minimum, Maximum and Desirable Clearance

Figure XVI

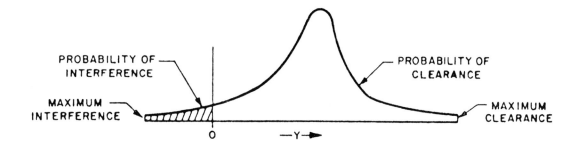

Frequency Distribution of Y

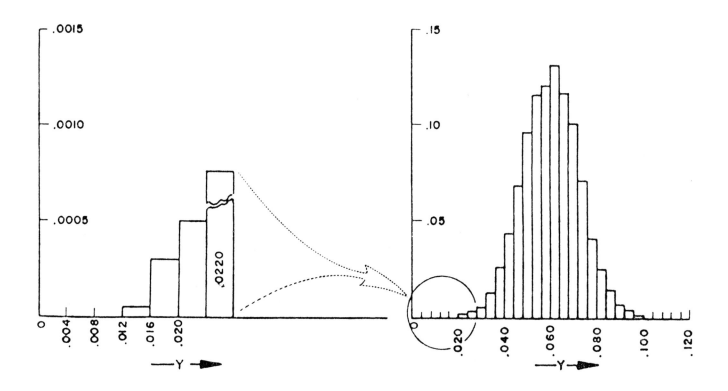

Histogram of Clearance Developed from Computer Simulation

Figure XVII

ENGINEER'S INPUT

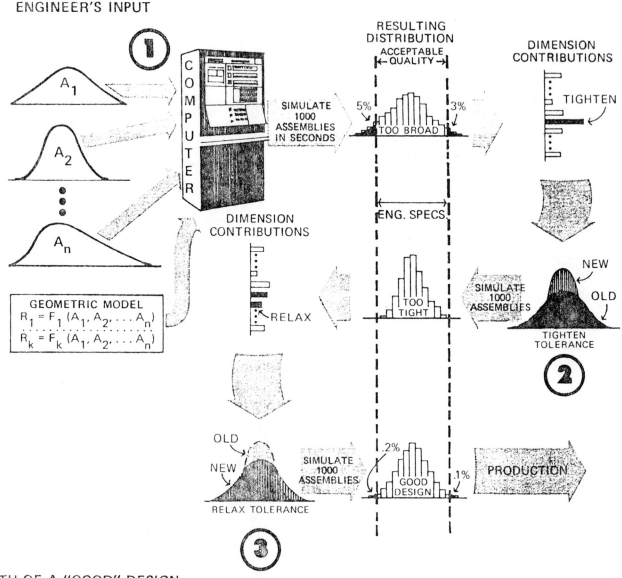

BIRTH OF A "GOOD" DESIGN

Providing the distribution patterns for component dimensions and a mathematical model (1), the engineer instructs the computer to simulate assembled units. Results are compared with acceptable quality limits. If a resultant distribution is too broad, he must tighten some tolerances (2). If the distribution is too narrow, he may relax some (3).

To know which tolerances to tighten or relax, the engineer examines the relative contributions of all individual dimensions to the variation of the resultant distribution. From this, he selects and adjusts the most critical tolerances to obtain the desired result. Tightening and relaxing continue until a good production design is born.

Figure XVIII

Prepared by Ernie Mazzatenta
Art by Dick Berube

SEARCH is published by the Technical Information Department, General Motors Research Laboratories, Warren, Michigan 48090.

Reprinted from *Tooling & Production*, February 1981

Tolerances Should Be Functional Not Impossible

By Edward Roth
Productivity Consultant

Some of the reasons given for decling productivity in the US deal with broad-scope problems that are beyond the control of any one company. These include factors such as lack of capital investment, an old-machine-tool base, inflation, less research and development funding, poor management, the 1974 depression, excessive governmental regulations etc. However, there are two productivity-improvement areas that can be addressed by any given manufacturing company.

One of these is inadequate product design, which has occurred over the last 10 or 15 years. Lack of formal training has caused mechanical engineers to tighten tolerances to make sure a product will work. In doing this they use unjustified tolerances and may end up with a product that cannot be built. I have noticed this tendency in my consulting work in plants.

A by-product of inadequate design is that datums are implied on drawings. A result of this is that the point of origin of all dimensions and therefore the point of origin of all measurements to be made is not specified.

The second problem is the retirement of personnel with expertise in surface plate inspection, tool and gage design, and tool and gage maintenance. Very few companies have their own tool and gage designers and toolmakers...they have to go elsewhere, which is a very uncomfortable thing.

The aforementioned reasons have far-reaching consequences on our productivity growth. Because of tight tolerances there are higher quotations being placed on products. Because we don't have the gage designers, many parts are not being gaged.

The Effects of Retirement

Evidence of poor tool design are apparent because the tools are too complex.

It's difficult to document this here, but they are much more complex than they need to be.

More coordinate measuring machines are being used as the acceptance criteria. In fact, I predict that in 10 years, there will be a substantial increase in their use. The problem with this is that coordinate measuring machines are more restrictive than functional gages in their acceptance criteria. Many don't allow parts to rotate and translate as they could when they're being assembled to their mating parts.

Engineers should be required to justify all tolerances tighter than $\pm 0.015"$. To do this, you must first make a detailed layout of the variations of size on parts and the operational characteristics that could occur when the various features that make up an assembly do vary at these two limits. If you do, therefore, need to have tolerances closer than $\pm 0.015"$, you should be able to justify it by this precise tolerance study. These are called design layouts. Designers should be taught to use design layout systems to establish limits of size, limits of locational tolerances and also to justify all tolerances tighter than $\pm 0.015"$ on a part.

We should also attempt to get design courses re-established in our educational institutions. For some reason, machine design and functional design courses are not required of graduating mechanical engineers, and the companies cannot provide them because of their experienced people are retiring. So the engineer is creating a tremendous problem in industry by designing products that, essentially, can't be built.

Apprenticeship programs should be re-established and all people that operate coordinate measuring machines and perform surface plate inspection, as well as tool and gage design, should have an apprenticeship program similar to those that were prevalent in this country 15 or 20 years ago.

As you know, coordinate-measuring machine operators now can be pulled off the street and taught to operate the machine in about 12 hours. This is attractive from a management standpoint but it's detrimental from a product acceptance standpoint. The retiring cadre of professionals, whether they be mechanical designers, tool and gage designers or toolmakers, should have the opportunity provided by their companies to write papers and record the state of the art, so that when they retire that knowledge does not go with them.

Process capability studies should be done on every machine tool at least once a year to assure that the coordinate measuring machine is maintaining its accuracy. We should also attempt to design more functional gages for the acceptance of products.

Last, I should like to see people in tooling and production stiffen their backs and throw it back over the wall if they can't build the product. When they get requirements from young mechanical designers that can't be achieved, or can't be understood, or with tolerances that are tighter than are required, they should take the design back and say, "Hey, tell me, why do you have to have these tolerances" or "justify these tolerances to me." Then, if the engineer is at a loss, help him to establish manufacturable requirements on the product line.

Presented at the SME Southeastern Conference, December 1976

The Role of the Independent Metrology Lab

By William C. Ledford
Ledford Machine & Gage Laboratory, Inc.

To appreciate the need for exacting methods, procedures and equipment for verifying the various dimensional elements of our gages, we need only note the requirements of MIL-C-45662 "Calibration Systems Requirements" supplimented by HDBK-52, MIL-STD 120 and similar quality documents. These requirements demand many procedures, training and equipment not practical to many medium and small manufacturing facilities.

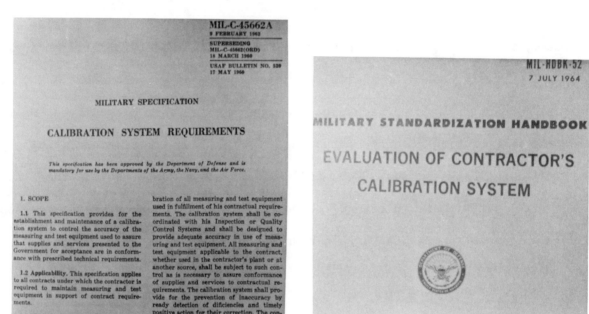

MIL-C-45662A
9 FEBRUARY 1962

SUPERSEDING
MIL-C-45662(ORD)
18 MARCH 1960

USAF BULLETIN NO. 520
17 MAY 1960

MILITARY SPECIFICATION

CALIBRATION SYSTEM REQUIREMENTS

This specification has been approved by the Department of Defense and is mandatory for use by the Departments of the Army, the Navy, and the Air Force.

1. SCOPE

1.1 This specification provides for the establishment and maintenance of a calibration system to control the accuracy of the measuring and test equipment used to assure that supplies and services presented to the Government for acceptance are in conformance with prescribed technical requirements.

1.2 **Applicability.** This specification applies to all contracts under which the contractor is required to maintain measuring and test equipment in support of contract requirements.

1.3 **Significance.** This specification and any procedure or document executed in implementation thereof shall be in addition to and not in derogation of other contract requirements.

2. APPLICABLE DOCUMENTS (None.)

3. REQUIREMENTS

3.1 **General.** The contractor shall establish or adapt and maintain a system for the calibration of all measuring and test equipment used in fulfillment of his contractual requirements. The calibration system shall be coordinated with his Inspection or Quality Control Systems and shall be designed to provide adequate accuracy in use of measuring and test equipment. All measuring and test equipment applicable to the contract, whether used in the contractor's plant or at another source, shall be subject to such control as is necessary to assure conformance of supplies and services to contractual requirements. The calibration system shall provide for the prevention of inaccuracy by ready detection of dificiencies and timely positive action for their correction. The contractor shall make objective evidence of accuracy conformance readily available to the Government representative.

3.2 **Description.** The contractor shall provide and maintain a written description of his calibration system covering measuring and test equipment and measurement standards. The portion dealing with measuring and test equipment shall prescribe calibration intervals and sources and may be main-

FSC MISC

Figure 1

MIL-HDBK-52
7 JULY 1964

MILITARY STANDARDIZATION HANDBOOK

EVALUATION OF CONTRACTOR'S CALIBRATION SYSTEM

Figure 2

The existing and on coming Consumer laws, the need for exacting record keeping of all inprocess gaging, military contract requirements, and state laws forcing measurements that are traceable to the National Bureau of Standards have overloaded many of the large product producing laboratories (captive labs). We need only send some gages in for calibration and experience the long return times to understand this problem.

The above noted reasons have brought into being the Independent Gage Laboratory. These facilities specialize in the specific service of gage calibration. The independent gage labs break down into these types of facilities;

1. Mechanical Standards
2. Electrical-Electronic
3. Combination of the above

The independent lab is designed to make available to manufacturing facilities to small to justify the cost of a fully equiped metrology lab. And because of the number of factories within the U.S. the National Bureau of Standards cannot calibrate and issue a traceable number to all. The cost of NBS calibration and time forced manufacturers to use gage manufacturers in the past. Another reason for the independent lab is engineering and technical information not available to the company without the full quality engineering department.

The independent is designed to furnish to the end user;

1. Traceability of Measurement
2. Proper environment
3. Appropriate measuring equipment
4. Standard measuring procedures
    consistent to the state of the art

Traceability is intended to assure that the accuracy of an instrument, process or measurement of length has as its base the national standard from the National Bureau of Standards.

MIL-C-45662 states under "Adequacy of Standards",
      Standards established for calibrating the
      measuring equipment used in controlling
      product quality shall have the capabilities
      for accuracy, stability and range required
      for the intended use.

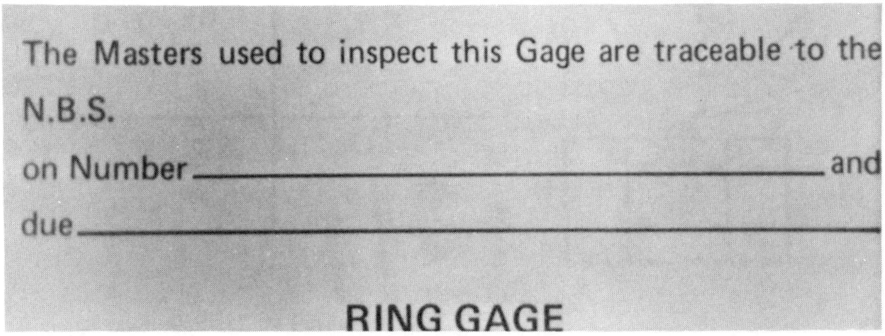

The Masters used to inspect this Gage are traceable to the

N.B.S.

on Number_____and

due_____

RING GAGE

Figure 3

Environmental controls for measuring and test equipment and gages shall be calibrated in "an environment controlled to the extent necessary to assure continued measurements of required accuracy giving due consideration to temperature, humidity, vibration, cleanliness and other controllable factors affecting precision measurement".

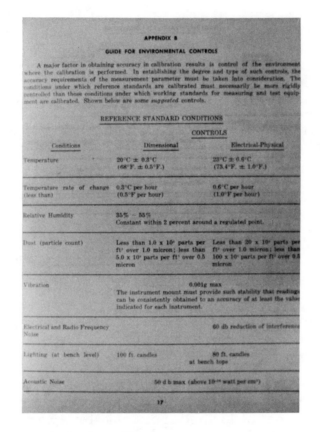

Figure 4

Measuring equipment varies within metrology labs, however certain basics are required.

1. Gage Blocks Grade 0.5 (AAA) ($\pm$.000001)
2. Gage Blocks Inspection (AA) ($\pm$.000002)
3. Gage Block Measuring Machines (.000001)
4. Basic Measuring Machine (.00001)
5. Thread Measuring Wires (set)
6. Gear Measuring Wires   (set)
7. Internal Measuring Machine (.000001)
8. Optical Projector

9. Lead testing Machine
10. Master Thread Setting Plug Gages
11. Electronic Amplifiers
12. Surface Plates (lab)
13. Indicator Tester
14. Master Square
15. Controlled Room
16. Digital Height Gages
17. Coordinate Measuring Machine
18. Other Support, equipment
19. Trained personel

The lab will maintain a set of standard procedurers showing the equipment required to calibrate, control requirements, a step by step procedure on how to make the actual calibration and disposition of gages.

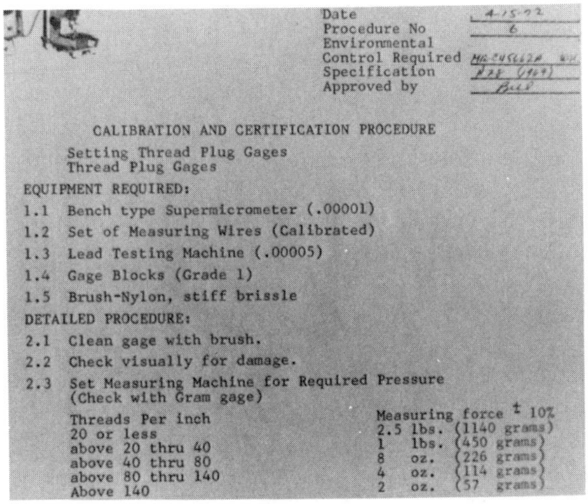

Figure 5

Most labs will assist customers in setting up a inhouse Metrology department. Because of the very broad experience of the independent, they are usually prepared to offer a very precise plan to the manufacturer doing job shop work. Most job shop labs are designed to calibrate the "common use" gage, leaving the masters to outside calibration. The labs will generally be clasified secondary standards labs, requiring less equipment and control than primary labs.

The independent lab is normally restricted to telling the customer the size and fuctional condition of his gage. Because of the varied specifications, Federal, Company, ISO, Society and others the independent really has no way of knowing your specification. The independent will in most all cases use the Federal Standards, IE: GGG-15, H-28, etc. For these reasons the independent will not "reject" the gage. This becomes the function of the user. This procedure also serves another purpose in the user will need to review the test reports furnished by the lab. Of course the customer can furnish to the lab a listing or copy of his spec's, where-by the lab can then make a determination of acceptance or rejection.

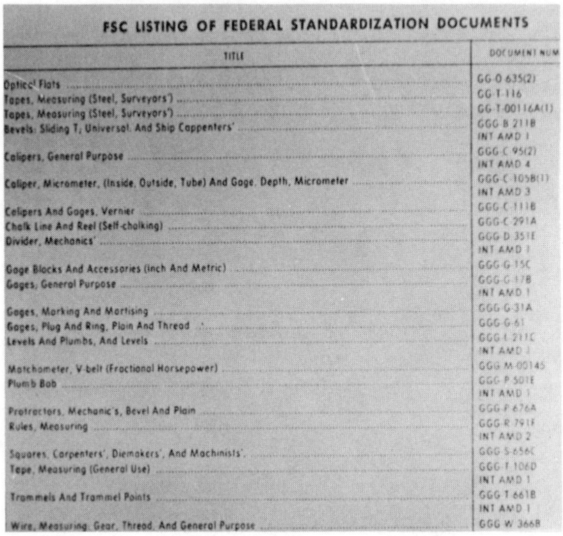

FSC LISTING OF FEDERAL STANDARDIZATION DOCUMENTS

TITLE	DOCUMENT NUM
Opticc' Flats	GG-O-635(2)
Tapes, Measuring (Steel, Surveyors')	GG-T-116
Tapes, Measuring (Steel, Surveyors')	GG-T-00116A(1)
Bevels: Sliding T, Universal, And Ship Cappenters'	GGG-B-211B
	INT AMD 1
Calipers, General Purpose	GGG-C-95(2)
	INT AMD 4
Caliper, Micrometer, (Inside, Outside, Tube) And Gage, Depth, Micrometer	GGG-C-105B(1)
	INT AMD 3
Calipers And Gages, Vernier	GGG-C-111B
Chalk Line And Reel (Self-chalking)	GGG-C-291A
Divider, Mechanics'	GGG-D-351E
	INT AMD 1
Gage Blocks And Accessories (Inch And Metric)	GGG-G-15C
Gages, General Purpose	GGG-G-17B
	INT AMD 1
Gages, Marking And Mortising	GGG-G-31A
Gages, Plug And Ring, Plain And Thread	GGG-G-61
Levels And Plumbs, And Levels	GGG-L-211C
	INT AMD 1
Matchometer, V-belt (Fractional Horsepower)	GGG-M-00145
Plumb Bob	GGG-P-501E
	INT AMD 1
Protractors, Mechanic's, Bevel And Plain	GGG-P-676A
Rules, Measuring	GGG-R-791F
	INT AMD 2
Squares, Carpenters', Diemakers', And Machinists'	GGG-S-656C
Tape, Measuring (General Use)	GGG-T-106D
	INT AMD 1
Trammels And Trammel Points	GGG-T-661B
	INT AMD 1
Wire, Measuring, Gear, Thread, And General Purpose	GGG-W-366B

Figure 6

The lab will furnish a test report for each gage calibrated showing measurable or visual condition on the gage. These reports should tell the user the actual measurement or master size (thread ring gages). The report will also reflect temperature, masters used and/or traceability number with due date of masters used. (See figure 7)

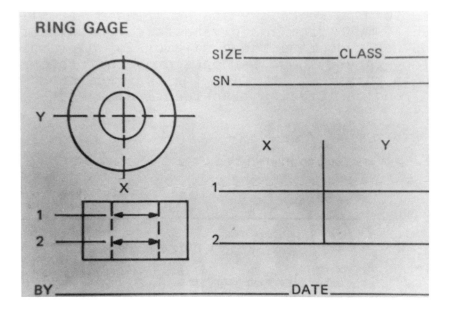

Figure 7

The services provided by the lab will be reflected in its printed brochure. The lab will provide calibration to any standard gage ie: gage blocks, thread gages UNF-UNC, thread wires, indicators, plug gages etc. Also most labs have masters for UNEF series threads. The lab is generally equiped to do coordinate measuring, plate measuring and other types. Some labs furnish laser interferometer measuring to calibrate NC machines and coordinate measuring machines. Many labs also do optical measuring machine calibration. All the services available cannot be listed in any single paper and the user must shop for or be available to the salesman to find all the services available.

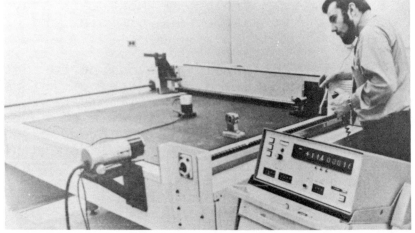

Figure 8

Some independent labs furnish the "Mobil Laboratory" designed to conform to secondary status. These facilities are designed to move from factory to factory and perform calibrations not requiring the conditions of a primary standards lab. A very high percentage of calibration can be finished in this facility allowing the gage to be available within good calibration practice time.

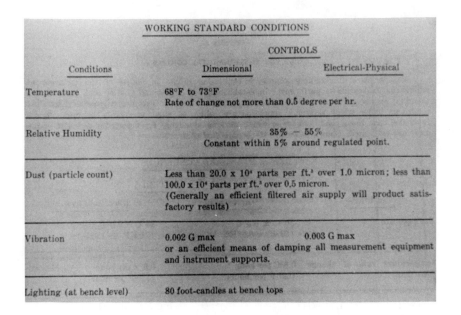

Figure 9

Because of the high degree of specalization of the independent lab the personel are highly trained engineers and technicians. Several machine tech programs in various schools provide basic training in Metrology. Most labs have in-house training programs to further the training and expertese of technicians.

The independent lab will only make replacement to defective gages on specific request, as they are not in business to sell gages. Normally the customer will make replacement from his local supplier. The lab can also do product testing with a total detachment as to product name. The Federal Government now uses several independent metrology labs for testing accuracy and wear life of measuring and test equipment.

In conclusion the independent lab is designed to furnish to the small manufacturer a service to assure traceability of their gages, maintain accuracy of measuring equipment used in controlling product quality. By using the independent lab the manufacturer unable to expend the full cost of an inhouse Metrology department can hold their cost down. Many small factories are finding this the most practical way to control gages and quality.

Reprinted from *Founding, Welding, Production Engineering*, October 1976

Presented to the S.A. Institute for Production Engineering

The Author

V. R. Burrows.
Technical Manager —
Metrology Division,
Alfred Herbert (South Africa)
(Pty) Ltd.

The Principles and Applications of Pneumatic Gauging

HISTORY

THE basic principle of blowing a jet from a nozzle against the surface of a workpiece to be measured is thought to have been known to scientists of the nineteenth century but the first application of the principle can be traced back through technical and potent literature as far as 1917.

In the application at that time the medium used was a liquid and it was not until about ten years later that a system using compressed air for engineering measuring systems was developed by a company in France. The first well-known application was in the standardising of jets for carburettors.

The pressing military demands of the 1940's gave increased impetus to the development of air gauging and by 1948 several systems were available. The United States industry was quick to appreciate the benefits which air gauging instruments had to offer and they were put to extensive use particularly in the automobile industry.

At the metrology division of the National Physical Laboratory in England study of the pneumatic method of gauging began during the 1939/45 war and subsequent theoretical and practical investigations were undertaken to obtain basic design data. Notes written by Evans, Graneek and Morgan have been published by the N.P.L. and are considered to be classic works on the subject.

Principles

By controlling the supply of compressed air to a system either the flow or pressure characteristics within the system can be interpreted to determine the dimensional relationship between a nozzle and the surface of a workpiece.

Simple flow or pressure indicators can be used to reliably display the dimensional relationship.

SIMPLE FLOW MEASURING CIRCUIT

A simple flow responsive system is shown diagramatically in Fig. 1a. In this system, compressed air at a constant closely controlled pressure is passed through a variable flow meter (i.e. a float in a tapered glass tube) and then to a suitable nozzle. The air from the nozzle impinges on the surface placed in front of the nozzle and if the surface

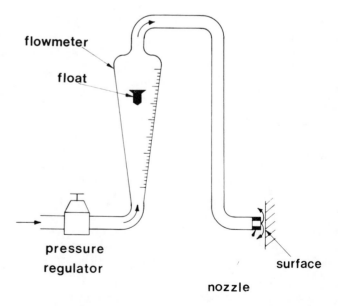

Fig. 1a – Simple Flow System

is moved towards or away from the nozzle the flow of air alters and the float in the tube moves accordingly.

Calibration can be carried out by applying known displacements to the surface and the graph shown in Fig. 1b shows a typical curve shape, relating air flow to the clearance between the nozzle and the surface.

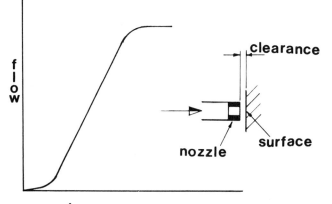

Fig. 1b – Flow/Clearance Curve

Simple Back Pressure Circuit

Fig. 2a shows a simple back pressure circuit. In this circuit, compressed air from the pressure regulator passes through a primary restrictor, commonly called the control orifice, before entering the cavity upstream of the escape orifice. Due to the presence of the primary restriction, changes in the restrictive effect of the escape orifice will give rise to changes in pressure in the cavity between the two orifices. Changes in the restrictive effect of the escape orifice are made by moving the surface towards or away from the nozzle face. Changes in pressure inside the cavity can be measured by a pressure indicator such as a Bourdon tube type pressure gauge. Again, calibration of such a system can be carried out by applying known displacements to the surface, and Fig. 2b shows a typical curve shape relating pressure in the cavity to the clearance between the nozzle face and the surface.

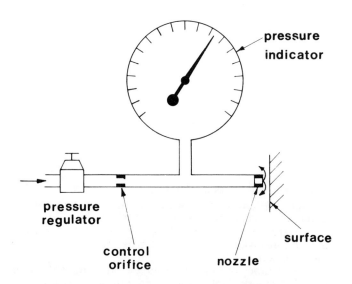

Fig. 2a – Simple Back Pressure Circuit

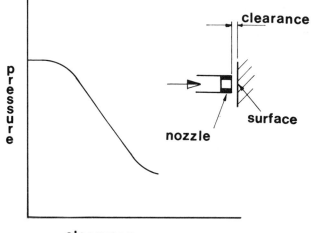

Fig. 2b – Pressure/Clearance Curve

Back Pressure Circuit Experimental Rig

An experimental rig with which a back pressure circuit can be studied is shown in Fig. 3a and the pressure/clearance curves obtained using a 0,55 mm dia. control orifice and nozzles of 1 mm, 1,5 mm and 2,0 mm are shown in Fig. 3b.

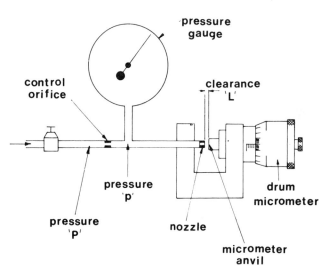

Fig. 3a – Back Pressure Experimental Rig

It will be clear that when the micrometer anvil is in contact with and sealing the nozzles, the pressure indicated on the gauge will be equal to the pressure at which the supply is regulated. Thus, the three curves have the same point of origin on the 'p' axis.

Other obvious conclusions which can be drawn from the three curves are:-

(a) The curves are not the same but are similar in shape.
(b) The range of measurement is fairly small, say, 0,125 mm in the case of the 1,0 mm dia. nozzle. (By means of indirect pneumatic gauging devices it is practical to achieve greatly increased measuring ranges. These will be discussed later)

Area of the Escape Orifice

Before any further theoretical study of air gauge circuits

is carried out, the area of the escape orifice should be considered as follows:-

When the clearance between the surface and the nozzle face is zero no air escapes from the nozzle and the area of the escape orifice is zero.

When the clearance between the surface and the nozzle face is very large, the area of the escape orifice is $\frac{\Pi D^2}{4}$ where 'D' is the diameter of the nozzle.

Between these extremes, especially where the clearance is small and where air gauging can be employed the area of the escape orifice is $\Pi D L$, that is, area of the curved surface of the cylinder shown in Fig. 4.

supply pressure = 280 kN/m²

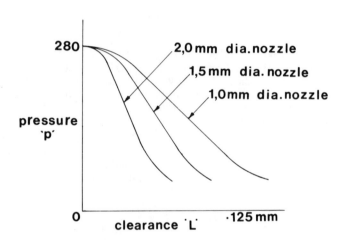

Fig. 3b – Pressure/Clearance Curves

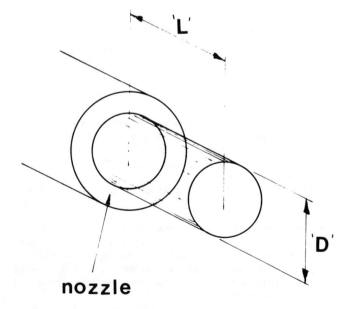

nozzle

Fig. 4 – Area of the Escape Orifice

Linearity and Sensitivity

If the experimental rig shown in Fig. 3a is used for more detailed study it can be shown that between certain limits 'p' is inversely proportional to the area of the escape orifice. In fact, between certain useful limits the pressure/escape orifice area relationship is linear within 1 %. If a lower order of linearity is acceptable, say within 2 %, measuring range is very usefully extended. These are conditions which are met by good commercial airgauge systems.

If the experimental rig is used to obtain figures for plotting pressure against escape orifice area using a number of different sizes of control orifice, it will be found that sensitivity increases as the diameter of the control orifice decreases i.e. for small control orifices the change in pressure is greater for a given change in escape orifice area. This is important since some commercial systems incorporate a variable control orifice the use of which enables sensitivity (magnification) to be set precisely.

Datum Control

If in the experimental rig, means is provided to change the pressure in the cavity using a variable bleed to atmosphere a datum or zero control is provided which varies the pressure 'p' when the escape orifice area remains fixed. This addition to a circuit provides means of accommodating the small differences which inevitably occur in the manufacture of gauge heads. Limited use of a datum control in the form of a bleed to atmosphere has an insignificant effect on linearity.

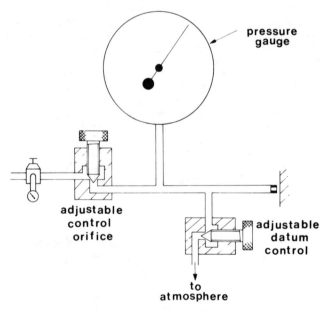

Fig. 5 – Practical Back Pressure Circuit

Practical Back Pressure Circuits

Back pressure circuits which are used in two popular airgauge units are shown in Figs. 5 and 6.

The circuit shown in Fig. 5 has a variable control orifice to adjust sensitivity and a variable bleed to atmosphere for datum setting.

The accuracy of systems which use the circuit shown in Fig. 5 depends highly upon the pressure regulator to maintain the supply pressure within very close pressure limits — the pressure regulator is a critical component in these circuits.

The circuit shown in Fig. 6 is a differential back pressure circuit in which accuracy is maintained regardless of some variation in the regulator performance which controls supply pressure.

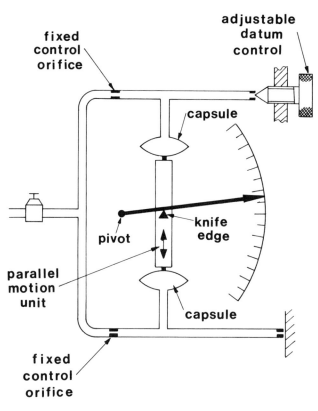

Fig. 6 – Differential Back Pressure Circuit

Practical Flow Responsive System

A practical flow responsive system which is in common use is shown in Fig. 7. This also employs variable orifices for sensitivity and datum control.

Air Gauge Unit Magnifications

The typical length of airgauge unit scales is 250 mm and these are commonly arranged to make the scale length represent between 0,10 mm and 0,010 mm giving reliable magnification ratios of between 2 500:1 and 25 000:1. Magnifications up to 100 000:1 are practical and available.

Applications

In addition to simple dimensional measurement, air gauging was developed to an extent where it was used to control precise and complex machining and other processes but from about 1960 onwards the development of electronic gauging had reached a stage where it began to seriously challenge air gauging.

Now, many applications for air gauging have been taken over by electronic gauging but some forms of measurement are still ideally performed pneumatically and it is these that we can usefully consider.

The Measuring Head

So far only air gauge circuits and indicators have been considered but it is the measuring head, by means of which the variable being measured is made to control the air flow,

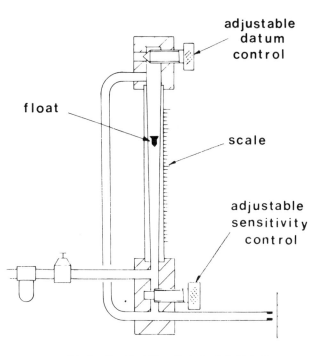

Fig. 7 – Practical Flow Responsive Circuit

which adapts the system to any specific problem of measurement.

Air Plug Gauges

The simplest form of airgauging head in common use is the air plug gauge. Many thousands of air plug gauges are in daily use providing industry with an economical and easily used facility for the precise comparative measurement of internal diameters.

The conventional open jet air plug gauge is an essentially simple device. Compressed air from the air gauge indicating unit is passed to the plug gauge and is allowed to issue from two or more jets in the periphery. The design of a two jet air plug is shown in Figure 8a.

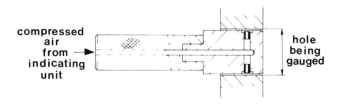

Fig. 8a – The Twin Jet Air Plug Gauge

When the air plug gauge is inserted into a hole, the air escaping from the jets is limited by the clearance between the jet faces and the hole. The small changes in clearance, arising when the air plug gauge is inserted in successive holes of differing sizes, produce changes in the flow rate or back pressure in the circuit.

The magnification and datum setting of systems with variable control orifices and zero bleeds is carried out with master holes, i.e. two setting rings, the diameters of their bores being known to a high degree of accuracy.

Only one ring master is required for datum setting in systems where the control orifice size is fixed.

Air plug gauges as small as 1 mm diameter can be produced at a price. A disadvantage of these is that their small jets provide only a small measuring range, say 0,010 mm.

Jet Recession

If we refer back to Fig. 3b it will be seen that when the surface being measured is very close to the nozzle face, equal increments of change in clearance do not produce equal increments of pressure change. The system is not linear under these conditions.

Because of this the faces of the jets on air plug gauges are ground below the body diameter of the plug as shown in Fig. 8b. This grinding back is called jet recession and it is the means by which the non linear portion at the high pressure (low flow) end of the pressure/clearance curves is avoided.

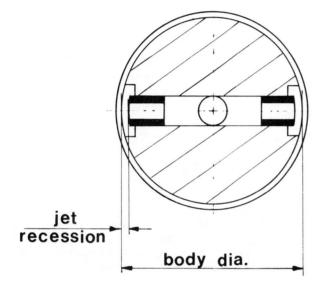

Fig. 8b – The Twin Jet Air Plug Gauge

Advantages and Disadvantages of Air Plug Gauges

Air plug gauges have the following advantages:-

(i)   They are of simple construction (no moving parts)
(ii) Their cost is relatively low.

They have the following advantages over solid GO/NOT GO plug gauges:-

(iii) No operator skill or judgement is necessary in their use since they are used in conjunction with an indicating instrument and any user will obtain the same indicated reading for a given size of hole.
(iv) Some clearance between the air plug gauge body and a hole at the low limit of size reduces wear on the gauge and limited wear has no significant effect on air plug gauge performance.
(v) Errors of form can be detected with an air plug gauge which are not possible with solid plug gauges.

The main disadvantage with air plug gauges is their limited measuring range which is usually restricted to 0,10 mm.
Some errors of form which can be detected with air plug gauges are:-

(i) Taper
(ii) Bell mouthing
(iii) Barreling

(iv) Ovality
(v) Lobing (using multi jet plugs)

Fig. 9 illustrates the use of air plug gauges in detecting these errors.

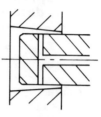

taper

ovality

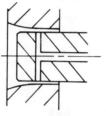

bell mouthing

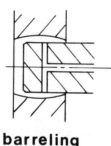

lobing

barreling

Fig. 9 – Errors of Form in Holes

Hole Straightness Gauging

Many modern mechanical assemblies demand that holes should be closely controlled for straightness as well as diameter. An air plug gauge for gauging hole straightness is shown in Fig. 10.

The Effect of Surface Roughness

If air from the jets of an air plug gauge impinges on the surface of a bore which is rough the size measured, is the size approximately mid way between the peaks and valleys of the surface roughness.

For the measurement of bores with rough surfaces, contact type gauging elements are interposed between the jet faces and the bore surfaces and the size of the bore which is functionally important is measured, i.e. the smallest diameter formed by the peaks of the surface roughness.

Contacting elements are also interposed between the jets

and the surfaces where the surfaces being measured are porous.

Indirect Pneumatic Gauging Devices

It has already been said that open jets have the disadvantage of small measuring ranges but air gauge indicators can be used for larger ranges if a gauging cartridge is used. Such cartridges employ a contacting stylus, the inner end of which is tapered and forms the restriction in an escape orifice. The position of the stylus and consequently the position of the taper in the orifice causes changes in the area of the orifice. Changes in the rate of the taper change the measuring range of the cartridge.

If necessary, measuring ranges up to 3 mm can be obtained easily with this type of cartridge.

Other Applications where Air gauging has Advantages

In addition to gauging hole size and geometry, air gauging has other applications where it is still a most suitable medium. For example, the non contact thickness gauging of soft and fragile materials such as plastic film and thin foils.

Another measuring problem for which air gauging has been used for some years is the gauging of the eccentricity of the conical seat and needle guide hole in diesel fuel injection nozzles.

Because the needle guide bore in these nozzles is quite small, in the order of 6 mm diameter, and the very small conical seat is at the bottom of a deep hole a small gauge probe without moving parts is desirable.

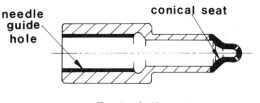

Typical Nozzle

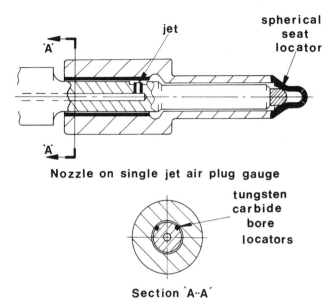

Nozzle on single jet air plug gauge

Section `A--A´

Fig. 11 – Injection Nozzle Gauging

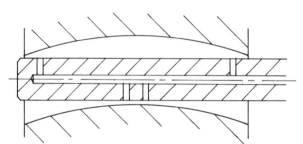

4 jet straightness air plug gauge at high pressure (low flow) position in hole

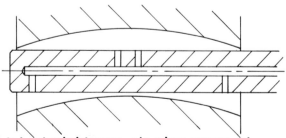

4 jet straightness air plug gauge at low pressure (high flow) position in hole

Fig. 10 – Hole Straightness Gauging

Fig. 11 shows a nozzle and an air plug gauge of special design which solves the problem of gauging seat/bore eccentricity. By locating the conical seat on a polished tungsten carbide sphere and the outer end of the bore on

polished tungsten carbide inserts in the air plug gauge, the eccentricity between seat and bore is detected by changes in clearance at the jet when the injector nozzle is rotated on the air plug. (The nozzle is usually rotated and biased onto the location points by a motor driven rubber roller lowered to contact the large outside diameter of the nozzle).

THE USE OF AIR/ELECTRONICS

Open jet, non-contact gauging has many advantages and a combination of pneumatic and electronic gauging systems is used to provide digital displays and switching for machine control.

The following examples of air/electronic gauging are in use.

Bore Grading:- The clearance between fuel injector nozzle bore and needles is limited to approximately 1,0 micron but it is uneconomical to hold nozzle bore and needle diameter tolerances closer than approximately 0,015 mm. This means that nozzle bores and needles have to be graded into 0,5 micron grades and appropriate grades of nozzle bore and needle diameter are assembled together. Because nozzle bores are small, approximately 6 mm diameter, an air plug gauge is suited to the conditions and accuracy of measurement required in their grading.

If a diaphragm is used instead of the pressure gauge in a circuit similar to that shown in Fig. 5, the deflections of the diaphragm will be proportional to hole size and the diaphragm deflection can be sensed by an electronic transducer. A voltage output from an electronic unit associated with the transducer can be used to drive a digital voltmeter and the thirty 0,5 micron grades within the 0,015 mm manufacturing tolerance allowed to produce the bores can be displayed on the digital voltmeter to provide a quick and reliable means of sorting nozzles into thirty grades.

Centreless grinding machine control:- As parts pass from the wheels of a centreless grinder, open jets can be incorporated in the discharge chute as gauging elements which do not impede the flow of parts from the wheels.

By measuring the deflections of a diaphragm in an air gauge circuit with an electronic transducer and electronic gauge the centreless grinding operation can be controlled by switched outputs from the electronic gauge.

If the grinding machine is not designed for automatic size control, the flow of parts into the machine from a feeder can be stopped in the event of the upper or lower limits of size being approached. If the machine is designed for automatic size control the control wheel can be moved in the appropriate direction when the upper or lower limits of size are approached.

Small spaces between the parts in the chute, chamfers on the ends of parts and undercuts in the parts will produce undersize signals but a timer can be incorporated in the low limit switching circuit which will prevent control action unless the undersize condition persists for some preset time.

Honing Machine Control:- Non-contacting open jets can be machined as an integral part of a honing machine mandrel and these can be fed with air from an air gauge circuit.

By sensing diaphragm deflection with an electronic transducer and electronic gauge, switched outputs from the gauge can be used to stop the honing operation at a predetermined hole size regardless of honing stone wear.

The advantage of non-contact gauging under the abrasive conditions is quite clear.

CONCLUSIONS

Pneumatic gauging is a reliable and precise system of measurement. It served industry well for a long period before electronic gauging became a reliable and economical workshop system.

There are many applications where pneumatic gauging is still very suitable.

CHAPTER 2

GAGING SYSTEMS

Economic Incentives for New Gauge Development

By Stephen A. Barre

Systems Research Laboratories, Inc.

ABSTRACT

The United States has traditionally shown great gains in employee product-ivity. In the last few years these gains have dropped off and during some periods productivity has actually declined. One of the reasons for this trend has been a reduction in the rate of application of explosive new tech-nology to the manufacturing environment. In addition to the practical difficulties in applying new technology in manufacturing, other environ-mental conditions contribute to this problem. This paper discusses some of the reasons for this reduction and some remedies from the point of view of the developer and supplier of measurement and control systems.

INTRODUCTION

The application of technology in manufacturing processes has for decades kept the United States ahead of other countries in individual productivity. A combination of factors has caused our increases in productivity to drop in recent years and during some periods we have experienced declining product-ivity. Changes in work rules, attitudes toward work in general, increased government regulation, and other economic factors contribute to the problem. We restrict our attention to the issue of applying advanced technology in the manufacturing process. It should be noted that the discussion presented here reflects the views of the developer of new industrial gauging and control systems and, therefore, includes whatever biases and limited vision this perspective may supply.

There is no question that new technological developments are occurring at a rapid rate. One has only to track the development of computer technology from the first large vacuum tube machines to today's powerful microprocessors to have an appreciation for the dramatic advances that have occurred in the last twenty years. Currently, the advances in computing power seem to be occurring at exponential rates. The potential for powerful computing capability in almost any corner of a manufacturing plant at a very reason-able cost is now a practical reality in terms of available hardware. What is not a practical reality in general is the effective application of the full potential of this power to the manufacturing process. To date, we have largely restricted our use of the microprocessor to simple replacement of more cumbersome computing schemes. The full potential of today's generation of microprocessors has not yet been conceived of by the developer or the user. Whole new approaches to control and manufacturing will arise as we begin to understand how to use the capability available to us in a creative way. And while we have not yet fully used the available power of the current gener-ation, microprocessor suppliers are busily developing even more powerful machines. When the full impact of this technology enters the industrial environment, it will revolutionize how we manufacture.

But computer technology is not the only area in rapid development. Fiber-optics, which has been used most prominently in laboratory and experimental settings, is finding its way into telephone transmission lines and high speed data links. Our own company uses a laser device for non-contact measurement of extrusions as they leave the extruding head. Accurate measurements of diameter while the extrudate is in a non-solid form was not possible before the advent of this type of non-contacting gauge. Everywhere we look one can see our understanding of the physical laws and the translation of those laws into devices occurring at an unprecedented pace. So the technology is there.

All of this rapid development notwithstanding, most manufacturing plants in this country are making little use of the potential available in such advanced capabilities. On the other hand, it is obvious that manufacturers have a great interest in increasing productivity and reducing cost. And if this new technology can meet this objective either by direct improvement in manufacturing processes or by providing better management information, or both, why is it not happening with the kind of enthusiasm we would like to see?

First, the process of bridging the gap between technology and practical application is not straightforward and is always difficult. We will discuss some key factors in this transfer process. Second, there are a number of factors present in today's economic climate which legislate against the creation of new high technology companies and the introduction of new gauging and control systems. These conditions were not present to such a degree in the late 1950's and early 1960's. The difficulty of the transfer process and these economic factors combine to give us a real challenge in our attempts to increase productivity by the application of new technology to the manufacturing process. Some fundamental changes may be required in the way suppliers of such technology and their customers do business, if we are to meet this challenge.

PRACTICAL ASPECTS OF NEW GAUGE DEVELOPMENT

Creating a successful new measurement and control system for any industrial application is a process which is not well understood by either gauge manufacturers or their customers. Indeed, most product developments do not succeed. Frequently they fail, not because it didn't work; but because of a number of complex, market-related factors which, taken together, may defeat even the most brilliant concept. The major factors which influence this process are: inventor understanding of plant needs, the unanticipated complexities often attendant with the innovative technology, the customer's technical strength, traditional buying patterns, and the risk of non-acceptance by user personnel.

Inventor Understanding of Plant Needs

The high-technology inventor frequently has significant difficulties in solving industrial problems. Very few of these technologists are directly involved in manufacturing processes. Scientists and engineers tend to think in terms of the technology they have developed rather than the benefits that technology provides to the customer. Few technologists look at the manufacturing process and think about what is needed for the process. They rather tend to come at it from the other direction, defining certain capabilities of their technology and forcing these capabilities on to the environment. Sometimes this is successful, but often it fails.

Technical specialists frequently fail to understand the subtleties of the environment in which they are dealing. The details of the process known to the operators by experience are not visible to the observer and are documented only in the minds of the plant workers. Careful and concentrated discussion with these people and observation of the operation is important to the development of a successful new gauge or control system. The scientist or engineer often does not speak the language of the process with which he is trying to deal. When we first involved ourselves with the extrusion industry our technical people didn't know the vocabulary, and considerable effort was expended simply learning how to communicate. The most successful developers of solutions to industrial problems will live in the plant and absorb not only the subtleties of the operation, but the work orientation and slang of the plant personnel. Development of this viewpoint is essential in designing equipment that "fits."

A second difficulty often arises when the technical specialist takes too sophisticated an approach and tries to do too much too soon. In gaining the confidence of the manufacturer and knowledge of the process, it is more important to do a useful thing well than to fail in an elegant attempt. In our extrusion line control systems, for example, we began by simply measuring and monitoring the variations in product diameter, comparing them with the traditional methods of diameter control. Through this process, both we and the manufacturer gained confidence in the reliability and accuracy of the gauging system before we added a direct control loop. Even with just the monitor, the individual operators were able to get a significant improvement in diameter control with a resulting reduction in the use of insulation on a wire line. Adding the automated control system gave a further improvement in performance. While not all applications of new technology allow this gradual confidence building process, it should be used whenever possible. In the medical electronics portion of our business where we design and manufacture automated lung testing systems, our approach was to automate the data collection and processing aspect of pulmonary function testing without asking the physician or medical technician to change any testing procedures or analysis technique. This approach probably had a more significant impact on the immediate acceptance of our system than we realized at the time.

Another difficulty often encountered by the scientist or engineer intent on applying his new idea to an industrial setting is that his commitment to his own approach may blind him to the real opportunity--the old NIH (Not Invented Here) Syndrome. For several years, we developed and tried to find an application for a distant indicating optical probe, (DIOP). Using light as the source and a photodetector as the receiver, we foresaw the use of this probe as a piece part measurement device in manufacturing. One of our engineers was working with this device on an assembly line and noticed the use of a laser to gauge the diameter of piece parts. Fortunately, he was sufficiently opportunity-oriented that he was able to see the value of a different concept. We made an agreement with the inventor of this device, worked with him to engineer it into a suitable product and found the application, not in piece part manufacturing, but in extrusion line control. Had our engineer fallen victim to NIH, it might have been several years before this most useful control system would have reached the extrusion industry.

Complex Technology

Most measurement and control problems involve a complex, interactive set of functions to be performed. Full control of an extrusion line does not simply result from a combination of black boxes and guages. There are many parameters of the line which need to be monitored, and control can be exercised through more than one function. To solve such a problem requires a combination of technologies of which the guage is only one part. Engineering of the system will require contributions from specialists in system design, control theory, digital electronics, computer software, optics, mechanics, and ultrasonics. The integration of this system with the extrusion line itself is a very challenging task. It requires very close coordination between the system supplier, plant engineering, and operating people.

The microprocessor offers new opportunities for data processing and control. The low cost and high computing power of these devices will allow redundancy and distributive approaches to be applied to control of plant processes. Individual control systems on each line will not only control the production process but also provide information to plant supervisors for management purposes. These will include alarms when a production line is down or producing bad product as well as production data on the quality and quantity of product produced. The distributive approach, as contrasted with the large central computer will allow individual processes to proceed even if any one of the systems goes down. At the same time, it allows data from these individual systems to be fed to a central point for collection and analysis.

Customer Technical Strength

Most small and medium-sized manufacturing plants and some very large ones have limited engineering staff. Besides tending to be small, they are frequently composed of mechanical and electrical engineers as opposed to electronic, optical or computer-oriented technical specialists. This is, of course, because the staffs were built predominantly to provide technical skills oriented to the process of production rather than to the guages and control systems required for production. This condition, however, presents a challenge not only in communication with the system designer; but also in acceptance of unfamiliar technologies in the plant environment. While this composition of plant engineering groups will change over time, it is unrealistic to expect a dramatic shift.

Customer Traditional Buying Pattern

Our experience in the extrusion industry indicates that the customer traditionally buys some controls and gauging as an incidental part of buying a new extrusion line. As years go by, he subsequently buys independent guages as retrofit items. These independent gauges are virtually always justified on a cost reduction basis. For an established gauging technology, this decision process is straightforward and the buyer usually has a choice of suppliers. A comparison of features and price can be made and a rational decision can be reached. With the introduction of new technology, this decision process is more complex. The potential buyer will doubt whether it will work, and may not know how to cost-justify the purchase.

Risk of Acceptance

The introduction of dramatic new technology into a manufacturing environment often requires changes in management perceptions, philosophies, and in production processes. When the economic benefit is equally dramatic as the technology, these changes are not as hard to induce. In one of our first installations of the laser gauging and control system, we were able to show a 25% reduction in material usage. Such a dramatic impact on the plant's profitability was a persuasive argument for the kind of changes necessary to accommodate the technology. This includes training people, reorganizing the production staff in some instances, and changing the attitudes of the work force. In many situations, however, the benefits of a control system are more subtle, harder to measure, and not so dramatic. In these cases, it is considerably more difficult to gain acceptance of a system which disrupts the status quo.

Another challenge to the introducer of a dramatic new gauging system is that in general, he will not be competing against a comparable system for the same function. Rather the competition is for other uses of the money within the plant. The market development job in the absence of directly competitive alternatives is more time consuming than a direct competition, and more likely to abort somewhere in the process.

Another risk is that the failure on the part of the system supplier and perhaps even plant personnel to understand the subtleties of the manufacturing process may cause an otherwise successful product to fail. One example where we experienced this was on a continuous vulcanization extrusion line in one particular plant. This is a situation in which the wire with the extruded insulation on it goes directly into a steam tube where it cures for many hundreds of feet before reappearing in the normal atmosphere. The approach to this gauging was to put a sight glass assembly in the steam tube close to the extrusion head and look through the steam to measure the diameter of the hot insulation. For some reason, the sight glass assemblies continued to deteriorate in unusual ways, giving the gauge at first erroneous readings and then no readings at all. It was only after many months of effort and analysis of the difficulty, that we discovered that the chemical conditioning of the steam before entry into the tube was working against us rather than for us. Obviously, we didn't make any profit on that transaction and our customer wasn't too happy either. Fortunately, this occurred in only one case; but if it had been a more general condition, and we had not found the problem, a potentially solid product would have gone down the drain.

ADDED CONSTRAINTS

All of the complexities and difficulties discussed above relate to the general problem of transitioning new technology into an industrial environment. For the most part, all of these factors are internal to the relationship of the system designer and the manufacturer. In addition to these, however, there are a number of external factors which increase the risk of the system designer, particularly for small high technology companies. They are: the availability of venture capital, the increase in legal liability, and the complexities of government regulation.

Venture Capital

In the late 1950's and throughout the 1960's, an inventor with a good idea could find ready backers. Many hundreds of small companies were started on a single idea of a new gauge or controller possessed by an individual inventor. This period was perhaps the heyday of the entrepreneurial inventor. Other companies that were a little further along also used venture capital as a major source of expansion resources. Today, and over the past ten years, the venture capital market has been considerably tighter. Certainly, some of this constriction came as a result of disappointment in many of the investments in high technology that were made earlier. But it is clear that the 1969 capital gains restriction severely restricted the entrepreneur. The rewards for risk capital were reduced to nearly the level of return for far less risky investments, such as tax-free bonds. In addition to this shortage of venture funds, it has also become much more difficult to leverage a small venture through borrowing money. When interest rates were low, the amount of income needed to service loans was nowhere near as demanding as when the rates run at eleven and twelve percent. These high-interest charges are a difficult financial burden for a young company.

Legal Liability

The increase in consumerism has led all products to come under greater scrutiny for potential hazards to human health and safety. While the cause of making products safe is a good one, some of the mechanisms have been more emotional than effective, and in any event, they create additional risk for the manufacturer of industrial as well as consumer products. As a result of the burgeoning number of multi-million dollar liability suits, insurance costs are extremely high if the insurance is available at all. In addition, from our experience, it is not quite clear how good our coverage is. If a company is forced to defend itself in a product liability suit, the cost in management time, regardless of the outcome, is substantial. These costs are especially difficult to bear or even predict for the smaller company.

Government Restrictions

The government's involvement in business has also tended to have a dampening effect on creativity and new products. In most areas we find increasing government certification of one kind or another required. For example, our laser scanner uses extremely low power and presents no hazard under normal, and most abnormal, uses; however, the regulations required us to add many dollars in unit cost and substantially increased our development cost. The Occupational Safety and Health Act, and its promulgation, defines acceptable human environments for many situations. In many instances, these are realistic. But when government regulations are dealing with new technology where there is little experience, extreme caution, regardless of cost, is the preferred bureaucratic viewpoint.

Many government agencies touch business in one way or another. The Internal Revenue Service, the Department of Labor, the Department of Energy, the EPA, and the Department of Transportation all have an impact on the way we do business. The requirement to deal with these many external forces is especially hard on the typical instrument or gauge manufacturer who tends not to have a lot of excess management resources around.

The three areas of constraint listed above have the effect of slowing down the infusion of new technology into new products and into the industrial environment. This is not to say that there are not other reasons why these constraints are valid, and I in no way intend to debate that question here. I simply point out that the effect on the process we are discussing is a negative one.

WHAT IS NEEDED

If we are to regain the record of increased productivity which has given our economy strength over many years, part of the answer is to do a better job of translating new technological advances into systems which will improve productivity. There are three fundamental factors which could significantly aid this process. They are: a partnership between the supplier of control systems and the using manufacturer, less government control and more encouragement, and the incentives to both the manufacturer and the system developer.

Partnership

The problems described under the technology transfer process could be helped considerably by a stronger relationship between the instrument/systems supplier and the using manufacturer. The exchange of information between plant people and system developers requires an intimate working relationship. In many instances, it should be feasible for the ultimate user to share some of the front end financial risk. Indeed, we have done this with some companies in the past. Another method of sharing the financial risk is a commitment process. All of these suggestions have their negative effects, too, but the complexity of the task we are trying to undertake requires approaches quite different from the old "show me what your gauge will do, and maybe I will buy one," approach. Staying with the arms length relationship will extend the time of technology transfer and, in come cases, kill it altogether.

Government Encouragement

Some changes have already occurred which we believe will help stimulate new investment in development. The latest capital gains tax rules should provide some improvement in the availability of risk capital. In addition, of course, the investment tax credit provides another incentive for investment on the part of the manufacturer. Perhaps a more aggressive approach to encouraging the introduction of new productivity improving systems is possible. Especially during a trial period of perhaps one year, special tax incentives and releasing the system from regulatory requirements could ease the burden on the creative small systems company as well as the manufacturer. While this seems unlikely to be politically acceptable, it is certainly true that government and industry in other countries, such as Japan, have a far more effective and dramatic partnership than is suggested here.

We are involved with the Air force's manufacturing technology activity. This seems to me, at least in special areas where the development might not otherwise occur, to be an effective way of making sure that new technology is at least proven to the feasibility level. To be more explicit the Manufacturing Technology Division of the Air Force Materials Laboratory has the specific mission to initiate, manage, and monitor new manufacturing methods and computer-aided manufacturing technology projects which satisfy current or potential generic Air Force production requirements. Such projects are based upon known and proven technology but require an up-front investment to prove manufacturability. In addition, it is a requirement that these demonstration projects involve technology or instruments which are not otherwise available, are beyond the normal risk of industry, and indicate a strong probability of providing tangible benefits to the Air Force. Specifically, under this program we have developed an improved ultrasonic inspection device for inspecting for cracks around bolts or rivets in wing surfaces. This device certainly has a limited market but a very important one. The potential savings to the Air Force, if the system continues to look as good as it does at this stage, are very great, and yet, no instrument manufacturer was likely to develop this system for such a restricted market. On the other hand, the technology, once developed, may well be applicable to other industrial problems. This is one way the government can constructively stimulate the development of technology for its own use which will also have payoff in other parts of the economy.

Incentives

The basic incentive for the transfer of new technology into industrial measurement and control systems is, of course, the potential economic gain for both the systems supplier and the manufacturer. While all of the conditions and risks discussed above exist, clearly this potential gain also exists or we wouldn't see any new gauge development of any kind. It is useful, however, to put into perspective the relative financial dynamics that at least one gauge manufacturer sees in pursuing this opportunity. The development of a typical extrusion control system costs approximately $150,000. However, development of the system itself is only part of the cost of having a new control system. An additional cost of at least equal size is the introduction of the product into the industrial environment and the development of the customer base to purchase the new system. If we assume that the product has a gross manufacturing margin (manufacturing margin is what is left over after subtracting labor, materials, and manufacturing overhead from revenues) of fifty percent, we can see that in order to have a twenty percent pretax profit on revenues, which is about the minimum an instrument manufacturer can afford, we have thirty percent to spend on sales, service, research and development, and general administration. Sales and service will cost about twenty percent of revenues. Since the general manager of any operation usually insists on being compensated, it's likely that the amount available for new product development will average five to six percent of revenues in a growth oriented company. Therefore, if we assume a research and development and product introduction cost of $300,000, a unit selling price of $10,000, and six percent of revenues available for research and development and new product introduction, we can see that at least 500 units will have to be sold to recover the research and development costs. Clearly undertaking such a commitment requires some clairvoyance and a good amount of faith on the front end of the development investment. The prospect of selling 500 units in the environment discussed throughout this paper is not an easy one. However, clearly we feel that it's not an impossible risk or we wouldn't be in the business.

Finally, incentive could be provided to the gauge or measurement and control systems creator if his return could be tied in some way to the improved productivity of the plant. In some of our extrusion control installations, the savings have been so significant that the customer clearly paid for the system in a couple of months. Obviously, he could have afforded to pay more for the benefit he received. On the other hand, it is extremely difficult to value price a guaging control system. But were a portion of the savings returned to the manufacturer of the equipment who made the savings possible, a considerably more aggressive development program would be encouraged.

CONCLUSION

The environment for the developer of a new measurement and control system is filled with opportunity and significant risk. The risks today are sufficient to have a restrictive effect on the speed with which new technological advance will be absorbed into manufacturing operations. No doubt the process will continue without added incentives. Some of the suggestions included here could aid the process considerably. The most important of all is the development of a real coordinative approach between the system developer and the manufacturer so that the knowledge and resources of each can be used to the benefit of both.

Presented at the SME Jigs & Fixtures Clinic, March 1980

Gage Designers Can Improve U.S. Productivity

By Edward S. Roth
Productivity Services

ABSTRACT

The decline in U.S. productivity growth has been repeatably documented in recent periodicals and the reasons given for this decline have been stated in terms that are probably impossible to resolve, because only political action is seen generally as the solution. There are, however, a number of solutions that do not require political action that can be lumped under the heading "work smarter, not harder." These are technical solutions that have been ignored by industry leaders who are mainly from law and accounting backgrounds. The technical reasons for our decline include poor product design practices by graduate engineers who have no formal training in design, an ever increasing number of engineering changes caused by these poor design practices, Material Review Board actions resulting in needless reject/rework also traceable to untrained engineers and non functional inspection practices, the use of Coordinate Measuring Machines as substitutes for functional receiver gages that represent mating parts, a decline in the use of the engineering team approach to integrated product design, and the retirement of over 50% of our skilled manufacturing personnel over the past ten years. Tool designers can provide a critically needed service to U.S. Industry by working with product designers and creating engineering teams when required to assure that product designs are manufacturable, functional, and have the largest tolerance possible.

INTRODUCTION

Three general approaches to product design exist today in U.S. Industry. These can roughly be characterized as 1) the Design Layout (DL) which is seldom used, 2) the Assembly Drawing (AD) which is sometimes called a Zero Drawing, and 3) the "Quick and Dirty" preparation of detail drawings with no DL or AD. Each of these approaches will be discussed in this paper in connection with check lists that the tool designer may follow in his design of the manufacturing process and while assisting the product designer. The tool designer is identified as the key person to improve U.S. productivity because he is a manufacturing engineer with the required indepth technical knowledge, and is in the interface position where he can most influence the product design.

TOOL DESIGN BASED ON A DESIGN LAYOUT

The DL is a unique assembly drawing. It geometrically defines the nominal size and the limits of size and location of all parts that make up the assembly. The most critical limits are usually shown in green (the MMC or "Go" gage size) and the least critical limits in red (the "Not Go" gage size). The DL is made at any required scale and complexity so that the designer can quantitatively determine what happens functionally when all parts are at their limits of size and location. In many cases, overlays will be required to look at these relationships. The DL enables the designer to utilize the largest tolerances of location and broadest limits of size possible before function is impaired. The following technical information should make up the DL.

- Datum identification, functional, qualification

- Critical dimensions, classification of characteristics

- Interchangeability parameters such as functional limits of
    size and location, allowances and clearances

- Materials and special processes

- Hardware and standard parts

The function of the DL is to enable the following:

- The preparation of sketches at the nominal, green and red limits so the DL can be functionally modeled in three dimensional hardware
- The preparation of all detail, support drawings and acceptance procedures
- Structural and stress analysis calculations
- The design of tooling to include jigs, fixtures and gages
- The design of life tests and materials capability analysis
- The determination of the manufacturability of the design

The first cut DL should be evaluated by all of the disciplines mentioned above which make up the engineering team and provide input to the designer. The key person in this engineering team is the tool designer, who must work closely with the designer and change the design if datums are inaccessible, tolerances are too tight or critical relationships and dimensions are not identified. Since even trained product designers cannot assess the manufacturability of a complex new product, the key person here is the tool designer.

In addition to the tool designer, the engineering team should also sign off on the DL when they are satisfied that it meets all their individual technical parameters. Many changes and iterations will occur at the DL phase before all are relatively satisfied with the design.

The key value of the DL is that it forces changes very early in the concept phase of the design where the cost of the changes is insignificant. Changes that occur after detail drawings have been completed, or later after the design has been released, become increasingly costly. Changes that occur after production has started are the most costly of all. Since changes are at least 50% of the design effort anyway, let's make them early so that most will not get into the Engineering Change (EC) system. Let's work smarter, not harder.

TOOL DESIGN BASED ON THE ASSEMBLY DRAWING

It should be obvious that the DL contains more pertinent information that effects tool design parameters than an assembly drawing. The AD is missing several key elements such as 1) datum definition, 2) datum targets, 3) assembly constraints (to include torque loadings), 4) critical dimensions and tolerances, and 5) the red and green limits of size and location. If no DL has been prepared, it would be advantageous if the tool designer could talk the designer into preparing one, or at least preparing partial DL's where critical relationships exist. If unable to do this, the tool designer must then use the following check list to extract the technical data required to prepare functional tooling.

- What are the primary, secondary and tertiary datums on each part?

- Will datum targets or form tolerances be used to achieve repeatability?

- Must the datums be tooled with special centering fixtures so they cannot translate, or can they be tooled with fixed size locators such that both rotation and translation can occur?

- If the dimensions are chained, which relationships are critical?

- Are the tolerances fixed in value, or may they vary with feature size?

- What product acceptance techniques will be used for each part; optical comparator, functional gage, CMM or surface plate?

- Are the tolerances and size limits functionally derived, or extracted from some ancient company standard by an anonymous author?

- If the parts are flexible (to the extent that they change measurably when released from fixtures and tools) can they be inspected on the machine or in the fixture?
- What is the batch size of the product?
- What reject or rework rate is acceptable?
- How much adjustment can be tolerated on the manufacturing line without seriously reducing production rates?

The above check list will enable the tool designer to assess the manufacturability of the assembly and the design knowledge of the engineer. If the designer cannot satisfactorily answer the above questions it would be mandatory that the tool designer provide additional assistance to the product designer via the engineering team.

TOOL DESIGN BASED ON DETAIL PART DRAWINGS

Parts defined using ANSI Y14.5, <u>Dimensioning and Tolerancing</u>, can be used directly for tool design providing the detail draftperson understands the direct implications imposed on tool design if this system is used. However, if the symbology is used without knowledge, or if no DL or AD is available due to the quick and dirty approach, the following check list will enable the tool designer to assess the adequacy of the detail drawing. (It is unfortunate that the least experienced draftpersons are used to make detail drawings when these drawings are used most by tool designers in determining hard tooling and gaging parameters. In most cases, no DL has been prepared and the inadequate drawings can cause unbelievable problems).

- Are the part datums accessible?
- If datum targets are specified, will they stabilize the part when clamped in the proper tooling?

- If datum targets are specified, will the part datum surfaces rock when clamped against machined surfaces or rails on the tooling?
- Are datums modified with either RFS or MMC?
- Are tolerances modified with either RFS, MMC, or LMC?
- Is the design such that only a single acceptance criteria is applicable?
- Must the tool designer make any personal interpretations?

Parts defined using standards that are not based on ANSI Y14.5 will probably be impossible to interpret and prove usable in tool design. Such drawings usually have implied datums, chain dimensioning and bilateral tolerances. The check list that would be required to extract the information needed would be a combination of the two lists above and probably would not even then give the tool designer adequate information because no DL was prepared. If the tool designer designs tools and gages, he will assume the role of the product designer, because his tooling will not be the mathematical equivalent of what is stated on the drawing.

TOOL DESIGN: ITS IMPACT ON THE ACCEPTANCE PROCESS

The retirement of skilled personnel has already created a shortage of gage designers in U.S. Industry. Pockets of gage designers still exist in key manufacturing areas and they usually work for "turn key" tool design and/or build shops. Since there are no extensive apprenticeship programs available, the retirees have not been replaced.

However, a substitute product acceptance technology is rapidly replacing these skilled gage designers and surface plate inspectors. The substitution utilizes Coordinate Measuring Machines (CMM's) manned by operators who do not understand surface plate inspection gage design or manufacturing. The concept of replacing skilled craftsmen with operators who can be trained in one day is very appealing to management, and these machines will probably be the inspection method of the future. The following table details the rough relative costs to accept mass-produced product based on the methods still in use today.

Surface Plate Inspection	Basic CMM with Operator	Advanced CMM with NC Computer & Touch Probe	Functional Gages
60	20	5	1

Functional gaging is still the most cost effective, but is losing ground to CMM's because the short-term cost replacement costs look attractive to management.

Unfortunately, CMM's are now used to inspect parts and calibrate tooling that is beyond their capability, if the standard of requiring an inspection process to be ten times as accurate as the part, and five times as accurate as the tooling, is still in effect. Process capabilities of current CMM's indicated that they usually have an accuracy of $\pm$ .0015 inches (to 3 sigma limits) when operated by trained metrologists. This indicates that parts with tolerances less than $\pm$ .015 inch and tooling with tolerances less than $\pm$ .0075 inch cannot be inspected or calibrated on these machines. Most CMM's however, are being used to inspect parts to tolerances of $\pm$ .001 inch and this probably occurs because the axis position indicators read to tenths! With the above data in mind, the tool designer should attempt to increase part tolerances at the DL or AD phases if the CMM's are to be used, or if that fails, require that the engineer justify all tolerances functionally that are less than $\pm$ .015 inch. If the designer can truly justify tighter tolerances, then the surface plate should be used.

If the surface plate is used, there are layout techniques[1] available that can take deviations shown on inspection reports and translate and rotate this data as if a three-dimensional gage were being used. Work is also underway to prepare software programs to functionally gage parts by computer. Product reject/rework evaluation using these techniques should be used by material review boards to reduce needless rework and rejects.

SUMMARY

The tool designer has been identified in this paper as the key person with the production knowledge to stop the decline in U.S. productivity growth. He is in the right organizational position to do this since he is the interface between that piece of paper called a product definition and the real world of costly hard tooling and gaging. He has the experience to immediately determine if the design and the designer are adequate, and to either meet with the designer, or instigate the formation of an engineering team, if required.

REFERENCES

1. Roth, Edward S., Functional Inspection Techniques, (Society of Manufacturing Engineers, Dearborn, MI 1967).

Presented at the Precision Machining and Gaging Conference, November 1974

Precision Gaging Applications

By J.R. Gibson
W. H. Rasnick
H.S. Schwartz
P.J. Steger
Union Carbide Corporation

Cartridge-type linear variable differential transformer (LVDT) transducers with captive, fully compensated, externally pressurized, linear air bearings offer exceptional accuracy and reliability of performance for precision gaging applications. In addition to the advantage of a low positive side-deflection characteristic, the air bearing provides, for all practical purposes, negligible friction. As a result, reliable operation at gaging forces as low as 0.1 gram (0.001 newton) is possible. Details of design and fabrication are discussed for several different geometries and sizes of transducers. Dynamic operational characteristics are presented along with the tests used to characterize performance criteria of the units. Application experience with these transducers used in several gaging systems concludes that the transducer can be effectively used to improve the accuracy of many gaging systems, while providing increased versatility and reliability of operation.

INTRODUCTION

For many years, cartridge-type LVDT (linear variable differential transformer) transducers have been used extensively at the Oak Ridge Y-12 Plant[a] and many other industries as a dimensional inspection tool. However, many of these transducers exhibit three characteristics which can cause serious limitations in their use: (1) shaft sticking, which requires unit replacement and involves much setup and recalibration effort; (2) high gaging forces of 20 - 30 grams, which can cause damage to soft metal parts, and (3) excessive shaft/bearing clearance (~ 20 μm), which can lead to an error of 140 microinches at 45 degrees.

In a conventional transducer, these factors are closely related because, in order to maintain accuracy, the bearing requires a close fit, which encourages sticking, and a high gaging force is required to overcome sticking.

Development of the air-bearing LVDT transducer has helped to overcome these problems to a large degree. Recent development effort has been concentrated on improvements to the original model as well as to other models for both general use and specific application. Fabrication, assembly, and testing procedures have been developed as well, and current efforts are being directed toward the acquisition of an acceptable commercial supplier.

LOW-FORCE CARTRIDGE TRANSDUCER

TRANSDUCER DESIGN

Five requirements were established in order for an air-bearing gaging transducer to be of value:

(a) Operated by the Union Carbide Corporation's Nuclear Division for the US Atomic Energy Commission.

1. Motion should be smooth and free of sticking.

2. The gaging force should be considerably less than the conventional model and preferably adjustable.

3. The bearing must be adequately stiff and free of lost motion.

4. The transducer must be of a size comparable to standard indicators and transducers.

5. The cost must not be prohibitive.

Satisfying the first of these requirements is an inherent quality of a hydrostatic air bearing, and meeting the first requirement is a necessary condition for meeting the second. The third condition is met by maintaining a small film clearance and the fact that a hydrostatic air bearing acts as a spring and technically has no lost motion. It was felt that to confine the unit to a 3/8 inch diameter would be possible but not economically feasible. However, a 1/2 inch diameter could economically contain all the necessary components and still be small enough to escape being unwieldy and to be adaptable to most situations using standard transducers. All components were designed to be fabricated on conventional precision machinery.

Bearing design was based upon approximate techniques which have been previously reported.[1] These techniques essentially consist of equating the Darcy equation of flow through a porous medium to the slot-flow equation for a recessed-pad hydrostatic gas bearing.

Shaft deflection (Δ) is considered as being made up of two components: beam deflection (δ_b) and film deflection (δ_f). General equations and Figure 1 are given for reference.

Total Deflection of Tip (Δ) = Beam Deflection (δ_b) + Film Deflection (δ_f),

where:

$$\text{Beam Deflection} = \frac{Fa^2}{2EI}\left[a-\left(\frac{a-3b}{3}\right)\right] , \tag{1}$$

$$\text{Film Deflection} = \frac{b+a}{b}\left(\frac{R_f}{2AS}\right)+\frac{a}{b}\left(\frac{R_r}{2AS}\right) . \tag{2}$$

Error due to side deflection caused by angular contact between the transducer axis and surface can be found by:

$$\text{Error} = \delta \sin \phi$$

In Equations 1 and 2:

      F   represents the gaging force, in newtons, N,

      E   the modulus of elasticity of the shaft material, in pascals, Pa,

      I   the moment of inertia of the shaft, in millimeters to the fourth power, mm^4,

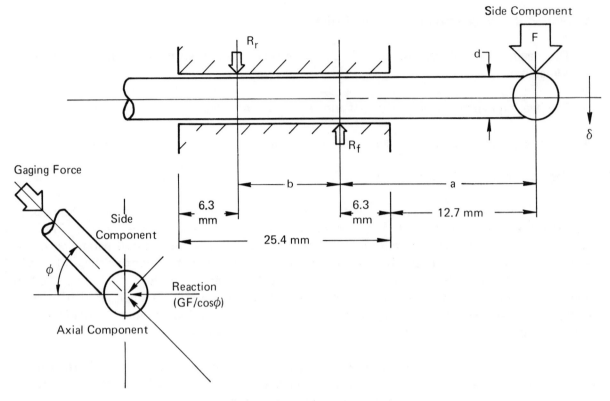

Figure 1. DESIGN PARAMETERS.

R the gaging force reactions, in newtons, N,

A the projected shaft areas, in millimeters squared, mm^2, and

S the film stiffness. $S = E_f \dfrac{0.35\,P_s}{h}$, where $E_f = 0.5$.

A cross-sectional view of the basic round-shaft-design transducer is shown in Figure 2. The bearing materials used have been either ATJ or EDA grades[b] of molded graphite, although other equivalent material could be used. Housing materials have been either brass or stainless steel. If brass is used, a shielding material should be placed around the LVDT per the recommendation of the coil manufacturer.[c] The shaft material most used has been aluminum (Type 6061-T6), although here again it is a matter of choice. A point to consider is that the lighter the material the better the time response, all other factors being equal.

The transducer can be operated using vacuum supplied at the back pressure bleed port to regulate back pressure low enough to obtain gaging forces lower than that obtainable with the bleed screw (< 3.5 grams force). In addition, vacuum can be used to retract the shaft for automatic or sequential operations. For applications where it may be desirable to gage from either direction, vacuum can be used to maintain a slightly negative gaging force.

(b) Union Carbide Corporation-Carbon Products Division, Cleveland, Ohio.
(c) Schaevitz Engineering Company, Camden, New Jersey.

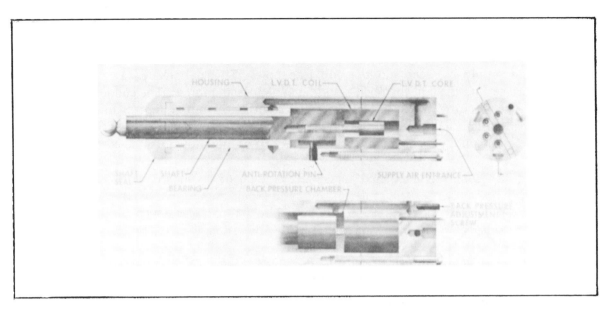

Figure 2. AIR BEARING LVDT TRANSDUCER.

FABRICATION AND ASSEMBLY PROCEDURES

All component parts are to be fabricated to the dimensional tolerances shown on the fabrication drawings using the materials specified. Special care is required during assembly in order to successfully meet the performance specifications. The assembly procedure followed at Y-12 is discussed in the sections that follow.

Graphite Bearing Assembly

The graphite (UCC grades ATJ or EDA, or equal) is to be machined dry using only ethyl alcohol for cleaning. Any use of oil or grease may clog the pores and inhibit air flow through the graphite. After machining, the graphite is to be cemented into the brass housing using Epon 828[d] epoxy (two parts) with Versamid 140 hardener[e] (one part), or an equivalent. Surfaces to be bonded must be slightly roughened (320-grit paper) and thoroughly cleaned before applying epoxy. Care must be exercised so that no epoxy enters the air distribution grooves, otherwise the air supply may be sealed off. This joint should be allowed to set up for eight hours.

The exposed ends of the graphite should be sealed with a Lucite[f]-in-methyl chloride solution to prevent the leakage of air through these areas. Care must be exercised to prevent excess Lucite from entering the bearing area as this will cause rubbing and is extremely difficult to remove.

The bearing assembly should be allowed to cure for two days at room temperature before machining to final dimensions.

(d) Shell Chemical Corporation, New York, New York.
(e) General Mills-Chemical Division, Kankakee, Illinois.
(f) DuPont, Wilmington, Delaware.

LVDT Coil Housing

A preliminary check of the brass cap should be made to insure that there is no breakthrough between the supply passages and the coil lead holes. This deficiency can cause leakage of high-pressure air into the coil chamber, resulting in high gaging forces. The coil leads are threaded through the holes in the brass cap and the coil is securely cemented in place to ensure that each lead is cemented into its hole. Epoxy cement has been used, but caution must be taken against using an excess amount which may prevent a correct fit or may flow into the core area.

The cross-drilled supply and exhaust ports must be plugged, as shown, using either small brass press-fit plugs or epoxy. Care must be taken not to block the internal passages. This precaution has been taken by inserting small Teflon[g] wires into the passages which can be removed after plugging. The single plug in the bearing assembly may be installed at this time.

The transducer may be optionally exhausted through an adjustable bleed screw or a vacuum line. If the vacuum line is to be used, the 0 - 80 tapped hole and associated cross drilling should be replaced by a 1.85-mm (0.73-in)-diameter x 0.3-mm (0.12-in)-deep counterbored hole. Short pieces of 1/16-inch-OD tubing are installed into the supply and exhaust ports as required.

Cabling, along with air lines, coil leads, sheathing, and an amphenol plug are to be assembled at this point.

Bearing/Shaft Fitting and Assembly

The bearing/shaft clearance should range between 0.000 to 0.005 mm (0.0000 to 0.0002 in) (diametral) as fabricated. These parts must now be lapped together to achieve a clearance of 0.005 to 0.010 mm (0.0002 to 0.0004 in) (diametral) so that the air film can be properly established. Lapping is performed using a 3-micrometer diamond grit in an alcohol vehicle and using alcohol for cleaning. Moderate air pressure should be applied to the air passages to blow out any material lodged in the bearing surface.

The LVDT core and threadstock are assembled to the shaft and positioned for the proper pretravel using small quantities of cement to maintain position. The antirotation groove (round shaft) should be checked for cleanliness and sharp edges. This antirotation provision of the transducer has been a major source of malfunction. The antirotation pin may be installed after insertion of the shaft into the bearing.

Unit Assembly

The bearing assembly with shaft and core are assembled to the cap, using the gasket and 0 - 80 screws, by observing the following directions: Tighten the screws securely and check the axial alignment of the two cylindrical components. Move the shaft and ensure that there is no interference between the coil and core or other components. Connect the supply port to a source of dry, filtered, oil-free, 400-kPa (60-psig) air, applying pressure gradually. Check the unit for high-pressure leakage of the gasket, plugs, shaft, and coil-lead penetrations. (Escape of

---

(g) DuPont, Wilmington, Delaware.

low-pressure air around the shaft is normal due to the air film established between shaft and graphite.) The dust cover is installed after these checks have been made. Upon completion of the assembly phase, the transducer is subjected to mechanical and electrical testing. The transducer is mechanically tested for gaging force and bearing film compliance. Electrical testing confirms LVDT output linearity with the range specified.

Operation and Performance Evaluation

Standard operating procedure for the transducer (refer to Figure 2) is to use dry, filtered, oil-free air supplied at 400 kPa (60 psig). The air is fed through the manifold, through the porous graphite, and into the bearing clearance where a pressurized support film is established. Air bleeding from the rear of the bearing is restricted by the adjustment screw to create a back pressure higher than ambient, which governs the gaging force. Screw adjustment can be made to vary the gaging force as required. To obtain lower gaging forces or automatic shaft retraction, useof vacuum is employed. A method of installation to use vacuum for both retraction and gage-force regulation is shown in Figure 3.

Important performance characteristics include gaging force, compliance, and time response. Gaging force is a function of shaft weight and chamber back pressure. Back pressure is adjusted to establish the gaging force required for a transducer in the orientation needed to satisfy a particular application. The gaging force is measured at the LVDT null point using a calibrated force gage applied axially to the transducer shaft. Transducer compliance is a function of both shaft bending and film deflection, and is measured with the transducer at the null position by applying a noncoaxial force at the gage tip and measuring lateral movement at this same point. A schematic arrangement of the component for setting the gaging force and measuring compliance

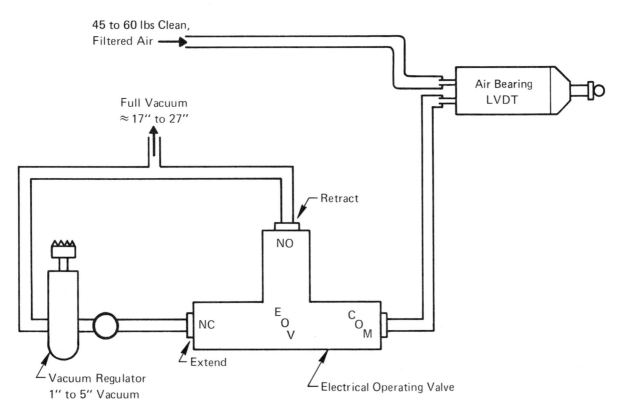

Figure 3. COMPONENT ARRANGEMENT FOR VACUUM REGULATION.

is shown in Figure 4. A comparison of the measured transducer compliance versus calculated is given in Table 1.

Time response or ability of an air-bearing LVDT transducer to respond to the traverse of an irregular surface is determined by the gaging force and the rate of traverse over some surface. If friction is considered negligible, the gaging force is only required to overcome inertia in order to respond to some surface irregularity (see Figure 5):

$$F = m \frac{d^2x}{dt^2},$$

where:

F   represents the gaging force in newtons, N, or kg-m/s^2, and

m   represents the transducer shaft mass in kilograms, kg.

Since $\frac{dx}{dt}$ and $x = 0$ when $t = 0$, then $x = \frac{Ft^2}{2m}$ (x is in meters), or $t = \left(\frac{2mx}{F}\right)^{1/2}$ in seconds (free, unrestrained response).

For the case of a periodically varying surface:

$$X = x_0 \sin \omega t,$$

$$dx/dt = x_0\omega \cos\omega t,$$

$$d^2x/dt^2 = -x_0\omega^2 \sin\omega t, \text{ and}$$

$$(d^2x/dt^2)_{max} = x_0\omega^2 \text{ (at } \sin\omega t = \pm 1) .$$

Therefore:

$$F = m\frac{d^2x}{dt^2} = mx_0\omega^2 \text{ in newtons, N, or kg-m/s}^2, \text{ and}$$

$$\omega = \left(\frac{F}{mx_0}\right)^{1/2} \text{ radians/s .}$$

Response frequency $(f) = \frac{\omega}{2\pi} = \left(\frac{F}{4\pi^2 mx_0}\right)^{1/2}$ in hertz, Hz.

The limiting velocity (v_{max}) at which some surface can be swept can be represented in terms of the response time (t) and the distance (L) over which some deviation (x_0) takes place, thus:

$$v_{max} = L/t, \text{ m/s .}$$

For a periodically varying surface:

$$v_{max} = Lf, \text{ m/s .}$$

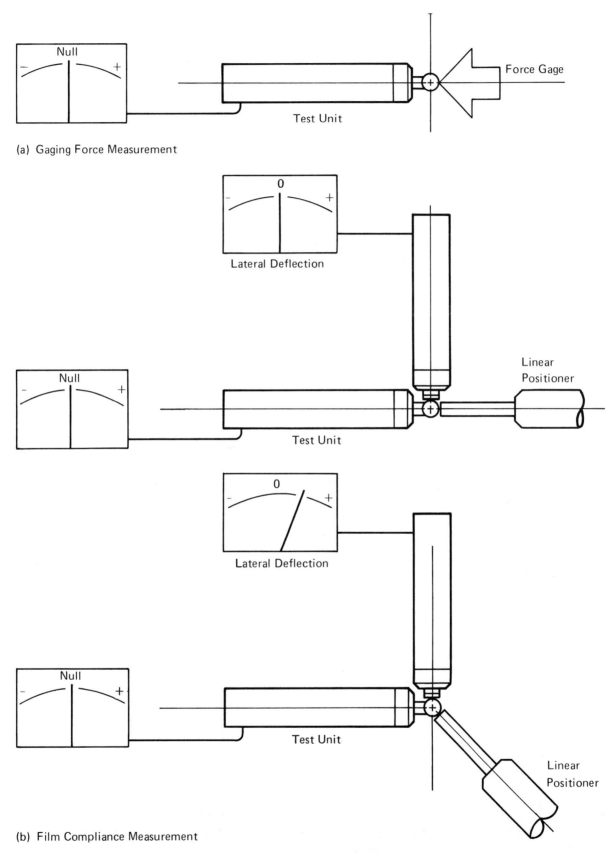

(a) Gaging Force Measurement

(b) Film Compliance Measurement

Figure 4. SCHEMATIC ARRANGEMENT FOR TRANSDUCER TESTING.

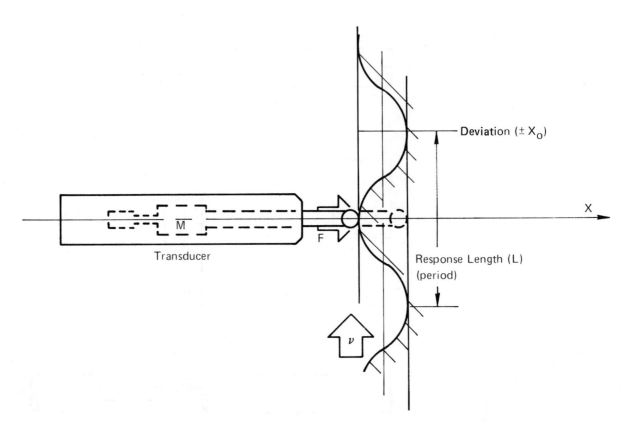

Figure 5. TRANSDUCER RESPONSE TO SOME SURFACE IRREGULARITY.

The transducer response testing was conducted using a motor/tachometer-driven 100-mm (4-in)-diameter offset circular cam to determine the limiting sweep velocity to which a transducer would respond. Tests were conducted using a conventional round-shaft transducer riding at two separate angles, as shown in Figure 6.

The surface velocity of the cam, above which the transducer no longer gave a true reading, is shown in Table 2.

Table 1

TRANSDUCER COMPLIANCE

Theoretical Compliance	Type Unit			
	Conventional			Miniature
	Round Shaft	"D" Shaft	Square Shaft	Square Shaft
$\mu in/g_f$	3.4	3.4	3.0	12.6
$\mu m/N$	8.9	8.9	7.7	32.8
inches/lb	1.55×10^{-3}	1.55×10^{-3}	1.36×10^{-3}	5.72×10^{-3}
Measured Compliance				
$\mu in/g_f$	5.22	5.86	$10.0^{(1)}$	18.6
inches/lb	2.38×10^{-3}	2.67×10^{-3}	4.55×10^{-3}	8.5×10^{-3}
Quantity Tested	29	2	8	2

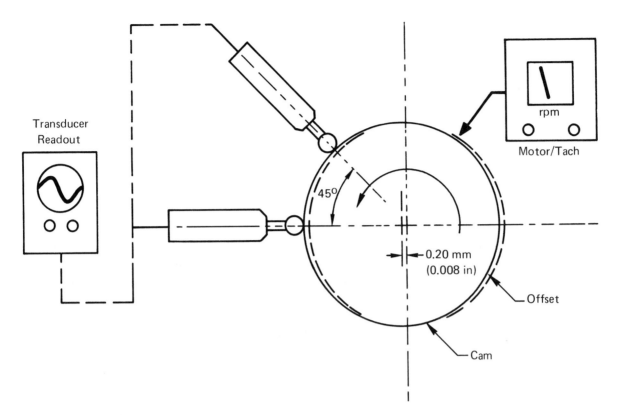

Figure 6. TRANSDUCER RESPONSE WHILE RIDING AT TWO SEPARATE ANGLES.

Geometry Variations

At the present time, three general types of air-bearing LVDT transducers, including several variations, have been or are presently being fabricated and tested. The three general types include: (1) the original round-shaft model, (2) a slightly larger square-shaft model, and (3) a miniature model. Variations on the original unit include an extended-range model and a "D"-shaft model, while variations on the miniature unit include both square and round shafts as well as two different gaging ranges.

Table 2

AIR BEARING LVDT TRANSDUCER RESPONSE LIMIT
(Round Shaft Unit, 0.049 N (5 gf) Gaging Force,
± 0.20 mm (0.008") Deviation, 4.5 g Shaft)

Horizontal Shaft Axis		45° Inclined Shaft Axis		Theoretical (Appendix B)	
rpm	m/s	rpm	m/s	rpm	m/s
1500	8.0	1200	6.4	2200	11.6

Each of these models is capable of the same small gaging force [< 0.01 N (1 gf, 0.04 oz)], using a 414-kPa (60-psig) air supply; and each uses basically the same design, fabrication, and assembly procedure. The differences have come about due to limitations in the particular models or needs for special-purpose units. Table 3 gives a summary of those units presently being evaluated.

The original model of the round-shaft transducer has been in use for about two years.[2] Most of the units being used were fabricated at Y-12; however, some attempt has been made to procure these transducers from industrial sources. Generally speaking, this transducer has been very successful and is now being used in the production areas. A major drawback with the original design has been the antirotation device which uses a stainless steel pin riding in a groove in the aluminum shaft. Galling has frequently occurred, increasing the drag and occasionally freezing the shaft travel. This problem has been lessened through the use of a nylon pin in place of the

Table 3
GEOMETRY VARIATIONS

Type Unit Unit Variation	Round Shaft			Square Shaft Conventional	Miniature		
	Conventional	"D" Shaft	Extended Range		Square Shaft	Square Shaft	Round Shaft
Shaft Size, mm (in)	4.66(3/16)	4.66(3/16)	4.66(3/16)	4.66(3/16)	2.54(1/10)	2.54(1/10)	2.54(1/10)
Cartridge Diameter, mm (in)	12.7(1/2)	12.7(1/2)	12.7(1/2)	15.9(5/8)	9.5(3/8)	9.5(3/8)	9.5(3/8)
Cartridge Length, mm (in)	65.0(2 9/16)	65.0(2 9/16)	89.9(3 17/32)	65.0(2 9/16)	58.5(2 5/16)	66.7(2 5/8)	66.7(2 5/8)
Gaging Range, mm (in)	± 0.51(0.020)	± 0.51(0.020)	± 2.54(0.100)	± 0.51(0.020)	± 0.25(0.010)	± 0.51(0.020)	± 0.51(0.020)

metal one. Other improvements have included the use of a hollow shaft to allow for in-place adjustment of the LVDT core, a more pliable gasket material (polyethylene) which has eliminated high-pressure air-leakage problems, and relief of the shaft/piston intersection which has eliminated shaft sticking at the fully extended position and eased shaft lapping.

The "D"-shaft variation (Figure 7) was designed to overcome the antirotation pin problem with the conventional unit. Fabrication and assembly is considerably easier than it is with the square-shaft unit, and operating characteristics are comparable to the conventional unit. The unit is identical to the conventional model except that the round shaft has a flat on one side, and the graphite bearing has a matching "D"-shaped hole broached through it.

The extended-range unit is identical to the conventional transducer with the exception that it uses a longer-range LVDT coil and core.

The square-shaft model (Figure 8) was the first major attempt to overcome the antirotation problem while adding some torsional stiffness to the transducer. Eight of these units have been fabricated and tested so far and can be considered only limited due to fabrication difficulties in maintaining proper shaft clearances. Bearing film stiffness for the initial six units was considerably less than that of the conventional units, which is attributable to the difficulty in maintaining squareness and flatness of the shaft and bearing. The last two units fabricated used unproved techniques in sizing and resulted in stiffness greater than the round shaft.

As far as antirotation is concerned, the square-shaft transducer is a definite improvement over the round-shaft unit both in terms of reliability and degree of rotation permitted. Most conventional transducers used at Y-12 have a 9.5-mm (3/8-in)-OD cartridge; and, subsequently, most existing fixturing is set up for this particular size. In an attempt to meet this need, two 9.5-mm (3/8-in)-diameter units (Figure 9) were designed, fabricated, and tested. Because of the small size, a miniature [± 0.25-mm (± 0.010-inch)] range LVDT coil was used, and air supply and exhaust passages were brought in through the side of the unit, eliminating the several long, small-diameter holes otherwise required. These modifications created some additional fixturing problems; however, a compromise will have to be accepted at some point. An advantage of this design is an allowance of a rear-end core adjustment provision which will allow adjustment during use if required.

Additional ± 0.51-mm (± 0.020-inch)-range units consisting of two round and two square-shaft miniature transducers are being fabricated for evaluation by the Y-12 Dimensional Inspection Engineering Department, as are all developmental units. These units will incorporate some improvements suggested by the earlier two transducers as well as providing a larger gaging range.

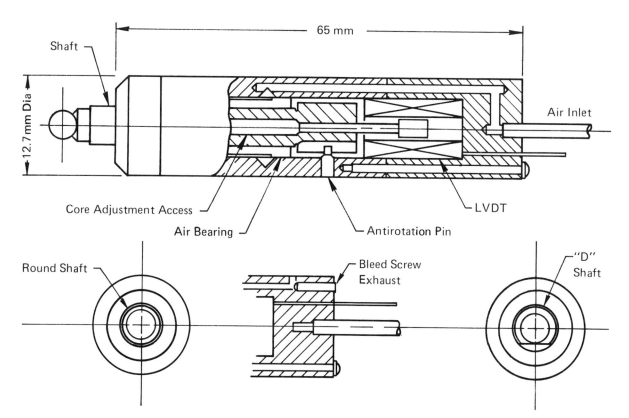

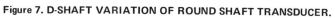

Figure 7. D-SHAFT VARIATION OF ROUND SHAFT TRANSDUCER.

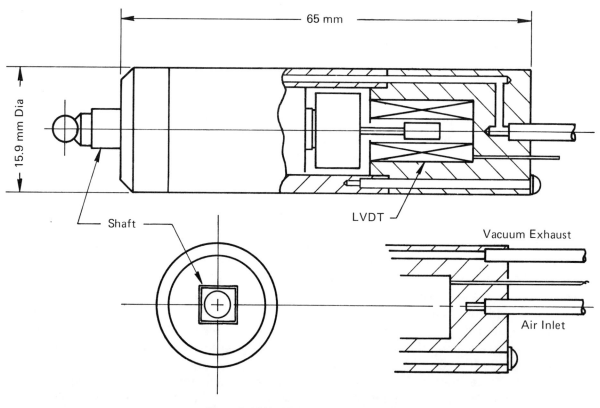

Figure 8. SQUARE SHAFT TRANSDUCER.

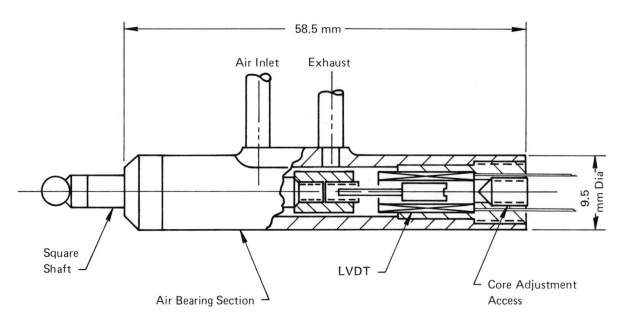

Figure 9. MINIATURE TRANSDUCER.

Because of the difficulties involved in fabrication and the poorer performance criteria, these units will be used only when it is not feasible to modify fixturing for the larger 1/2-inch-diameter units.

All of the air-bearing transducers developed thus far have met with some degree of success and acceptance; however, development effort will continue to follow the use of these units in an effort to overcome the remaining limitations in size, operating characteristics, and fabrication techniques. Current major efforts include attempts to provide an outside source of supply and the evaluation of the miniature units.

Transducer Application

One particular application of these LVDTs is on a Universal Y-Z measuring machine. Measurements are made in a latitude fashion or point-to-point mode at angles of ± 45 degrees from normal to the part surface. Three hundred and sixty deviation readings are taken and the data are fed into a minicomputer where various manipulations are made and part disposition is decided (Figure 10).

Close proximity of the transducers to the part demanded that the LVDTs have retractable shafts. For this application, the normal air bleed port was replaced with an outlet for vacuum supply, and retractability of the shaft was accomplished by full vacuum being supplied through the normally open port of an electrically operated valve (EOV), as shown in Figure 3. During measurements and therefore shaft extension, the same vacuum was routed through a vacuum regulator and the normally closed port of the EOV. By using the vacuum regulator, the vacuum supply to the LVDT was adjusted until the desired tip pressure was achieved at the end of the shaft. In moving along the contour of the part, the gage force resulted in a force vector which varied over 90 degrees. This variation caused the force normal to the shaft to vary from 70% of the gaging force, then to zero, then to 70% of the gaging force in the opposite direction, as shown in Figure 10. For this application, the advantage of the LVDT was a low gaging force and a predictable deflection characteristic, which could be compensated for if required.

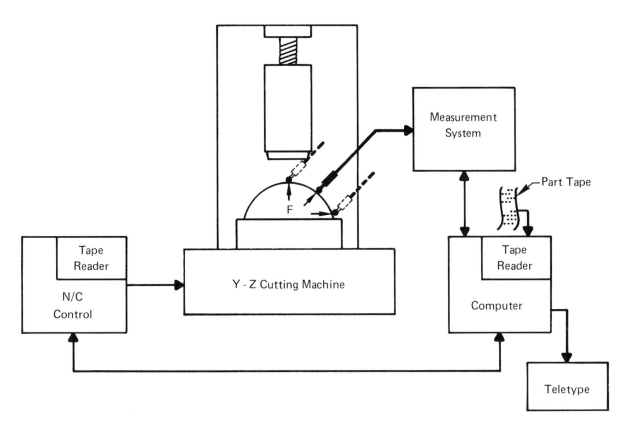

Figure 10. COMPUTERIZED Y-Z MEASURING MACHINE.

By far, the major problems with the conventional LVDTs have been with "stickshun" (erratic operational friction characteristic) and radial play. When the measurement angle of the LVDT is other than normal to the part, the allowable radial-play tolerance becomes much tighter. Unfortunately, the tighter the allowable radial play, the more chance the shaft of the LVDT will stick due to friction and atmospheric dirt. Another inherent problem of the conventional transducer is its difficulty in achieving tip pressures less than 15 grams.

In one particular application, the transducers were maintained at 45 degrees to the part, and tip pressures less than 15 grams were a necessity. To further add to the upset, the LVDT shaft needed to have a reliable low friction due to the dangerously high possibility of part galling. First attempts to inspect these parts with conventional LVDTs were truly disastrous. Measurements with various different types of shaft tips were made (tungsten carbide, ruby, nylon, and Teflon). Use of the most common type (tungsten carbide) was impossible due to surface finish damage and galling. The ruby tip proved to be most successful; but, even then, great care against part damage had to be taken. Additional inspections were required to insure against erroneous data due to "stickshun". Constant lubrication of the transducer had to be maintained throughout the entire part run. With all possible precautions taken, the LVDT failure rate was still quite high; and, worse, the possibility of a part "kill" was ever present.

Upon replacement of the transducers with air-bearing LVDTs, all of the problems just enumerated disappeared. They are indeed "stickshunless" and can be maintained at tip pressures much less than 15 grams. Since the initial installation of these transducers in mid 1972, several hundred part runs have been made, and the air-bearing LVDTs have yet to be replaced due to failure during normal usage. The only two actual replacements that have been necessary were due to machine mishaps which damaged the LVDT shaft and caused transducer failure.

In addition to dimensional inspection applications, the low gaging force possible with this transducer makes it ideal for referencing diamond turning tools on numerically controlled machine tools. The edge of the single-crystal diamond tool used in specular surface metal turning is extremely fragile because it is only a fraction of a microinch in thickness. Any sudden force applied, such as due to stickslip, would generally damage the edge of the tool.

REFERENCES

(1) Steger, P. J.; *Practical Design Considerations for the Application of Gas Bearings to Machining Processes*, Y-DA-4541; Union Carbide Corporation-Nuclear Division, Oak Ridge Y-12 Plant, Oak Ridge, Tennessee; September 18, 1972.

(2) Rasnick, W. H.; *Air Bearing LVDT Transducer*, Y-DA-4839; Union Carbide Corporation-Nuclear Division, Oak Ridge Y-12 Plant, Oak Ridge, Tennessee; November 14, 1972.

APPENDIX

SAMPLE CALCULATIONS
(Reference: Figure 1)

Deflection (Δ) = Beam Deflection (δ_b) + Film Deflection (δ_f)

$$\Delta = \frac{Fa^2}{2EI}\left[a - \left(\frac{a-2b}{b}\right)\right] + \frac{b+a}{b}\left(\frac{R_f}{2As}\right) + \frac{a}{b}\left(\frac{R\nu}{2As}\right)$$

	Round Shaft	Square Shaft	Miniature Square Shaft
Shaft Diameter (in)	0.1875	0.1875	0.100
Material, 6061-T6 Aluminum, $E = 12 \times 10^6$ psi			
Side Force, F (lb)	0.011	0.011	0.011
Moment of Inertia, I (in^4)	6.08×10^{-5}	1.03×10^{-4}	8.3×10^{-6}
Bearing Area, A (in^2)	0.09375	0.09375	0.050
Film Modulus, S (lb/in^3)	4.375×10^4	4.375×10^4	4.375×10^4
$\left(S = E_f \dfrac{(0.35\,Ps)}{h}\right)$			
Beam Deflection, δ_b (μin)	5.66	3.35	41.5
Film Deflection, δ_f (μin)	11.4	11.4	21.5
Total Deflection, Δ (μin)	17.0	15.0	63.0
Compliance, $C = \Delta/F$ (in/lb)	1.55×10^{-3}	1.36×10^{-3}	5.72×10^{-3}

Transducer Time Response

An example of the time response of the conventional round shaft transducer with the following characteristics is (refer to Figure 5 and Page 7):

$$F = 5 \ g_f \ (49 \times 10^{-3} N)$$

$$m = 5 \ g \ (0.005 \ kg) \text{ aluminum shaft and } 0.020 \ LVDT \text{ core}$$

$$x_o = \pm 0.010 \text{ in } (0.25 \times 10^{-3} m)$$

$$L = 10 \ mm \ (0.01 \ m)$$

$$\text{Free Response, } t = \left(\frac{2mx_o}{F}\right)^{1/2} = \left(\frac{2 \ (0.005)(0.00025)}{0.049}\right)^{1/2} \doteq 7.13 \times 10^{-3} \ s$$

$$v_{max} = L/t = 0.01/0.0073 = 1.37 \ m/s$$

$$\text{Periodic Response, } f = \left(\frac{F}{4\pi^2 m x_o}\right)^{1/2} = \left(\frac{0.049}{4\pi^2 \ (0.005)(0.25 \times 10^{-3})}\right)^{1/2} = 31.5 \ Hz$$

$$v_{max} = Lf = (0.01)(31.5) = 0.315 \ m/s$$

The first units listed as popular units; the units in parenthesis are International System (SI) units.

Presented at the SME Precision Machining Workshop, May 1977

Linear Air Bearings for Precision Applications

By P. Donald Brehm
Pneumo Precision Inc.

ABSTRACT

Linear air bearings can be used effectively in the design of machine tool and gaging slides where accuracy and repeatability are important. The basic design theory of hydrostatic linear air bearings and typical precision linear applications are presented.

INTRODUCTION

Equipment for precision machining and gaging require rotary and linear movements of the operating components to the highest degree of accuracy. The application of air bearing spindles for precision machining has been well documented. The purpose of this paper is to discuss the similar application of air bearing slides.

For many years the only type of linear machine tool bearing used was the conventional sliding way system with scraped cast iron members and hydrodynamic lubrication - sometimes with low pressure positive oil supply. Variations use cast iron against hardened steel ways, and various non-metallic materials against cast iron and steel ways. All of these systems exhibit to varying degrees the undesirable characteristics of "stick-slip" (break away friction higher than running friction) and "floating" (inconsistent lift depending on slide speed, amount and viscosity of lubrication, loads, settling time, etc.) Also since sliding contact can take place, wear can occur, with resulting loss of accuracy.

To overcome these limitations, antifriction linear bearing systems with ball or rollers - either in cages or recirculating - acting against steel ways are sometimes used in precision machines and gages. Although friction levels are much improved, the rolling elements can cause "noise" (vibration due to size and geometry variations of the rolling elements and ways) and shock loads can cause "brinneling" (denting of the ways by the rolling elements).

Linear hydrostatic bearings answer many of these design limitations. Although hydrostatic bearings can use fluids as well as air, this paper will deal specifically with air bearings.

DISCUSSION

Hydrostatic air bearings drive their load carrying ability and stiffness through the pneumatic compensating force developed within the bearing in response to an external load. The bearing must receive a constant supply of pressurized air, this air being then distributed within the bearing as to support the load before it escapes to the ambient atmosphere around the bearing. Figure (1) shows the cross section of a typical round pad restriction type bearing. By equating expressions (2) for the air flow through the restrictor and through the bearing gap the bearing performance can be predicted mathematically. A typical set of performance curves derived in this manner is shown in figure (3). These curves are based on the single sided (single directional thrust) bearing shown in figure (1) - if a preloaded double-sided (bi-directional thrust) bearing is analyzed, it will be found that the stiffness is much greater and the yield curves are linear - figure (4). For this reason most precision applications require the preloaded design. Since air is compressible, the various bearing parameters must be chosen with great care to avoid pneumatic instability within the bearing. With optimum design, high stiffness can be obtained - for example a thrust bearing with a diameter of 10" O.D. x 4" I.D. (400mm x 100mm) has a stiffness of 4,000,000 pounds per inch at 60 psig supply pressure. This means that the deflections for precision application are usually in the microinch range.

The characteristics of hydrostatic air bearings most important to precision linear motion applications are:

A. Low friction - due to the low viscosity of air (2000 times less than oil) an air bearing slide moves with very low friction and no "stick-slip". The starting friction is zero and friction increased linearly with speed. At the slow working speeds of most linear applications friction can be considered negligible in any calculation of moving force required.

B. Accuracy and repeatability - Since the linear air bearing has only one stationary member and one moving member there is no error contribution from intermediate balls or rollers. The pressurized air film maintains a uniform slide position regardless of speed or direction and the air film averages many of the residual form errors in the bearing components. Straightness accuracy can be held to .000,010 (0,25u) or less per 12" (300mm), and short term errors can be in the range of 1 microinch (0,025u) or less. Repeatability of the linear motion path will be within 1 microinch (0,025u) or better.

C. <u>Low</u> <u>vibration</u> <u>and</u> <u>noise</u> - because there is no rolling or sliding contact and any residual roughness in the bearing parts is averaged by the air film, linear air bearings have a very smooth signature.

D. <u>Long</u> <u>life</u> - Absence of contact means linear air bearings have unlimited wear life. The broad bearing area when air is either on or off protects from damage from shock loads and shipping/handling.

E. <u>Absence</u> <u>of</u> <u>contamination</u> - Air is a clean lubricant that does not have to be collected and recirculated.

The disadvantages of air bearings for precision linear applications are:

1. They usually have to be custom designed for each application for optimum performance.

2. Although plain thrust stiffness is very good, overhung or moment loads require special design consideration.

3. A source of clean air is required - although restrictor type air bearings (and to a lesser degree, porous bearings) have a remarkable tolerance to air line contamination and with proper materials choice, rust from air line moisture can be controlled; air bearings do require the added expense and complexity of a filtered air source.

4. Tolerances and alignment requirements are very critical both in manufacture and mounting, since total bearing

clearances are from .0001 (2,5um) to .001 (25um).

DESIGN VARIATIONS OF AIR BEARING LINEAR SLIDES

Many design variations are possible in applying linear air bearing technology. The basic configurations are shown in figure (5). The choice of the optimum design for a given application will depend on many considerations and constraints, including:

A. Accuracy/manufacturing consideration - high accuracy requires designs that have a minimum of surfaces requiring precision finishing, and designs that have the ability to be mounted and aligned to close tolerances. Bearing in mind that the air film averages out small errors in the bearing parts, but not long path errors (straightness, twist, etc.) and that the most critical surface of the bearing from an accuracy consideration is the surface opposite the member that has the air restrictors, the best design can then be chosen.

B. Load/stiffness consideration - the magnitude and direction of the external forces on the linear bearing will determine the best design. In general the air bearing surfaces will require more space than ball, roller or plain bearing designs to carry the same load. Many times creative design of the total machine or gage can minimize the loads, and particularly the offset loads that the linear bearings must support.

ACCURACY VERIFICATION

Accuracy checking of linear slide for precision applications requires very careful techniques. Straightness displacement accuracy of short travel slides can be checked with an optical flat reference master and high magnification gage head - amplifier - strip recorder. If short term excursions from true travel are important, then the set up must have sufficient frequency response to see all movements of the gage head relative to the optical flat.

For longer travels, a granite straightedge can be used in place of the optical flat, however the grain structure of the granite will affect the high frequency readings. In either case the reference master error can be separated from the linear slide error by displacing the master and reversing the master, and observing the difference in the recording chart traces.

Optical auto collimators and laser alignments interferometers can be used very effectively to check linear straightness accuracy, including pitch, roll, and yaw errors, as well as displacement errors.

Any of these methods must be used with measurement taking place along the functional path of the machine - i.e. at tool height or if this is not possible, by translating mathematically the known displacement, pitch, roll, and yaw errors to the proper working point or path.

APPLICATIONS

We have applied linear air bearings to the design of straightness gages to give a straightness path accuracy of .000,060 (.,5um) in 60 inches (1520mm), and .000,001 (0,025 um) per inch

(25mm).   See figures 6 and 7.

We have used linear air bearing slides in the design of our MSG diamond lathes and flycutting machines. Surfaces of metal optical mirrors have been turned and flycut on these machines to finishes of 0.2 microinch AA (0,05um) and geometry (Flatness) of .000,010 per 12 inches (0,25 um per 300 mm). These linear applications have included slides to support both tools and workpieces, with the slides supporting up to 1500 pounds (680 kg). See figures 8 and 9.

CONCLUSIONS

With proper selection and application linear air bearings can be applied to meet the highest demands of precision machining and gaging devices.

REFERENCES

1. Brehm, P. D.; "Use of Air Bearings in Gaging Equipment", 1Q76-803, S.M.E.

2. Rasnick, W. H.; Arehart, T. A.; Littleton, D.E.; Steger, P. T. "Porous Graphite Air-Bearing Components as Applied to Machine Tools", MRR74-02, S.M.E.

3. Gerth, H. L.; "Improving a Production Lathe to Machine Optical Components High Energy Lasers" MR74-942, S.M.E.

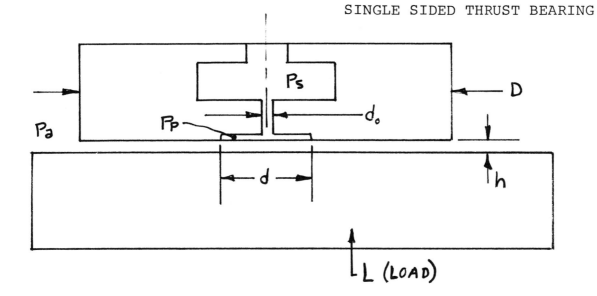

Ps = Supply Pressure
Pp = Pad Pressure
Pa = Ambient Pressure
D = Bearing Diameter
d = Recess Pad Diameter
d_O = Orifice Diameter
h = Air Film Thickness
L = Applied Load

AIRFLOW THROUGH ORIFICE d_O

$$= \left[C_d \, \pi \, \frac{d_o^2}{4} \, P_s \sqrt{\frac{2K}{(K-1)RT} \left[\left(\frac{P_p}{P_s}\right)^{2/k} - \left(\frac{P_p}{P_s}\right)^{\frac{K+1}{K}} \right]} \right] \ \#/_{SEC}$$

AIRFLOW THROUGH BEARING AIR
FILM h =

$$= \left[\frac{\pi h^3}{12 \mu RT} \right] \left[\frac{P_p^2 - P_a^2}{\ln D/d} \right] \ \#/_{SEC}$$

EQUATING THESE FLOWS AND SOLVING FOR
FILM THICKNESS:

FILM THICKNESS (h) $= \left[\dfrac{3\mu RT \ln D/d}{P_p^2 - P_a^2} \, C_d \, d_o^2 P_s \sqrt{\frac{2K}{(K-1)RT} \left[\left(\frac{P_p}{P_s}\right)^{2/k} - \left(\frac{P_p}{P_s}\right)^{\frac{K+1}{K}} \right]} \right]^{1/3}$

FIGURE 1. AIR BEARING DESIGN CALCULATIONS

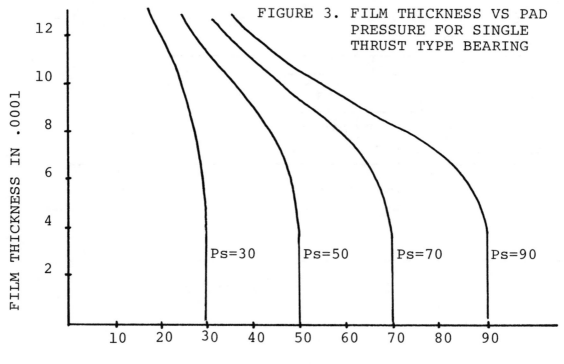

FIGURE 3. FILM THICKNESS VS PAD PRESSURE FOR SINGLE THRUST TYPE BEARING

FILM THICKNESS IN .0001

Ps=30   Ps=50   Ps=70   Ps=90

PAD PRESSURE - PSIG

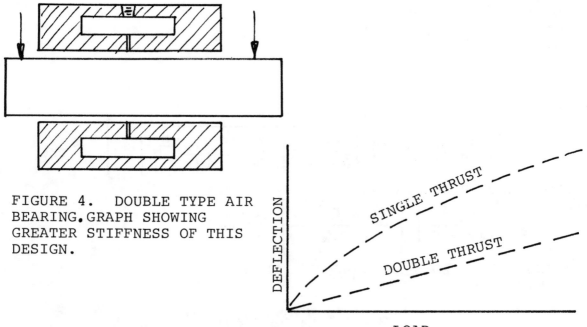

FIGURE 4. DOUBLE TYPE AIR BEARING. GRAPH SHOWING GREATER STIFFNESS OF THIS DESIGN.

SINGLE THRUST

DOUBLE THRUST

DEFLECTION

LOAD

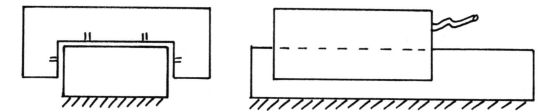

A.) AIR SUPPLY TO MOVING MEMBER

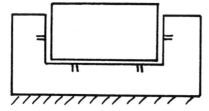
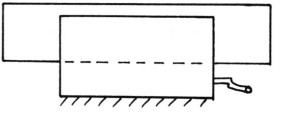

B.) AIR SUPPLY TO FIXED MEMBER

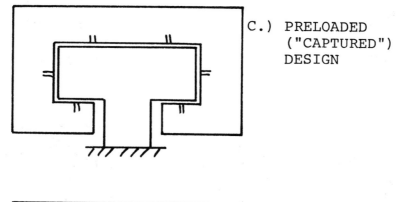

C.) PRELOADED
    ("CAPTURED")
    DESIGN

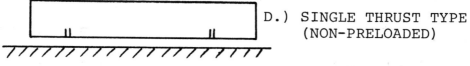

D.) SINGLE THRUST TYPE
    (NON-PRELOADED)

FIGURE 5. LINEAR AIR BEARING DESIGN VARIATIONS

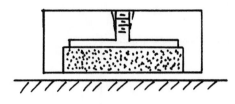

E.) POROUS TYPE

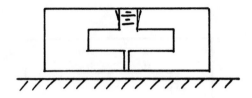

F.) RESTRICTOR (ORIFICE)
TYPE

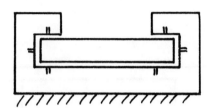

G.) RECTANGULAR TYPE

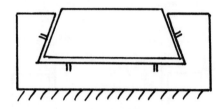

H.) ANGULAR TYPE

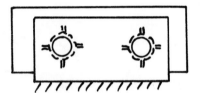

I.) ROUND WAY TYPE

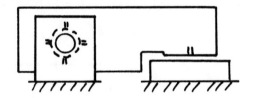

J.) ROUND AND FLAT

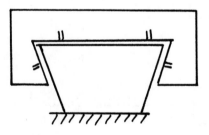

K.) FIXED PADS

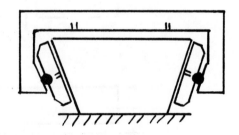

L.) SELF-ALIGNING PADS

FIGURE 5. (Continued)   DESIGN VARIATIONS

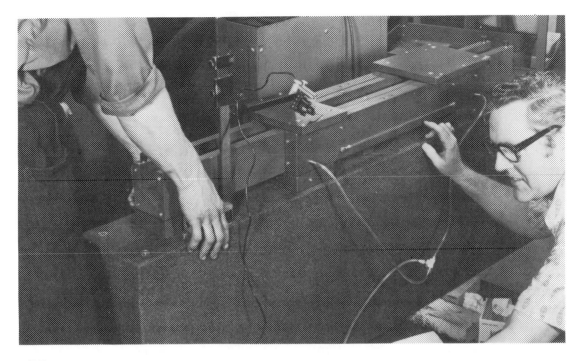

LINEAR AIR BEARING FOR MONOCHROMATOR

LINEAR AIR BEARING FOR CAM GRINDER

FIGURE 6.   LINEAR AIR BEARING APPLICATIONS

FIGURE 7. LINEAR AIR BEARING GAGING SLIDE APPLICATIONS -
MAX TRAVEL (TOP PHOTO) OF 60".

FIGURE 8.   AIR BEARING WORK SLIDE ON DIAMOND TOOL
FLYCUTTING MACHINE.   SLIDE TRAVEL 8" WITH MOVEMENT
BY AIR/HYDRAULIC DRIVE.

FIGURE 9.   FULLY ENCLOSED AIR BEARING TOOL SLIDE

ON MSG-300 DIAMOND LATHE.   TOTAL TRAVEL OF 18" BY

BALL SCREW/SCREW SERVO MOTOR DRIVE.   STRAIGHTNESS OF

TRAVEL OF .000,010 per 12".   CAPABLE OF FINISHES TO

0.3 MICROINCH AA.

PAPER LAYOUT GAGING

Paper layout gaging is a direct and inexpensive technique for making an immediate functional check of inspection results, permitting the adjustments possible with functional three-dimensional receiver gaging. It provides a method of determining if a part can be reworked, and the most economical rework required. It is also a useful way to evaluate tooling, indicating the resetting required to produce an acceptable product. Moreover, paper layout gages never wear out nor require storage space, as do receiver gages.

Paper gaging is especially useful in reconciling conflicting inspection results, where drawings with implied datums have forced inspectors to select arbitrary setup datums. For example, one inspector might use a hole or holes to establish datums; another might use part edges. The inspection reports would probably differ in their results, but a paper gage could be used to determine if the part is functionally acceptable, regardless of the setup used.

APPLICATION

The first step in applying the "paper gaging" technique is to decide when to use it. An examination of the inspection report will yield certain essential information about design specifications and inspection procedures, indicating whether or not paper gaging is necessary or feasible.

Parts That Cannot Be Paper Gaged

A part with coordinate datum dimensions, all originating from a single specified datum reference frame, does not lend itself to paper gaging, regardless of the tolerancing method, because the part features are fixed in relation to the datum reference frame (see Fig. 1-8).

Special Cases

There are some parts that can be paper gaged, but the procedure is more trouble than it is worth. It is better and easier to analyze the inspection results directly. A prime example of this is inspecting a part with concentricity requirements.

Fig. 3-1 shows a drawing of a part that must be concentric about a datum axis. Part concentricity, specified as a positional tolerance, can be partially checked by rotating the part about its datum axis and measuring surface runout with an indicator.

There are numerous ways to determine the axis of a datum feature. In this functional case, a gage pin, 0.7015 in. diameter, just fits datum feature A. The setup, using this pin inserted into a precision chuck, is shown in Fig. 3-2.

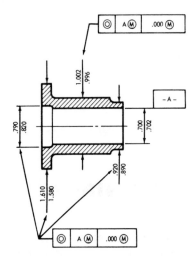

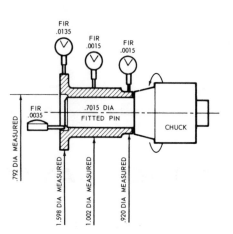

Fig. 3–1. A concentricity requirement.

Fig. 3–2. Setup for inspecting concentricity.

Since zero MMC tolerancing is specified, the concentricity tolerances allowed depend on the finished size of each diameter, and even the datum diameter can influence them. Because the datum diameter is 0.7015 in., or 0.0015 in. larger than MMC, all concentricity tolerances can be increased from 0.000 in. to 0.0015 in. to begin with (see Table III-1).

Note that the FIR (full indicator reading; same as total indicator reading) checks in Table III-1 can be deceiving, because they include elements of runout,

Table III-1. Inspection Results

Measured Feature Diameters (in.)		Datum Tolerances (in.)		Concentricity Tolerances (in.)	FIR (in.)
0.792	+	0.0015	+	0.002 (0.792 − 0.790)	0.0035
1.598	+	0.0015	+	0.012 (1.610 = 1.598)	0.0135
1.002	+	0.0015	+	0.000 (1.002 = MMC)	0.0015
0.902	+	0.0015	+	0.000 (0.920 = MMC)	0.0015

out-of-roundness, measurement axis error, and so on. A detailed form inspection should be made along with all FIR measurements if eccentricity is to be segregated.

Parts That Can Be Paper Gaged

Parts with independent patterns of axial features (holes or pins, for example) related only to a primary datum, which would ordinarily be gaged with a feature relation receiver gage, can be more effectively paper gaged.

Dimensioning to allow hole-pattern independence on a part is shown in Fig. 3-3. This type of part creates problems for the gage designer when he attempts to create a design to gage hole 4, because hole 4 is erroneously considered to be a datum hole. Many gage designers therefore gage this hole with a tapered pin, which unduly restricts the relationship of hole 4 to the other holes in the pattern.

PAPER GAGING PROCEDURE

Separate layouts are made for a pattern of part features, one of measured axis locations, and the other of positional tolerance zones. The tolerance zone

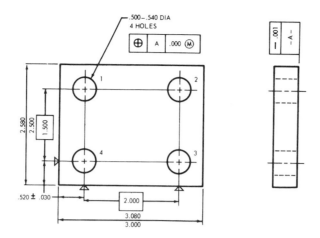

Fig. 3-3. Independent hole pattern dimensioning.

layout is superimposed over the measured axis layout, and rotated and translated to determine if any single orientation allows all plotted axis points to fall within their respective tolerance zones. (The procedure is *analogous* to drawing an extremely large part exactly as described on the inspection report and comparing it to another equally large drawing of the gage to see if the part fits [see Fig. 3-4]. Obviously this particular procedure is impractical because the scale factor required to make the tolerances visible would create drawings to fill a room.)

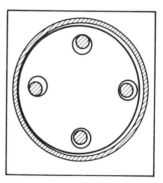

Fig. 3-4. Gage/part drawing method of paper gaging.

The part shown in Fig. 3-3, which will be used to illustrate the procedure, can be inspected using two outside peripheral surfaces or any two holes as secondary and tertiary datum features. In any case, datum surface A is used as the primary datum. The results of one inspection procedure were shown in Table II-2, based on Fig. 3-3. Two outside surfaces were used as secondary and tertiary datums, and the part was set up as shown in Fig. 2-1. The (A) modifying a measurement in Table II-2 merely indicates that this was the coordinate location of a hole-center point nearest datum surface A.

Scaling Tolerances

Tolerances are generally in the order of magnitude of thousandths of an inch. For this reason, tolerance zones and measured axis variations must be scaled up so that they can be seen and accurately plotted and evaluated.

Scaling Dimensions

If dimensions are scaled up by the same factor as tolerances, the layout will be too large to handle. The same scale factor is therefore not used for dimensions unless the actual paper gaging operation indicates a marginal part.

Format

Experience has dictated that the layout format should resemble the geometry of the part. Unsuccessful attempts have been made to use prepared plots of concentric circles about a common axis, but this technique has led to confusion and many mistakes in plotting and evaluation.

Material

Plastic materials such as Mylar can be used in place of paper for layouts. Mylar is somewhat more stable than paper, and is reusable, since pencil lines on it can be easily erased. Several Mylar sheets and a grid can be used over and over again, with copies of each gage made as a permanent record.

Plotting

The method of layout and plotting is determined by pattern complexity. A simple pattern of 4 to 8 features can be plotted on an 8 1/2 by 11 in. sketch-pad sheet laid over a standard metric grid. A complex pattern of 12 to 36 features is laid out on a larger sheet of paper, using a drafting board and regular drafting instruments.

Inspection Results Layout

The measured locations of points at both ends of each feature axis are plotted in relation to the X and Y axes established by the datums specified on the inspection report. The two perpendicularity points plotted for each hole axis can be joined by a line to indicate that they are points for one hole axis. This procedure creates the effect of a three-dimensional gage.

Grid Plot. Each square in a preprinted metric grid is assigned a value, perhaps 0.001 in. (Table III-2 gives tolerance values that can be assigned to

Table III-2. Positional Tolerance Values (10 × 10 grid)

Positional Tolerances (in.)	Equivalent Grid Square Value (in.)
0.002 − 0.005	0.0001 or 0.0002
0.005 − 0.030	0.0005 or 0.001
0.030 − 0.060	0.002
0.060 and over	0.005

the squares of a 10 by 10 grid.) Since each square is 1 mm, or 0.039 in. on a side, this would give a scale factor of 39:1. A piece of transparent or translucent sketch paper is laid over the grid, fixed so that it will not shift, and the basic locations of each axis plotted with a compass point for accuracy. As an aid to quick identification of part geometry, the center lines may be joined in a rectangular grid pattern.

The difference between the basic axis location and each measured axis point location is translated into metric grid squares, and counted off in the X and Y directions from each basic axis location plotted on the layout.

Fig. 3-5 shows the inspection plot for Fig. 3-3 and Table II-2.

Drafted Plot. The basic axis locations of a bolt circle or similar simple circular pattern can be laid out with a drafting machine and a large beam compass. The basic locations of complicated or rectangular patterns can be laid out with a drafting machine.

The measured axis locations are readily plotted from each basic location; plotting the X and Y measurements is made convenient by the fact that the drafting instrument moves in the X and Y axes.

An appropriate scale factor for plotting the measurements is selected and monitored on the drafting machine scales, perhaps 1/16 or 1/32 in. for each 0.001 in. of actual variation.

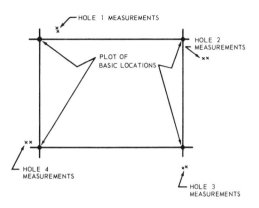

Fig. 3–5. Plot of hole center deviations from inspection report.

Tolerance Layout

It is important that the basic axis locations of the inspection results layout be plotted identically on the tolerance layout, which is the second layout required in paper gaging. This can be done by redrafting, or laying the tolerance layout sheet over the inspection results layout and transferring the basic locations directly through with a compass point.

The positional tolerance zone diameters taken from the inspection report are scaled up by the same factor selected for plotting the measured axis locations, and drawn in with a compass.

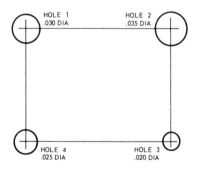

Fig. 3–6. Plot of positional tolerance zones.

A tolerance layout for Fig. 3-3 and Table II-2 is shown in Fig. 3-6. The positional tolerance zones are plotted directly from the "Tolerance Diameter Allowed by MMC" column on the inspection report.

Combining Layouts

Since the sizes of the tolerance zones vary, the results of paper gaging will be entirely nullified if the tolerance zones are not placed over their properly corresponding axis locations. It is thus advisable to mark or number each hole location on both layouts of a complex pattern, particularly if the pattern is symmetrical.

The completed tolerance layout is placed over the inspection results layout, and slightly shifted and/or rotated if necessary to determine if all sets of measured axis location points fall simultaneously within their respective tolerance zones, thus graphically demonstrating whether or not the part being gaged will assemble with its mating part.

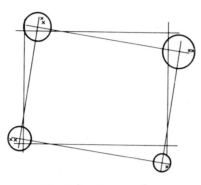

Fig. 3–7. Paper gaging.

Fig. 3-7 shows the completed paper gage for the part of Fig. 3-3 and Table II-2. The tolerance layout has been rotated and shifted, much the same as a receiver gage or mating part would be adjusted for assembly. All four sets of axis points (top and bottom readings) fall within their respective tolerance zones, demonstrating that the part is acceptable.

To complete the inspection analysis, holes 1, 3, and 4 are checked to see if they meet their 0.520 in. (±0.030) location specification.

Measurements taken from any setup used for the part in Fig. 3-3 will result in plots of hole centers and tolerance zones that have the same relationship to each other. The tolerance zone plot (Fig. 3-6) is uniformly the same, and the finished hole sizes and their perpendicularity remain constant for the same part. Thus identical results are obtained with paper gaging.

ALLOWANCE FACTORS

The overall accuracy and validity of paper gages are affected by such factors as the accuracy of the layouts themselves, the accuracy of inspection measurements, and the completeness of inspection results. Also, allowances must sometimes be made for gage tolerances and gage wear if paper gaging is to be used in conjunction with functional receiver gages. Reducing the diameters of the tolerance zones on the paper gage layout by a fixed percentage can serve as an overall safety factor to compensate for inaccuracies and gage allowances.

Layout Accuracy

The material on which the layouts are made introduces error equivalent to its coefficient of expansion and contraction. If the same material is used for both inspection and tolerance layouts, this type of error cancels out.

The grid used for a plot can introduce error if it is not a perfect grid (and a printed grid can be assumed to be somewhat imperfect). Using the same grid in an identical manner for making both layouts cancels out this error.

There is likely to be at least 0.010 in. error in positioning lines, points, and diameters. This error is directly minimized by the scale factor selected, thus:

$$\frac{\text{Positioning Error}}{\text{Scale Factor}} = \text{Actual Error}$$

In making the layouts, a 0.010 in. width of pencil line, with a 100:1 scale factor, can cause a 0.0001 in. error, 0.00005 in. on each side of the line (0.010/100). This error can be minimized by working to one side of the line.

Inspection Measurements

Obviously a certain amount of error is inherent in inspection measurements. However, since open-setup inspection is used to calibrate tooling, fixtures and gages, it can be assumed that carefully made measurements will not produce an error factor greater than 5 per cent in the data on an inspection report.

Incomplete Inspection Results

Sometimes an inspection report is incomplete in that it does not contain information about setup or hole perpendicularity, yet a paper gage must be made from it. Therefore, some reduction in the size of the tolerance zones can be made to compensate for the uncertainty of incomplete inspection results. This reduction may be as much as 25 per cent and not be unreasonable. Such a reduction is useful when the inspection report contains an uncertainty factor, as previously discussed, since the reduction can be directly applied to modify the tolerance zone layout.

Gage Allowances

Reducing the size of each tolerance zone can also serve to include allowances for gage tolerances and wear. This is important if the paper gage is intended to accept or reject to the same degree as a receiver gage. Such would be the case when paper gages are used for in-process checks, or at the beginning of a tool run when receiver gages are not yet available but will be used.

No allowance should be made on the paper gage for gage tolerance or wear if no receiver gage will be used for the part or parts being inspected. This would only lead to rejection of otherwise acceptable parts. Also, no reduction of tolerance zone size is necessary to accommodate virtual size; the plotting of both ends of the axis to establish perpendicularity automatically takes care of this, because both points must fall within the tolerance zone. This also gives the paper gage a three-dimensional effect.

Gage Policies

The question of allowing for gage tolerances and wear brings up the problem of differing policies. No standard yet in general use presents an unequivocal interpretation of go and no-go gage tolerances, nor is there a commonly accepted standard percentage of wear to be allowed before a gage is taken out of service to be reconditioned. It is extremely important for users and suppliers (or design, manufacturing, and inspection departments) to arrive at a common understanding on these matters before production starts.

WHO PREPARES THE PAPER GAGE?

The paper gaging operation can be performed by the inspector, a quality control engineer, a draftsman, or the designer. The inspector or the quality control man is able to make a paper gage, but the final decision on acceptance is not his, it is the designer's. Particularly in a small company, where the designer is probably also the draftsman, he might be expected to want to make his own gages. It would not be unrealistic for a large company to employ a layout man specifically for the purpose of making paper gages. His salary would be paid many times over by the cost of acceptable parts saved from rejection.

ANALYZING RESULTS

It is in analyzing a completed paper gage that the true power of this technique becomes clear. Not only does it indicate functional acceptance, it also can be used as the basis for determining the feasibility of reworking a part, the nature of the most economical rework required, and what tooling changes might be required to bring parts into acceptance. A series of paper gages made during a production run can be used to monitor the rate of tool wear, and predict accurately when tooling should be replaced or reworked.

Rework Determination

Paper gage analysis, including scaling plotted axis locations in relation to the perimeter of their respective tolerance zones, can be of great value in determining rework.

Suppose the hole centers lie outside their respective tolerance zones; if one or more tolerance zones can be increased in diameter (which is accomplished by reaming out the corresponding holes to a larger diameter), the part can be accepted. Measurement of the paper gage relationships between axis locations and tolerance zone perimeters can be used to determine the amount of rework, and when and where it is required.

Whoever makes the paper gage can indicate on the inspection report the rework required to bring the part into acceptance, and indicate a provisional "OK" in the results column for the particular feature or features to be reworked.

Ordinarily the inspector does not judge acceptance of a part; he merely records inspection measurements on the report, and the designer determines acceptance. However, if he makes the paper gage, the inspector can indicate rework on the report. Of course it is necessary that the part be reinspected and perhaps paper gaged again after rework to make sure that it is then acceptable.

Tooling Check

Succeeding parts made on the same tooling will look much the same as the first part, with only minor variations. Thus necessary changes in tooling can be determined on the basis of one paper gaging operation, and perhaps additional perfunctory checks made during production.

Paper gages can also be made on a regular basis to conveniently monitor tool wear so that the frequency and occasion for change and rework of tooling can be accurately predicted.

COMPLETING THE INSPECTION REPORT

Paper gaging is actually a supplement to variables inspection that completes the inspection operation.

The paper layout gage can be considered a part of the inspection report, and should be attached to it. Once the paper gage has been verified (therefore "calibrated") by a second independent check to catch any incidental errors, it can take on the status of any other functional gage.

If any number of parts meets essentially the same dimensional requirements, and is therefore acceptable by the same paper gage, the individual inspection reports can: (1) reference the report that includes the paper gage, or (2) include a copy of the original paper gage.

PAPER GAGES COMPARED TO OTHER FUNCTIONAL GAGES

Paper layout gaging offers advantages comparable to those of three-dimensional receiver gaging and optical comparator gaging, plus additional advantages that are unique.

The receiver gage, however, has one distinct and unique advantage: being a three-dimensional object, it receives a part exactly as a worst mating part would, thus giving direct evidence of function. The optical comparator and paper gage, being two-dimensional representations of the receiver gage, are one step removed from a true functional gage.

Compensation for the value of the missing third dimension sometimes can be achieved on an optical comparator if the part is mounted so that the primary datum surface is perpendicular to the collimated light beam, and the part moved so that an edge throughout the depth of the part can be examined in focus to ensure that it meets its locational requirement. Obviously the part should be thin enough to be conveniently checked. The paper gage will include the three-dimensional effect of a receiver gage so long as three-plane datum setup and axial perpendicularity information is included on the inspection report and included in the layouts.

One disadvantage of the receiver gage (in addition to cost and lead time required to obtain one) is the necessity for gage tolerance and wear allowance, which cuts into its acceptance rates. Neither the paper gage nor the optical comparator need include these, unless they are being used as preliminary inspection devices prior to receiver gaging.

An important and unique advantage of the paper gage is that there is no significant time factor involved in making one; it can be developed quickly and easily for a single part with little planning and can be as quickly verified. Another advantage of the paper gage is flexibility of application. It is impractical to design and construct receiver gages or chart gages for only one or two parts, but paper gages can be applied conveniently to 1,100, or 10,000 parts. In the case of large numbers of parts that look alike, only a few paper gages may be required to check questionable parts. Finally, inspection report comparisons are more meaningful when paper gages are used.

BENEFITS OF PAPER GAGING

Manufacturing Requirements

Current dimensioning and bilateral tolerancing techniques usually employ a hedge factor that cuts tolerances in half to make sure that manufacturing stays within the necessary limits. Paper gaging the first tool-made sample part in a production run can tell a manufacturer immediately how well his tooling meets the design specifications during production. This might encourage him to relax his tolerance and size limits or establish new tooling "nominals," thus increasing potential acceptance rates.

Material Review Boards

Material (or manufacturing) review board (MRB) meetings are post-mortems that examine part rejections. Under current evaluation techniques, these meetings are frequently based largely on conjecture, and the conclusions influenced more by engineer rank and persuasiveness than engineering information and techniques.

Paper gages can serve as an excellent basis for resolving MRB disagreements. A manufacturer's representative, for example, can produce paper gages made from production parts to support an argument for changing design specifications that he considers unrealistic. This can also help persuade the product designer to perhaps employ the zero MMC concept if feature sizes go beyond certain limits. A properly verified paper gage can give members of a material review board a solid basis for their decisions.

AUTOMATED PAPER GAGING

In place of manually constructed paper gage layouts, a computer program can be written to process the X and Y coordinate measurements and tolerance zone diameters on inspection reports, "rotating" and "translating" this data in the computer to determine functional acceptance. The computer would print out data indicating acceptance or rejection — or produce the data in graphical form, resembling manual layouts.

The program could be written so that data indicating rejection would include specific instructions for the most economical rework required for acceptance.

FURTHER APPLICATIONS

Any method of tolerancing can be incorporated into paper gages—even bilateral tolerances and RFS positional tolerances.

For bilaterally toleranced parts, the tolerance zones on the layout can be any geometry the drawing describes. RFS positional tolerances would simply not vary in size.

MMC positional tolerances for slot features are represented by rectangular, not circular, tolerance zones that vary as slot sizes change.

RFS-to-datum specifications are reflected in paper gages that can rotate slightly but not shift (see Figs. 3-8 and 3-9). Paper gages used for MMC-to-datum specification may shift as well as rotate slightly, if the datum features are not at their MMC size (see Figs. 3-10 and 3-11). Such a paper gage has an additional tolerance zone for the datum feature. It is used to gage parts that would otherwise be gaged with a three-dimensional shake gage.

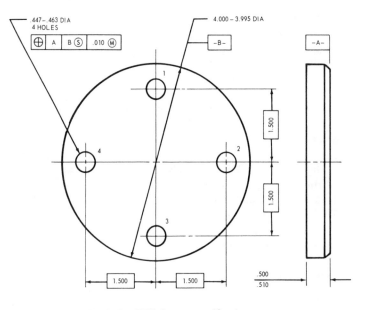

Fig. 3-8. RFS datum specification.

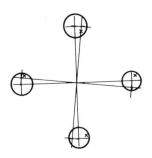

Fig. 3-9. Paper gaging RFS datum.

154

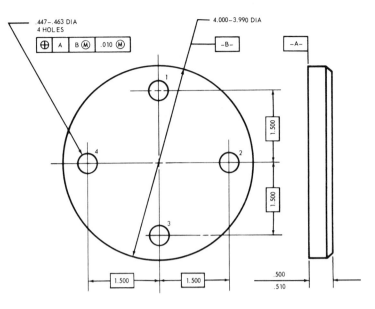

Fig. 3-10. MMC datum specification.

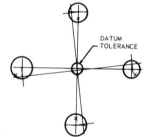

Fig. 3-11. Paper gaging MMC datum.

Reprinted from *Tooling & Production*, June 1978

How to measure rundness

by William E Drews
Rank Precision Industries, Inc
Des Plaines, IL

Life in our industrial world depends on machines with all types of rotating devices. From watches to automobiles to power generators—all have rotating parts in the form of shafts, bearings, gears, bushings, wheels etc. The common element in all of these parts is that they are round. But how round is round?

It can seldom be assumed, without measurement, that components are round enough for their intended purpose. In order for journal bearings to carry high loads and resist wear, the journal must be round. Hydraulic valve spools and bodies, as well as diesel fuel injector plungers, must be round to reduce leakage to a minimum and still allow assembly and operating clearance. Surfaces that run in contact with seals must be round to avoid leakage and reduce wear.

Modern manufacturing requires closer tolerances and more accurate measurement of out-of-roundness (or of roundness, as it is more conveniently called). Instruments are now in use for measuring roundness and it is important for anyone connected with manufacturing—designers, machine tool operators and inspectors—to have a basic understanding of the operating principles, capabilities and limitations of these instruments and how to interpret the results they provide. Not only can these re-

sults determine acceptance or rejection of round parts, but they can also be used to monitor machine tool performance, detect tool wear and reveal the effects of poor operating procedures.

What is roundness?

Generally, a part is considered to be round in a specific cross-section if there exists within that cross-section a point (the center) from which all other points on the periphery are equidistant. The cross-section is, therefore, a perfect cir-

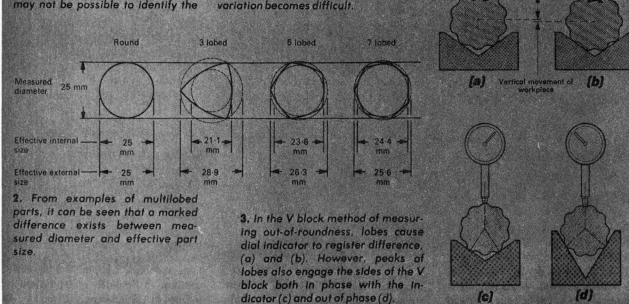

1. A perfectly round cross-section (left) has all points on the periphery equidistant from the center. In a symmetrical shape (center), out-of-roundness is specified as r_1 minus r_2. In an irregular profile (right), it may not be possible to identify the center, thus measuring profile variation becomes difficult.

2. From examples of multilobed parts, it can be seen that a marked difference exists between measured diameter and effective part size.

3. In the V block method of measuring out-of-roundness, lobes cause dial indicator to register difference, (a) and (b). However, peaks of lobes also engage the sides of the V block both in phase with the indicator (c) and out of phase (d).

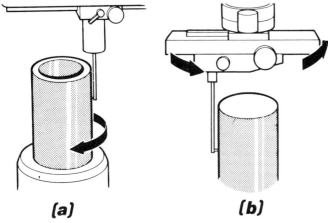

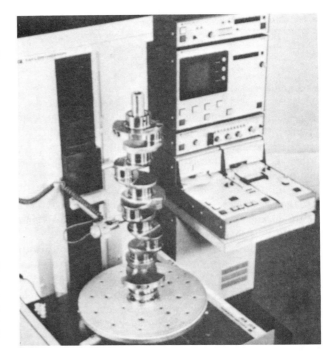

4. *Two basic types of roundness measurement instruments—turntable type (a) in which the workpiece rotates; rotating pickup type (b) in which pickup rotates around workpiece. Among the features of the Talycenta system shown in the photo is automatic centering and leveling capability accomplished by pushbutton once the component has been brought within range of the pickup. Roundness and straightness functions are built into one system to minimize setup time and, in this case, accommodate crankshaft measurement.*

cle. When dealing with a cross-section which is not a perfect circle but is a symmetrical shape, the out-of-roundness is specified as the *difference* in distance of points on the periphery from the center, as seen in **Figure 1**. Specifying out-of-roundness of an irregular profile in the same manner is only possible if we can find a center from which to make measurements. Finding the center is an important part of roundness assessment.

Roundness must not be confused with diameter. The two are measured by different methods and with different instruments. To measure roundness, rotation is always necessary. Diameter is a static measurement made, for example, with a micrometer. Although roundness and diameter are two distinct parameters, roundness does have a practical effect on the measurement of diameter and can make such measurement very misleading.

Measured size vs effective size

Measured size means "diameter" as measured between a pair of parallel faces such as those of a micrometer. The true effective size of a part, however, takes into consideration the out-of-roundness condition of the part. The effective size of a shaft or OD part is the measured size plus out-of-roundness. For a bore or ID workpiece, the true effective size is the measured size minus out-of-roundness. It can be seen by the examples in **Figure 2** that the shape (roundness) of a part can affect its size as measured in the conventional way.

Because out-of-roundness cannot always be detected or measured by a micrometer, some other method must be adopted which will involve workpiece rotation.

A simple method for measuring out-of-roundness is to place the part in a V block with a dial indicator mounted above, and in contact with, the part. Rotate the part slowly and carefully by hand, taking care not to disturb the V block or gage stand and making sure the part rests on the two arms of the V at all times. If the part is truly round with negligible irregularity, the pointer of the indicator will not move. If, however, the part is out-of-round, the irregularities in contact with the sides of the V block will cause the part to move up and down. The irregularities will also displace the plunger of the gage as they pass under it, as seen in **Figure 3**. It can be seen from this and similar three-point methods of checking for out-of-roundness that they suffer from the limitation that results will vary according to the angle of the V and the spacing of the irregularities.

A more accurate method of checking for roundness is to rotate the part between centers, but this is only applicable to parts which have been machined on centers or have sufficiently accurate center locations. Although this method is sensitive to all lobing patterns, there are a number of factors which could contribute to inaccuracy. These include roundness of both male and female centers, alignment of both sets of centers, use of dial indicator, rotation of part by hand.

Causes of out-of-roundness

Undulations and lobes do actually exist on nominally round parts; they are by no means mere theoretical concepts. These deviations from roundness are an important—yet possibly unsuspected—attribute of all machining operations.

Out-of-roundness of machined parts could be due to poor bearings in the lathe or grinding wheel spindle, or to deflections of the workpiece as the tool is brought to bear on it. Shafts, ground between centers, can be out-of-round due to poor alignment of the center or deflection of the shaft.

Lobing is particularly difficult to avoid in many machining processes. A round bar or ring-type part held in a 3- or 5-jaw chuck is compressed at the points of contact. Even if the part is turned or ground perfectly circular on the machine, when it is removed from the chuck the stress in the metal will be relieved, causing three or five lobes.

Measuring roundness

Roundness measuring instruments are of two basic types, **Figure 4**. In the turntable type, the workpiece rotates; in the rotating pickup type, the workpiece is stationary and the pickup revolves around the workpiece. Each type has its advantages and is more suitable for certain types of measurement; the choice depends largely on the measurements to be made and the size and shape of the parts to be measured. The photo in **Figure 4** shows a turntable instrument with an automatic centering and leveling system.

The turntable type lends itself to the measurement of concentricity and alignment because the pickup is not associated with the spindle. Transferring the pickup from an external to an internal surface has no effect on the reference axis. Also, there is more freedom in positioning the pickup in difficult-to-reach areas. On the other hand, the weight of the component and turntable is carried on the spindle bearing, and this imposes some restrictions on workpiece size that can be measured.

In rotating-pickup instruments, the spindle has only to carry the comparatively light and constant load of the pickup. As a result, these instruments are capable of greater accuracy. The worktable can be of substantial construction and the weight of an offset workpiece load is not a limiting factor.

Polar graph interpretation

The trace produced by a polar graph-ing instrument is simply a graphical record, suitably magnified, of the displacement of the stylus of the measuring elements, as either the stylus or the part rotates on the axis of the precision spindle. As discussed earlier, the problems of center location relate to out-of-roundness in that out-of-roundness is expressed as the difference between the greatest and least distance of the profile from the center. To assist in the measurement of this distance, or rather to help identify the maximum peak and valley, it is usually necessary to make use of a reference circle or pair of circles which can be drawn, recorded or otherwise superimposed on the graph. The position of the reference circle and its center is not arbitrary, but is chosen to fulfill certain conditions. Only in this way can an unambiguous and repeatable value for the out-of-roundness of any particular profile be obtained. There are four ways in which such a center can be chosen:

1) Minimum radial separation (MRS); also known as minimum zone circles (MZC).
2) Least squares circle (LSC).
3) Maximum inscribed circle (MIC).
4) Minimum circumscribed circle (MCC).

5. *Profile filters are incorporated in the electronics of a roundness measurement instrument to clarify the profile trace for visual assessment. In polar graph (a), the basic profile of the workpiece is recorded with 1 to 500 upr (undulations per revolution) filter. Graph (b) is the same profile using a 1 to 15 upr filter. In this case, the closely spaced undulations are suppressed and the form of the profile can be more clearly seen. A 15 to 500 upr filter is used in (c). The form error is suppressed and the graph is centered.*

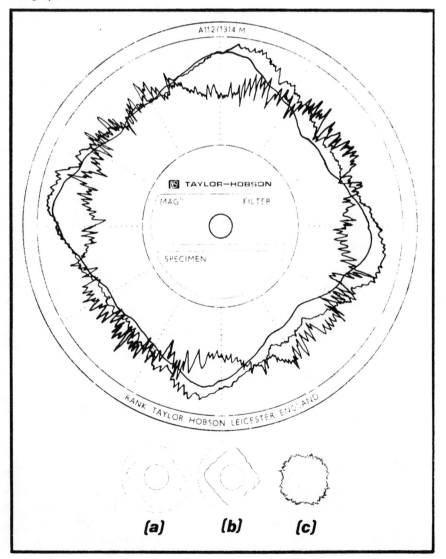

(a)   (b)   (c)

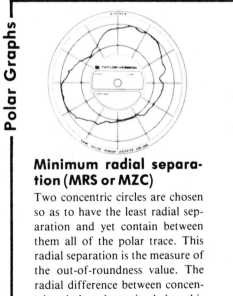

Polar Graphs

Minimum radial separation (MRS or MZC)

Two concentric circles are chosen so as to have the least radial separation and yet contain between them all of the polar trace. This radial separation is the measure of the out-of-roundness value. The radial difference between concentric circles determined by this method is numerically unique, in that by definition a smaller value cannot exist.

Each of the four types are detailed in the accompanying box entitled "Polar graphs."

A template for rapid manual assessment of the graph is supplied with most instruments. It consists of some transparent material on which is engraved or printed a series of concentric circles matching the circular ordinates of the chart. To use it, the chart is placed under the template and then moved about to determine the position of the reference circles. It can be used to find the circumscribed or inscribed reference circles and the minimum zone circles.

Profile filters

If all of the radial deviations of a circular cross-section were fully represented by a measured profile, the presence of surface irregularities of high frequency could mask the lobing condition or the form of the profile. Since the lower frequency surface irregularities may be of greater importance to the part function than the higher frequency irregularities, or vice versa, electronic filters are incorporated in the electronic unit to clarify the profile trace for visual assessment.

The filters perform two functions: either they suppress the closely spaced (high frequency) undulations to reveal

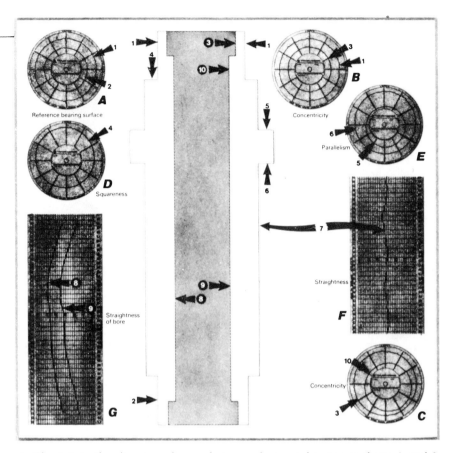

6. *Chart A in the diagram of a workpiece references bearing surfaces 1 and 2. Chart B shows good concentricity of internal surface 3 with bearing surface 1. Chart C shows the eccentricity of the top of main bore 10 compared with 3. Chart D shows the amount by which shoulder 4 is out-of-square with the axis. The lack of parallelism between faces 5 and 6 can be seen in Chart E. With an instrument equipped to measure straightness, it can be seen that the external surface 7 is generally straight and parallel to the axis. Chart G shows that the bore (surfaces 8 and 9) is curved but its diameter is constant throughout.*

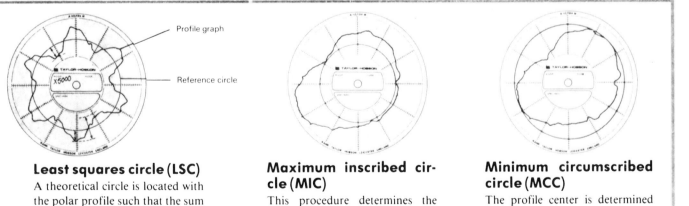

Least squares circle (LSC)

A theoretical circle is located with the polar profile such that the sum of the squares of the radial ordinates between the circle and the profile is a minimum. The out-of-roundness value would be determined by the sum of the maximum inward and maximum outward ordinates divided by the proper chart amplification factor.

Maximum inscribed circle (MIC)

This procedure determines the center of the polar profile by the center of the largest circle which can be fitted inside the profile. From this circle the maximum outward departure of the profile denotes the out-of-roundness.

Minimum circumscribed circle (MCC)

The profile center is determined by the smallest circle which will just contain the measured profile. From this circle, the maximum inward departure of the profile can be measured; this maximum departure is the out-of-roundness.

the form of the profile more clearly, or they suppress the widely spaced (low frequency) undulations and allow the magnifications to be increased to reveal the closely spaced irregularities. The effects of filters can be seen in **Figure 5.**

Although manual methods are widely used, instruments—in the form of accessory units—are available which greatly simplify evaluation of results obtained by roundness measurement instruments. One such accessory is the Taylor Hobson Reference Computer 5, an electronic centering device. The unit computes the least square reference circle of the profile and controls the recorder so that the circle is drawn on the chart, correctly superimposed on the graph, either in a centered or offset mode.

Another accessory is Talydata, Taylor Hobson's module that simplifies and speeds-up analysis of departures from roundness by providing a visual display of workpiece profile and any of the four reference circles. It also includes a digital display of values of departures from the selected reference circle. This accessory renders a recorded graph unnecessary, but if one is required for permanent record, it can be provided.

More than roundness

A single roundness graph is often of limited value because it can give only part of the information needed to determine the overall functioning of a part. However, several related roundness and straightness measurements can give a much more complete picture. In other words, the *whole* geometry of the part must be considered, as far as it can be assessed on a single instrument. In addition to roundness, other parameters can be assessed with a roundness system. These parameters are concentricity, alignment, squareness, parallelism, straightness and conicity.

The workpiece shown in **Figure 6** is an example of the ability to measure several parameters. The part is centered and leveled to bring the bearing surfaces 1 and 2 concentric with each other and the axis of the part in alignment with the axis of rotation of the spindle. This means that all measurements made by the instrument also relate to the axis of the part.

In an automotive engine, one of the most highly stressed parts is the crankshaft. It is subjected to continually vary-

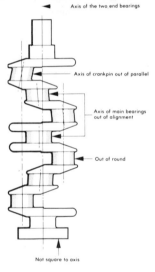

7. Alignment parameters of an automotive crankshaft are shown in the sketch. Photo shows a twin caliper pickup for measuring parallelism of crankshaft journals.

Axis of the two end bearings

Axis of crankpin out of parallel

Axis of main bearings out of alignment

Out of round

Not square to axis

ing bending, torsional and shearing loads. In operation, it is essential that no extra strain or vibration is imposed by the geometry of the bearings. The main bearings must be in line, round and straight along their length as seen in **Figure 7.** The crankpins must also be round and straight with their axes parallel to the axis of the main bearings. All these measurements can be made on a suitable roundness instrument.

Large roundness measuring systems are available with load-carrying capacities of 1000 lb or more. These have centers of gravity within the worktable, allowing complete engine blocks to be measured. Piston bore roundness and straightness and crankshaft and camshaft bores can be checked for alignment and roundness.

Roundness measuring systems are available with special features to accommodate special situations. For example, a special piston measuring system measures concentricity of the piston's crown, skirt and grooves. Squareness of the grooves to the piston axis is also recorded on such an instrument.

Special instruments, such as the Talynova Processor, reduce time and increase versatility of the measuring instrument. Its standard program provides: Reference center functions with

the reference circles displayed or recorded on the polar graph automatically centered; auto-ranging with the profile centered and displayed at highest magnification possible depending on out-of-roundness of the workpiece; harmonic analysis with amplitudes and phase relationships of lobes displayed; cylindricity where measurements made and stored at a number of cross-sections along the cylinder give least squares reference axis; plus others.

Commercially available roundness measurement systems can accommodate the most complex measurement situations, either as standard instruments or modified to suit specific needs.

FUNCTIONAL GAGING WITH SURFACE PLATE EQUIPMENT

Surface plates can serve as the base for specially constructed functional gages, which are partially or completely simulated by standard surface-plate inspection accessories and fixtures. Such gages are carefully constructed setups of fixed geometry that resemble receiver gages, and can be assembled with almost any degree of accuracy. They are clamped together with mechanical or magnetic force, and can be calibrated in place when required.

These surface-plate gages can include wear allowances and can even be toleranced against the part if they are used as emergency measures until actual gages become available. Surface-plate gaging should be considered whenever any or all of the following conditions exist:

1) A number of parts, made from the same drawing, must be inspected.
2) The parts do not require variables inspection data; that is, simple go or not-go attributes are sufficient.
3) Schedules are tight and an immediate in-process or acceptance gage is needed.
4) Open-setup inspection of variables on the surface plate would be time-consuming because: (a) parts are difficult to set up, or (b) part surfaces are hard to contact.
5) The parts do not lend themselves to paper gaging analysis.

GAGING AN MMC HOLE PATTERN

Fig. 4-1 shows a part containing a pattern of clearance holes dimensioned from part datum surfaces A, B and C. Each hole has a variable tolerance of location allowed by the MMC callout (0.010 in. tolerance for a 0.510 in. diameter, 0.017 in. tolerance for a 0.517 in. diameter, etc.).

The setup gage (Fig. 4-2) is complex and requires careful planning. If the part is nonmagnetic, the gage can be held in place by magnetic force instead of clamps. The four pins are all 0.500 in. diameter (0.510 in. hole at MMC minus the 0.010 in. tolerance allowed at MMC), and their basic center locations are 0.750 and 1.750 in. from the angle plate and the parallels. The part is acceptable if the holes fit over the form gage pins and surfaces B and C contact the angle plate and parallel plates. It should be possible to position the primary datum surface (A) of the part parallel to the surface plate, which supports the surface-plate gage items in the plan view. The 0.510/0.520 in. hole size limits must be checked separately to complete the inspection.

Breakout

Fig. 4-3 shows a method of locating a hole pattern on a part where only the breakout distance from the hole to the part periphery is critical. (The holes might be positionally toleranced—see Figs. 4-1 and 4-2.) Fig. 4-4 shows a not-go

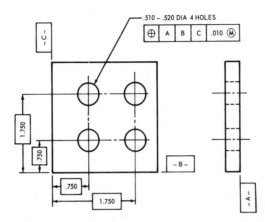

Fig. 4–1. Drawing for a clearance hole pattern.

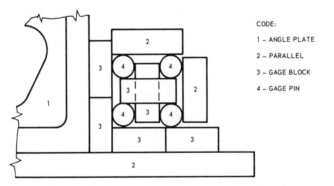

CODE:

1 – ANGLE PLATE

2 – PARALLEL

3 – GAGE BLOCK

4 – GAGE PIN

Fig. 4–2. Surface-plate hole pattern gage (plan view).

gage setup for each of the four holes. The part periphery — rotated for each hole — is placed on the gage-block stack, and the pin, which is clamped to the V block, should *not* enter any hole located a proper distance from the part periphery. Holes should be checked at each part surface, or eight times in all. The pin should be slightly smaller than the minimum hole size specified.

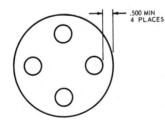

Fig. 4–3. Specification for minimum hole breakout distance.

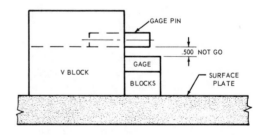

Fig. 4–4. Surface-plate setup for gaging minimum breakout.

GAGING OTHER POSITIONAL TOLERANCES

Symmetry

Fig. 4-5 shows a part with a symmetry requirement (which might also be specified by a positional tolerance) that is difficult to inspect by conventional means. The gage stack shown in Fig. 4-6 is a true go functional gage. The gage

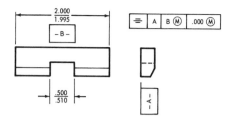

Fig. 4–5. MMC specification for symmetry.

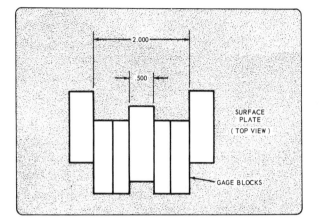

Fig. 4–6. Surface-plate setup for gaging symmetry.

setup has a 2.000 in. gap between the outer gage blocks to accommodate part datum feature B at MMC (2.000 in. on the finished part) and a 0.500 in. slot gage (the slot is allowed no symmetry tolerance when it is 0.500 in., and up to 0.010 in. symmetry tolerance when it is 0.510 in.). The entire gage stack is placed on a surface plate, which must contact part datum surface A during gaging. If the part enters the gage, the inspector need only inspect the 0.510 in. maximum slot width and the 1.996 in. minimum datum-B width to complete the inspection.

Concentricity

Fig. 4-7 shows a concentric part that is difficult to inspect. The single functional-gage setup shown in Fig. 4-8 will relieve the inspector of several calculations and individual setups involving the actual size of datum feature A. The

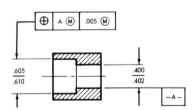

Fig. 4–7. MMC specification for concentricity (MMC datum).

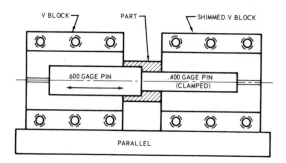

Fig. 4–8. Surface-plate setup for gaging concentricity (plan view).

0.600 in. diameter gage pin (0.605 in. MMC hole minus the 0.005 in. concentricity tolerance allowed at MMC) and the 0.400 in. diameter pin (the MMC size of the datum hole) should be coaxial. If both pins enter and touch bottom in the holes, all the inspector need do to complete the inspection is check the 0.605, 0.610, and 0.402 in. limits.

GAGING FORM TOLERANCES

Perpendicularity

Fig. 4-9 illustrates a part to be measured against a variable perpendicularity tolerance. When the hole is finished to a 0.505 in. diameter, the perpendicularity tolerance is 0.005 in. FIR; when the hole is finished to a 0.510 in. diameter, the perpendicularity tolerance increases to 0.010 in. FIR. (All finished hole sizes in between have corresponding perpendicularity tolerances.)

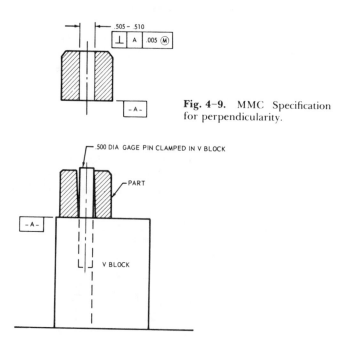

Fig. 4–9. MMC Specification for perpendicularity.

Fig. 4–10. Setup for gaging perpendicularity.

Fig. 4-10 shows the functional gage setup for measuring variable perpendicularity, consisting of a 0.500 in. diameter gage pin (0.505 in. MMC hole diameter minus the 0.005 in. perpendicularity tolerance allowed at MMC) clamped in a V block. If the part fits over the pin and datum surface A makes flush contact with the V block, the part is acceptable — providing the hole diameter is within the 0.505 to 0.510 in. limits.

Straightness

Fig. 4-11 shows a pin that must be straight within a 0.002 in. FIR when it is at MMC (0.400 in. diameter) and can deviate within a straightness tolerance of 0.004 in. FIR when it is at 0.398 in. diameter. Instead of measuring each pin to its exact diameter and then computing the exact straightness tolerance required (that is, a 0.3992 in. diameter pin is allowed a 0.0028 in. FIR) the inspector can inspect a series of these parts more rapidly with the setup shown in Fig. 4-12.

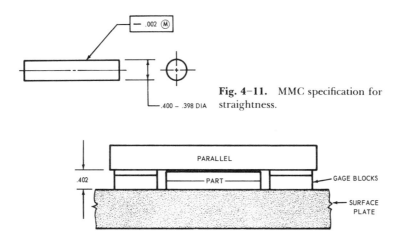

Fig. 4-11. MMC specification for straightness.

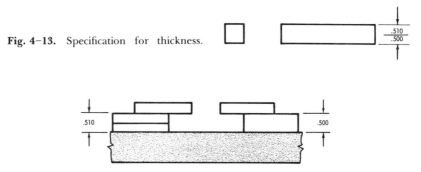

Fig. 4-12. Surface-plate setup for gaging straightness.

The surface-plate gage in Fig. 4-12 has a 0.402 in. diameter opening (0.400 in. diameter plus 0.002 in. straightness tolerance at MMC). The pin need only roll under the bridge to be acceptable, providing other measurements have shown that it is no larger than 0.400 in. diameter and no smaller than 0.398 in. diameter.

Thickness

Fig. 4-13 shows a simple thickness requirement that is readily gaged with the go and not-go gage stacks shown in Fig. 4-14. These same methods can be used to check the pin limits in Fig. 4-11 if other gages are not available.

Fig. 4-13. Specification for thickness.

.510
.500

Fig. 4-14. Surface-plate setup for go and not-go gaging of thickness.

.510   .500

Contour

Fig. 4-15 shows a contour tolerance of 0.010 in. FIR normal to the basic contour. (The series of basic coordinate contour-locating dimensions have been omitted for clarity.) The inspection of such a requirement is time-consuming with normal setup procedures. Fig. 4-16 shows a go-gage setup for inspecting contour tolerances consisting of standard surface-plate equipment. The pin sizes are determined by drawing a greatly magnified layout of the contour and fitting a series of circles (representing gage pins) to the contour so that the circles contact the maximum (go) contour. This kind of gage can *monitor* at a rapid rate production of parts too large for optical projectors.

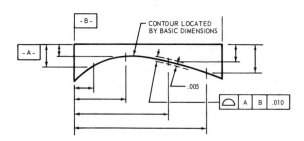

Fig. 4–15. Contour tolerance specification.

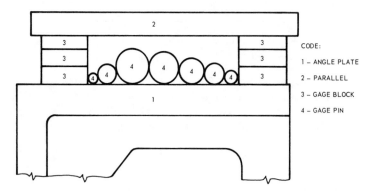

Fig. 4–16. Setup for gaging contour.

Reprinted from *Tooling & Production*, July 1979

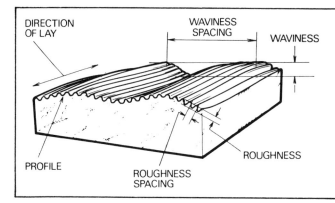

DIRECTION OF LAY
WAVINESS SPACING
WAVINESS
PROFILE
ROUGHNESS SPACING
ROUGHNESS

The components of surface texture measurement are: (1) Roughness—the irregularities in the surface texture which are inherent in the production process, but excluding waviness and errors of form; (2) waviness—that component of surface texture upon which roughness is superimposed. Waviness may result from such factors as machine or workpiece deflections, vibrations, chatter, heat treatment or warping strains. Profile is the combination of roughness and waviness, or total texture. Each pattern is characterized by the lay (the principal direction of the predominant surface pattern), the spacing of the principal crests and, in height, its departure from a reference line.

Surface texture—
a synopsis of instrumentation and methodology

by **William E Drews**
Manager, Metrology Products
Rank Precision Industries, Inc
Des Plaines, IL

Surface measurements provide a link between manufacturer and product function. This opens up new challenges to and responses from surface metrology assessment and application.

In recent years there has been a growing awareness of the importance of surface texture measurements. However, not many of us realize the effect surface roughness plays in our everyday lives.

The abrasiveness of different dentifrices on human dental enamel is monitored in several laboratories around the world by measuring surface finish. The test consists of measuring the abrasive loss of enamel by comparing a surface measurement made before the test, and one made after 1000 strokes on a special brushing machine. This insures that the abrasive loss of enamel during toothbrushing is clinically insignificant.

Many workers in the field of total hip replacement have stressed the need for rigorous quality control of the prosthesis during its manufacture. Stylus measuring instruments have many applications in quality control and wear studies for these artificial body replacements and are currently used by prosthesis manufacturers and research centers.

In the dairy industry, surface finish is helping to insure sanitation by facilitating appropriate cleaning procedures. During inspection, most dairy surfaces are visually checked for sanitation, because pits, cracks and inclusions are a breeding ground for bacteria. Grinding and polishing these surfaces is necessary to permit (and pass) these visual inspections.

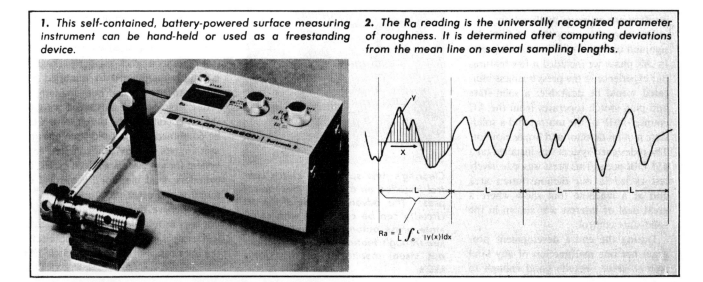

1. This self-contained, battery-powered surface measuring instrument can be hand-held or used as a freestanding device.

2. The R_a reading is the universally recognized parameter of roughness. It is determined after computing deviations from the mean line on several sampling lengths.

$$R_a = \frac{1}{L} \int_o^L |y(x)| dx$$

Clearly, surface finish measurements affect us in these and various other industries, and in many areas of our lives. Health and welfare, electronics, and metalworking industries all rely on surface measurements for quality products.

Typical stylus

A typical stylus type of instrument is illustrated in **Figure 1**. The ANSI-B 46.1.1978 surface texture standard mainly addresses itself to surface roughness averaging using stylus-type instruments. It establishes definitions of surface texture and requirements for instrumentation, along with roughness specimens. Because this standard deals almost exclusively with averaging-type stylus instruments, these instruments are the most widely used by industry.

Prior to 1978, the American Standard specified surface roughness measurements as arithmetical average (AA) deviation from the mean surface. The mean surface is the perfect surface formed if all the roughness peaks were cut off and used up in just filling the valleys below this surface. Today, this measurement is known as the R_a reading. It is the universally recognized parameter of roughness, determined as the mean result of several sampling lengths, **Figure 2**.

The type of stylus to be used for R_a measurements on a new instrument can have a radius of 10 micrometers (400 microinch) ± 30 percent. After an instrument has been in use, the tip radius is allowed to vary ± 50 percent, which is equivalent to 5 to 10 micrometers (200 to 600 microinch). However, an exception is an instrument used to calibrate precision reference specimens. This should have the 10 micrometers nominal size radius ± 20 percent. The stylus should be cone-shaped with a spherical tip. Other radii or shapes should be used only in unusual circumstances and must be specified.

Reading the meter

To maintain uniform interpretation, the standard specifies the method of reading the meter on a meter-type averaging unit. The reading that is considered significant is the mean reading around which the needle tends to dwell. Although extremely high and low momentary readings often occur, these extremes do not represent the average surface condition, and such readings should not be used in determining average roughness.

Several instrument manufacturers have now made systems with digital readout, which helps the inspector by removing the annoying problem of meter fluctuations.

Several surfaces have the same Ra

R_a readings serve well for surface finish control in most instances. Because this is an average type of reading, however, several different surfaces can be obtained with the same average.

Although the R_a readings have approximately the same values, the surfaces will function quite differently. In preparing part surfaces by means of manufacturing operations, it is the final function of the part that we strive for.

For example, if a part is to be prepared to a certain size and painted, a key function of that part surface is to help the paint adhere to the surface in a smoother manner. Should the end function be oiling or wiping of another surface, as with cylinder liners, a different pattern of surface finish on the part is required.

The same pattern in one case would not function in the other, yet in using R_a readings as a sole method to determine surface finish, both patterns could have the same average reading even though they differed widely in suitability for their intended uses. Therefore, in some applications, more knowledge of the surface texture is required.

Profiling

It was because of this need that instrumentation employing recorders capable of graphing the profile of the part became increasingly useful, **Figure 3**. Profile systems using skids can

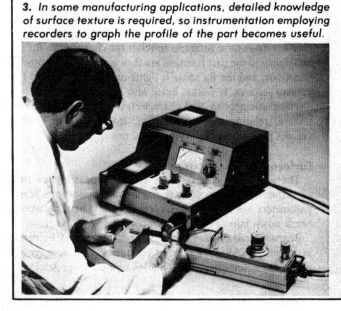

3. In some manufacturing applications, detailed knowledge of surface texture is required, so instrumentation employing recorders to graph the profile of the part becomes useful.

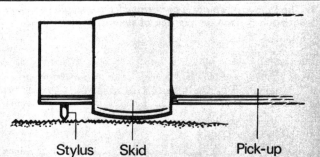

Stylus   Skid   Pick-up

4. The pick-up is a variable reluctance transducer which is supported on the surface to be measured by a skid, a curved support projecting from the underside of the pick-up in the vicinity of the stylus. As the pick-up traverses across the surface, movements of the stylus relative to the skid are detected and converted into a proportional electrical signal. The radius of the curvature of the skid is much greater than the roughness spacing, so it rides across the surface almost unaffected by the roughness, thus providing a datum representing the general form of the surface.

...continued

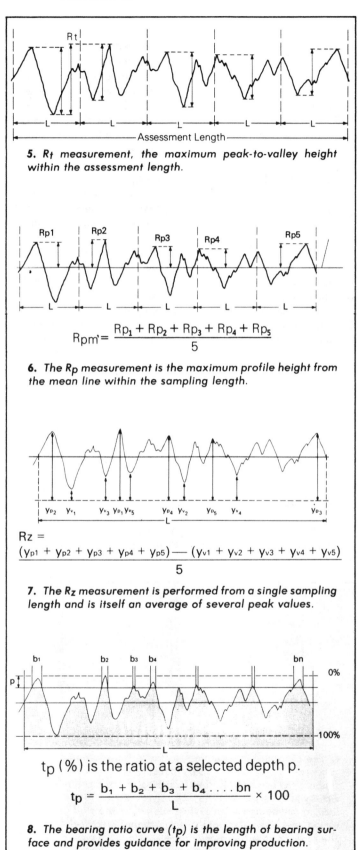

5. *R_t measurement, the maximum peak-to-valley height within the assessment length.*

$$Rpm' = \frac{Rp_1 + Rp_2 + Rp_3 + Rp_4 + Rp_5}{5}$$

6. *The R_p measurement is the maximum profile height from the mean line within the sampling length.*

$$Rz = \frac{(y_{p1} + y_{p2} + y_{p3} + y_{p4} + y_{p5}) - (y_{v1} + y_{v2} + y_{v3} + y_{v4} + y_{v5})}{5}$$

7. *The R_z measurement is performed from a single sampling length and is itself an average of several peak values.*

t_p (%) is the ratio at a selected depth p.

$$t_p = \frac{b_1 + b_2 + b_3 + b_4 \ldots bn}{L} \times 100$$

8. *The bearing ratio curve (t_p) is the length of bearing surface and provides guidance for improving production.*

plot roughness profiles, **Figure 4**. With built-in datum references, total profile can be recorded.

Once the profile graph is completed, consider what measurement to make to better define your surface. The most common measurement on these graphs is a total roughness-height evaluation, also known as the peak-to-valley measurement, because it is quickly and easily made. Other evaluations such as the frequency of peaks or the slope angle from peak to valley are most difficult. This is because the typical random characteristics of most surfaces give peaks of varying height and varying frequency, as well as a variety of slope angles.

Another difficulty arises when an attempt is made to ascertain the R_a value of a surface from this profile graph. Theoretically, this value could be calculated; in fact, many people suggest dividing the apparent peak-to-valley height by four to arrive at the R_a figure. At best, this will result in an *approximation* of the R_a value. Don't use this method for making accept or reject decisions, or as a reading to be compared with that obtained by averaging instruments.

First, the theoretical mean line through the texture must be found. Just how accurately can this line be established on a typical profile graph? How many points along the profile graph must be considered and measured to come up with a sufficient number of accurate readings to establish, with some degree of certainty, the R_a value? How do you apply the cutoff value to these readings? In short, the results of such an attempt are an approximation of the R_a value.

Of course, this profile ability leads to many word drawings. For example:

> "The roughness trace should be composed of plateaus and valleys. The plateaus provide good bearing area and the valleys on the trace show surface porosity which provides lubricant feed from the bushing interior to the shaft-bearing interface. Peaks, if present, should not project more than 250 microinches above adjacent plateau areas."

Many word drawings exceed 10 pages in length and include typical traces illustrating good and bad varieties.

In a laboratory, profiling techniques require the time and patience to make profile graphs, charts and illustrations. Time and skill are needed to interpret these recordings. The production floor demands fast answers with minimal inspection time. In many cases these attempts to relate the character of the surface finish to the part function are then ignored by the production floor, and the R_a value is tightened to cover up and thus prevent failures. It works, but it also increases cost.

These attempts to find characterization factors of surface texture, relating to part function, led directly to forming various parameters requiring special readouts.

Surface characterization

The wide range of parameters available opens up new fields of surface measurement, assessment and application. Several parameters are defined in **Figures 5** to **8**. These parameters break down into two groups: *amplitude* and *spacing*.

Measurement of peak heights, for instance, may be needed in painting (SAE J911 already includes peak count requirements for automotive steel), plating and glass applications to

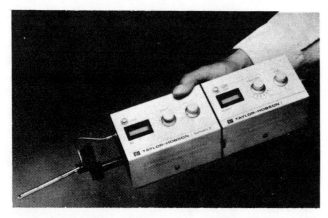

9. *This hand-held roughness measuring instrument provides surface measurements in six parameters.*

control the pitting of gear teeth, improve seals between surfaces and increase the stiffness of press fits.

R_t measurement is the maximum peak-to-valley height within the measurement length. It is valuable for analyzing finish to provide guidance for planning subsequent metalcutting operations. For example, if a honing operation is to follow, the R_t value will indicate how much material can be removed in honing before part size reaches a particular limit.

The R_p measurement, defined as the maximum profile height from the mean line within the sampling length, is the amount of material to be removed from a workpiece to obtain a 50 percent bearing area.

The R_z measurement is also known as the ISO 10-point height parameter. It is measured over a single sampling length and is itself an average of several peak values. It comes in handy when surfaces are short and the R_a value with a lower cutoff will not include irregularities that could cause problems.

Irregularity spacing information is useful for measuring the wearing in of surfaces in relative motion, and assessing the electrical and thermal conductivity between surfaces in contact.

Skew will show whether porous, sintered and cast iron surfaces will yield a meaningful R_a measurement. Also, the criterion for a good bearing surface is that it should have a negative value of skew.

Measurement of the bearing ratio curve (t_p) on a surface subject to wear is the length of bearing surface (expressed as a percentage of the length of measurement) at a depth below the highest peak and made before and after a life test. This measurement helps improve production processes.

Irregularity spacing and height parameters used in combination are valuable for sheet-steel applications and for friction and lubrication studies.

Variety of instruments

There is a wide variety of instruments available, from battery-powered portable shop instruments that can have parameter units added to them, **Figure 9**, to profiling systems with multiparameter readouts, **Figure 10**, all designed to help define surface characterization.

Minicomputers have made available complete statistical *multitrace* systems measuring several places over a given area,

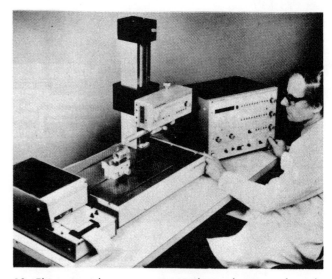

10. *This system has a processor with a wide range of roughness and waviness parameters for industrial or research applications. Other modules shown include a traverse unit, a linear recorder, a motorized recorder and stand.*

and can be coupled with computers to give standard deviations and average over area-type readings. These systems lend themselves to research applications where specialized programming can achieve auto-correlation, power spectrum analysis and peak curvature.

These other parameters of surface texture represent a major breakthrough, in that the determination of the relationship between proper characterization of a surface and its function could save untold dollars in industries where surfaces either fail to operate properly or are over-specified because they are improperly characterized.

It is the proper application of all of these surface texture parameters and their precise measurement that permits manufacturers to make products that more efficiently fulfill their functions. Perhaps another way of saying this is simply that the correct use of surface measuring techniques is the manufacturer's best tool for achieving optimum quality control of his end product.

Reprinted from *Production Engineering*, Copyright, Penton/IPC Inc., March 1980

Gaging With Light

Optical methods can boost your productivity with fast, noncontact, on-line measurement and inspection.

By NORMAN N. AXELROD
Principal
Norman N. Axelrod Associates
New York

Linear dimensions—like length, diameter, gap, and displacement—are among the most common manufacturing parameters that production engineers have to measure every day. And a variety of optical techniques that employ linear diode arrays, lasers, and diffraction phenomena are ideally suited for making these measurements right on your production line.

An optical measurement system consists of a light source, optical components, photodetectors, and an electronics unit which converts the light into an electrical signal for control and/or display. Such systems typically are used for au-

171

tomatic inspection, but can also be used for manual inspection and bench testing.

Because the dimensional information is carried by the light beam, the way in which it is coded and decoded is a major factor in determining system accuracy. And, therefore, the circumstances under which the measurements will be made, and the characteristics of the objects to be measured, are crucial to deciding which approach is suitable for a specific application.

Linear diode array

A linear diode array consists of photodetector diodes regularly spaced along a straight line. The spacing between the diodes is constant and provides the scale or ruler for the measurement. An image of the illuminated object is formed on the linear diode array and the electronics unit counts the number of adjacent, illuminated diodes.

The length of the object is directly proportional to the number of diodes counted, with a correction for the magnification of the image. For example, with a 0.001-in. center-to-center diode spacing, and an image with a magnification of one-half, then 20 illuminated diodes correspond to an object length of 0.04 in. The accuracy of the measurement depends on the fidelity of the image (which depends on lens performance and image focus) as well as on the spacing of the diodes and the uniformity of their response.

The only diodes that provide information about the object's length are those which define the difference between light and dark at the ends of the object's image. The other diodes in the linear

diode array permit measurements to be made without concern for exact positioning of the object and with no mechanical adjustment for objects of moderately different lengths. If only GO/NO-GO gaging is required, then it is only necessary to determine whether the distance between the two ends of the object falls within acceptable limits, and one simplified on-line system uses only four diodes. Two of these detectors signal presence of one end of the object between them, and the electronics unit then determines whether the other end of the

object is between the other two detectors at that same instant. This system discriminates between objects that are acceptable, too short, or too long. And two additional detectors can alert the operator to trends in object length before tolerances are exceeded.

Scanning laser beam

The scanning laser beam employs a rotating mirror to sweep a

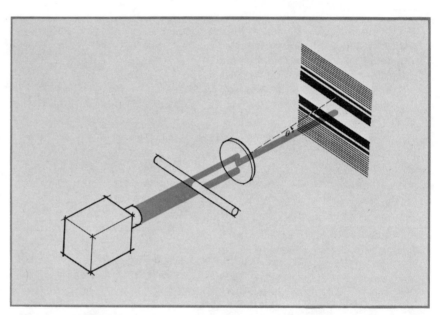

Diffraction patterns can be used to measure small dimensions. The parallel coherent light in the laser beam is diffracted by the wire, and the resultant pattern focused on the screen by the lens. At the focal plane of the lens, the angle of the first minimum of light intensity, θ, is equal to the wavelength of the light divided by the diameter of the wire ($\theta = \lambda/d$). The distance in the focal plane from the lens axis to the first minimum is equal to the angle θ multiplied by the focal length of the lens ($k = \theta f$). With $\lambda = 1$ micron and a lens with $f = 10$ in., a 0.0004-in. diameter wire yields $k = 1$ in., which is easily measured with a linear diode array (not shown).

172

single beam of laser light across a planar area. The laser beam is reflected from the rotating mirror at its axis of rotation. And the axis of rotation coincides with the focal point of a lens. Therefore, because all rays of light originating at the focal point of an ideal lens are parallel after they emerge from the lens, the laser beam is continuously translated parallel to itself, generating a continuous plane of light.

A second lens focuses the parallel rays of laser light onto a photodetector located at the focal point of the second lens. If there is no object in the planar light beam, the photodetector provides a continuous output signal. An object placed between the two lens partially blocks the plane of light and the length of the object is measured by timing the interval over which no laser light reaches the photodetector.

Practical scanning systems use low-mass mirrors mounted on rugged galvanometers, and are capable of dimensional measurement accuracies of ±0.001 in. or better. Systems using massive rotating-polygon mirrors are capable of even greater accuracy.

In both the scanning laser beam and linear diode array techniques, misalignment and lens imperfections can blur the light defining the edges of the object. Hence, an adjustable threshold of sensitivity is necessary. They can, however, measure an overall linear dimen-

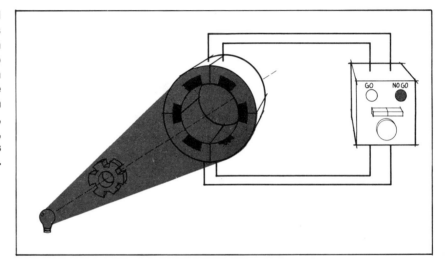

Inspection can be greatly simplified by taking advantage of the subject's rotational symmetry properties. An image of the subject is projected onto a ring-shaped sensor made up of an appropriate number of separate detectors. Any deviation from symmetry, like a missing gear tooth, causes an uneven illumination, which the electronics unit interprets as a defect.

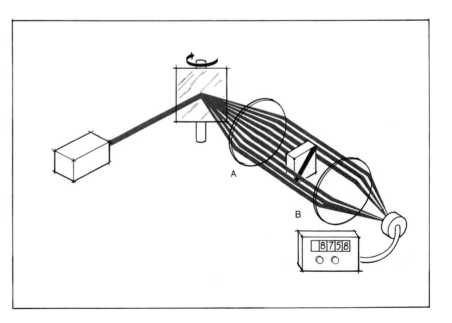

Scanning laser beam measures linear dimensions by measuring the total time that a laser beam—sweeping across the subject at a known rate—is cut off. Because the rotating mirror is located at the focal point of lens A, the laser beam is swept parallel with the axis common to both lenses and unless cut off by the subject, impinges on the detector located at the focal point of lens B. The electronics unit calculates the subject's length from the sweep rate and the total cutoff time.

sion of an object that includes a gap—e.g., the diameter of a washer—with suitable electronics units.

Diffraction and symmetry

Small dimensions can demand accuracy beyond the capabilities of practical linear diode arrays and laser beam scanning systems. And length gaging is often performed indirectly, by measuring the displacement of a table or carriage on which the object is mounted. Diffraction effects offer practical methods for dealing with both of these situations.

Also, an optical property of the areas containing the edges to be used in gaging can sometimes be used to discriminate between the features of interest and other features that interfere with simple gaging. For example, many common objects have rotational symmetry (e.g., spur gears, grommets, buttons, and multipin electrical connectors) which permits gaging for dimensional deviations by comparing one part of the object with another part of the same object.

Diffraction effects are well-suited for gaging objects like small-diameter wire, and also for measuring narrow gaps. A narrow beam of light incident on a small-diameter wire is spread due to diffraction, with the spread inversely

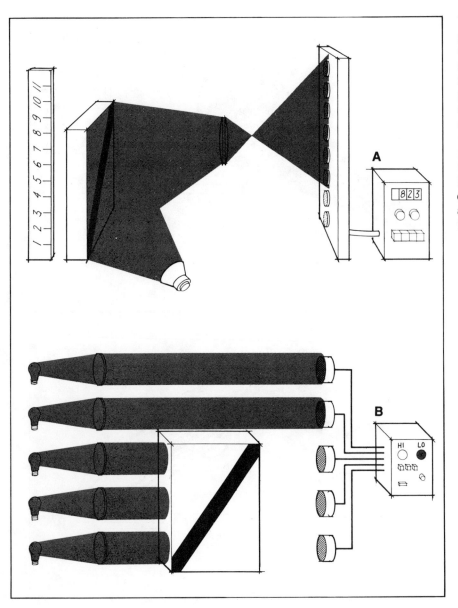

The linear diode array is optical gaging's analog to the ruler. Casting an image of the subject on the array (A) illuminates a string of diodes. The number of diodes illuminated is proportional to the length of the subject, so the electronics unit has only to count the number of diodes illuminated and apply the proportionality factor to display the length of the subject directly. An inexpensive variation on the linear diode array (B) is useful for simple applications like discriminating between cartons of different heights.

proportional to wire diameter.

The diffracted beam can be focused on a screen using a suitable lens, resulting in the typical diffraction pattern of alternate light and dark bands. The linear distance between these bands can be measured using a linear diode array. Furthermore, it bears a precise mathematical relationship to the wire diameter, wavelength of the light used, and focal length of the lens. The latter two are known constants of the system, hence the wire diameter can be calculated directly from the diffraction pattern measurement.

This method is independent of the position of the wire in front of the lens, because all parallel light rays incident on a lens at a given angle are focused on one point. Therefore, the diffraction pattern is independent of the location of the wire as long as the diffracted light is collected by the lens. And finally, this method is equally useful for measuring narrow gaps because the diffraction pattern from narrow gaps is the same as the diffraction pattern from opaque strips of the same diameter.

Another diffraction technique, which is commonly used to measure displacement along a straight line, is the method of Moire fringes. Moire fringes can be produced by using two similar, transmission diffraction gratings placed nearly face to face and with the grating lines nearly parallel. Light transmitted through both gratings produces a pattern of dark and light bands or fringes. The fringes are quite visible to the naked eye even though the grating-line spacing is too fine for visual discrimination. The distance between adjacent dark fringes depends on the grating-line spacing and on the angle between the lines on the two gratings.

If one grating is moved relative to the other, the pattern of fringes moves perpendicular to the direction of relative motion. The ratio of fringe pattern displacement to grating displacement is a function of the angle between the two gratings, and can be as great as 1,000. Hence, relatively simple, low-resolution optical components can be used to count the fringes and still yield high-accuracy measurements. And microinch displacements can be measured easily using a linear diode array to count the fringes.

Objects with rotational symmetry can often be inspected by using one part of the object as the comparison standard to check other parts of the same object. This method provides a simple gaging technique for situations where dimensional variations are infrequent and can be regarded as defects.

In practice, an image of the object is cast on a multidetector sensing array which has the same symmetry as the object itself. Deviations from acceptable tolerances cause uneven illumination of the detectors, which the electronics unit interprets as an unacceptable object.

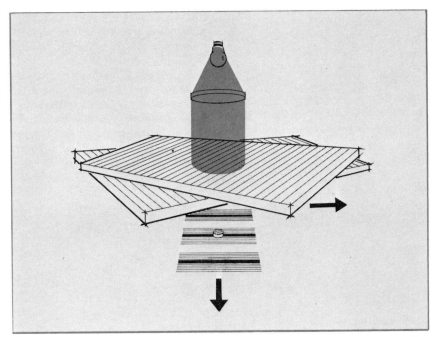

Moire fringes are alternate light and dark bands produced by light passing through two diffraction gratings oriented with their respective grating lines nearly parallel. Moving one grating with respect to the other causes the fringe pattern to move in a direction perpendicular to the relative movement of the gratings. The movement of the fringe pattern can be as much as 1,000 times greater than the actual relative movement of the gratings, allowing relatively simple measurement of microinch displacements.

Presented at The American Society of Mechanical Engineers'
Design Engineering Conference and Show, April 1981

A Computer Simulation of a True Position Feature Pattern Gauge

J. R. O'Leary

Assistant Professor,
Civil Engrg. Dept.,
Ill. Inst. of Tech.,
Chicago, Ill.
Mem. ASME

Described in this paper is an approximate technique for the simulation of a mechanical feature pattern gauge (a go/no-go gauge). The procedure involves the unconstrained minimization of a judiciously constructed response function. The formulation of this function as well as the development of the associated algorithm are presented. Moreover, the technique is demonstrated on an assortment of sample problems.

NOMENCLATURE

CPT	composite positional tolerancing
FRT	feature relating tolerance
PLT	pattern relating tolerance
PPMT	positional plus-minus tolerancing
C_i	$\cos(\overline{\theta} + \theta_i)$
$\underset{\sim}{D}E$	gradient of E
$\underset{\sim}{D}^2E$	the Hessian matrix for E
$k_i^{(1)}$	
$k_i^{(2x)}$	proportionality coefficients
$k_i^{(2y)}$	
ℓ_i	magnitude of $\underset{\sim}{R}_i$
M	number of hole centers
n	hardening exponent
$\underset{\sim}{p}_i$	location of hole center i
$\underset{\sim}{q}_i$	location of PLT zone center
$\overline{\underset{\sim}{R}}$	location vector for the rigid body
$\underset{\sim}{R}_i$	location vectors for FRT centers
S_i	$\sin(\overline{\theta} + \theta_i)$
$\underset{\sim}{u}_i$	vector from FRT zone center to PLT zone center
$\underset{\sim}{v}_i$	vector from FRT zone center to hole center
V_{ix}	$v_{ix} + \ell_i C_i$
V_{iy}	$v_{iy} + \ell_i S_i$
$(\overline{x}, \overline{y})$	cartesian components for $\overline{\underset{\sim}{R}}$
$\varepsilon_1, \varepsilon_2$	convergence parameters
δ	the variation symbol
$\overline{\theta}$	rotation of the rigid body
θ_i	direction of R_i
$\underset{\sim}{\chi}$	configuration space

INTRODUCTION

In order to control tolerances on mechanical components, the use of the American National Standard for Dimensioning and Tolerancing (1) has in recent years achieved wide spread acceptance by manufacturers from both the defense and the private sectors. The former being more so effected than the latter. One particular aspect of this standard is the location of feature patterns. Normally one determines the acceptability of a particular pattern of holes by the employment of a mechanical gauge (often referred to as a "go/no go" gauge). These gauges are manufactured by skilled tool makers to a high precision, hence their cost may be a nontrival factor. This cost multiplies as the number of such patterns increases and hence the use of such gauges on limited run production parts may become prohibitively expensive. An alternative to the use of a mechanical guage is to employ an optical scanner to determine the actual hole locations, then to construct a template and inspect the pattern graphically.

Contributed by the Design Engineering Division of The American Society of Mechanical Engineers for presentation at the Design Engineering Conference and Show, April 27-30, 1981, Chicago, Illinois. Manuscript received at ASME Headquarters January 27, 1981.

Copies will be available until January 1, 1982.

This procedure also has a distasteful expense associated with it aside from the question of graphical accuracy. This leads one to consider a computer simulation for such an inspection gauge. If such a simulation can be developed which is fast and requires limited storage, then a mini-computer coupled to an optical scanner can be employed by quality control inspectors to give real time inspection.

As will be shown in this paper, an "exact" simulation for such a gauge is actually a repetitive exercise in constrained optimization. The object of this paper is not to simulate the gauge exactly, but to obtain a close approximation using unconstrained optimization. The attractiveness here being that two levels of iteration can be avoided and hence large patterns can be inspected substantially faster using this procedure rather than an "exact" simulation. It will be demonstrated that this approximate technique is quite accurate and if it should error, it is a conservative error, i.e. a bad part will never pass the simulated gauge, however, there is a remote possibility that a good part may fail.

The discussion will begin with a review of the fundamentals of feature pattern location tolerancing in the true position setting. After which will be detailed the construction of a response function so as to produce the desired results when subjected to unconstrained minimization. The mathematical formulation for the above minimization is then presented. The final section of this paper will outline the implementation of such an algorithm into a computer program and will present several sample problems.

FEATURE PATTERN LOCATIONS

In the design of mechanical components the location of a hole center relative to other holes within the same pattern is often more critical than the location of the pattern as a whole. Recognizing this fact, procedures have been developed for specifying two independent sets of tolerances. One of which stipulates the tolerance on a hole center when measured relative to the other centers within a pattern. The resulting regions, wherein a hole center may acceptably be located, are defined as the feature relating tolerance zones (FRT zones). The regions generated by the other specified tolerances are defined as the pattern locating tolerance zones (PLT zones), and their function is to tolerance a hole center relative to the component's global datum. In accord with (1) there exists two acceptable schemes for specifying these zones, i.e. composite positional tolerancing (CPT) and positional plus minus tolerancing (PPMT).

The specification of a CPT scheme as one would typically find on an engineering drawing, is shown in Fig. 1a. Such a scheme is a double application of a true position dimensioning, generating two independent systems of circular tolerance zones as shown in Fig. 1b. The PLT zones (the larger circles) are fixed and centered at the theoretical (basic dimensioned) locations for the hole centers. The FRT zones (the smaller circles) are also related to one another by the basic dimensions, however, these zones are free to float as a pattern so long as their locations relative to one another are fixed. The criteria for the acceptability of a pattern of hole centers is first that each hole center be within its corresponding PLT zone and secondly, that there exists a configuration for the pattern of FRT zones such that each hole center is located within its respective FRT zone.

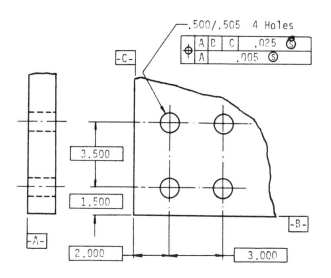

Fig. 1a   Example of Composite Positional Tolerencing

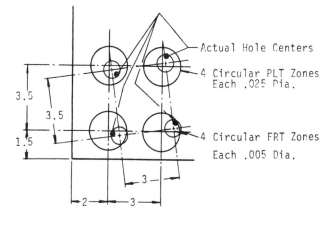

Fig. 1b   Interpretation of Composite Positional Tolerencing

Fig. 2a shows a representative example of the specification for a PPMT scheme. This is a hybrid method where global locations are given in the traditional plus-minus form, while relative locations are specified true positions. Due to the hybrid nature of this scheme, zone geometry for the two independent systems are different. The PLT zones are fixed and centered at the theoretical locations for the holes, however, due to the fact that they are specified via plus-minus tolerances they are rectangular. As is the case for the CPT scheme the FRT zones are circular and free to float as a fixed pattern. Aside from zone geometry the major difference between the two schemes is the criterion for acceptability. For the CPT scheme one merely needs to show the existence of a configuration

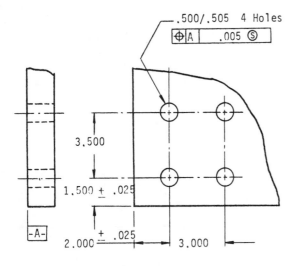

.500/.505 4 Holes

⊕ | A | .005 Ⓢ

3.500

1.500 ± .025

-A-

2.000 ± .025   3.000

Fig. 2a   Example of Positional Plus Minus
Tolerencing

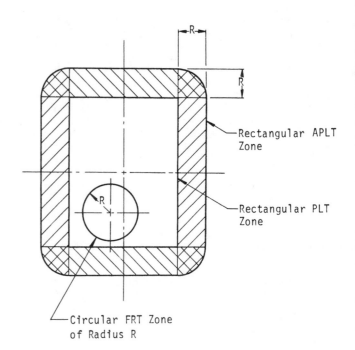

Fig. 2c   Typical PPMT Scheme Zones

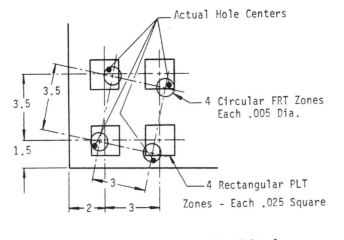

Actual Hole Centers

3.5

3.5

1.5

4 Circular FRT Zones
Each .005 Dia.

4 Rectangular PLT
Zones - Each .025 Square

2   3

Fig. 2b   Interpretation of Positional
Plus Minus Tolerencing

so that each hole center lives within the intersection
of the respective zones. For the PPMT scheme one
must show the existence of a configuration for which
first, each hole center lies within its FRT zone and
secondly, that the centers for these FRT zones lie
within their corresponding PLT zones, a graphical
interpretation of this scheme is shown in Fig. 2b.
One notes that by this scheme it is possible for a
hole center to be located outside of its PLT zone,
Fig. 2c shows the region wherein a particular hole
center must reside, we will define this region as the
APLT zone. It should also be noted that it is this
criterion of acceptability which makes the PPMT scheme
more difficult to simulate then the CPT scheme.

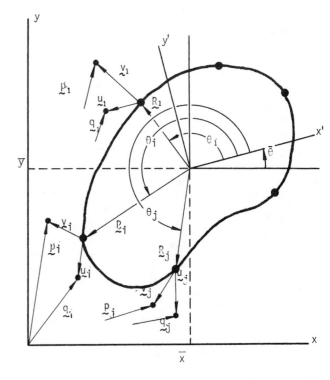

Fig. 3   Rigid Body Model

A RESPONSE FUNCTION

The fundamental approach which will be followed, is to consider the FRT zone centers as points on a rigid body so as to require that distances between these rigid body points, the corresponding PLT zones centers and hole centers, satisfy the criterion outlined above. This model is shown in Fig. 3 wherein we define a reference coordinate system (x, y) with base vectors $(\underline{i}, \underline{j})$ in addition, a convected coordinate system (x', y') embedded in the rigid body is also introduced. These two coordinate systems are related by the vector $\overline{R} = \overline{x}\underline{i} + \overline{y}\underline{j}$ and the rotation $\overline{\theta}$. Hence all motion for the rigid body is defined by the prescription of $(\overline{x}, \overline{y}, \overline{\theta})$. The set of all such motion is defined as χ i.e.

$$\underline{\chi} = \underline{R}^2 \times [o, 2\pi] \qquad (1)$$

On this set is defined the norm

$$\|\overline{\underline{x}}\|_{\chi_p} = (|\overline{x}|^p + |\overline{y}|^p + |\overline{\theta}|^p)^{\frac{1}{p}} \qquad (2)$$

In addition, point (i) on the rigid body is located relative to the convected coordinates via the polar coordinates (ℓ_i, θ_i) or equivalently by the vector $\underline{R}_i$.

The actual hole centers and the PLT zone are respectively identified with the vectors $\underline{p}_i$, $\underline{q}_i$, which in turn allows for the definition of the displacement vectors $\underline{u}_i$, $\underline{v}_i$, viz.

$$\underline{v}_i = \underline{p}_i - (\underline{R}_i + \overline{R}) \qquad (3)$$

$$\underline{u}_i = \underline{q}_i - (\underline{R}_i + \overline{R}) \qquad (4)$$

As a point of notation, a Latin subscript will denote a vector component with respect to the reference coordinate system, for example $v_{kx} = \underline{v}_k \cdot \underline{i}$. Moreover, the symbol $\|.\|$ is reserved for the usual euclidean norm whereas $|.|$ signifies absolute value.

One approach to the PPMT problem is to proceed in an iterative fashion where for the i-th iteration one minimizes some measure of the vector $\underline{v}_i$ while constraining $\underline{v}_j$ $j = 1, \ldots i-1$ to lie in their respective PLT zones and then to test if $\underline{v}_i$ is less than the FRT tolerance. This leads at each iteration to a problem of constrained optimization and hence a relatively laborious algorithm. Rather than following this exact procedure, presented here is the construction of a measure for the total of vectors $\underline{v}_i$, the minimization of which will lead to a near optimal positioning for the rigid body. In this approach the constraints on $\underline{u}_i$ are trivally satisfied in CPT scheme; however, for PPMT scheme they will prove to be problem for which an additional trick must be incorporated. An approach of this nature involves no iteration, only a single unconstrained minimization.

As a first attempt one may seek to minimize the vectors $\underline{v}_i$ in a least squares sense. Applied to our model, this is analogous to attaching linear springs between each point on the rigid body and the corresponding hole center and then to seek the resulting equilibrium position. For most problems this will yield a valid approximation; however, when applied to a problem of the type shown in Fig. 4, the technique fails. The explanation for this becomes apparent when the spring analogy is considered viz., the force in the spring at position 4 needs to equilibrate the vector sum of forces at positions 1, 2, 3, hence in the equilibrium configuration $\|\underline{v}_4\| \geq \|\underline{v}_i\|$ i=1,2,3.

What is needed is to minimize some measure of $\underline{v}_i$ as above, but to also maintain their norms at near equal value while doing so.

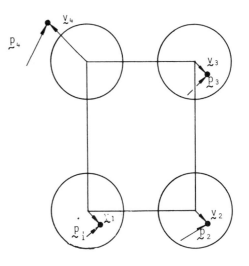

Fig. 4   Equilbrium Position for

A Linear Spring Model

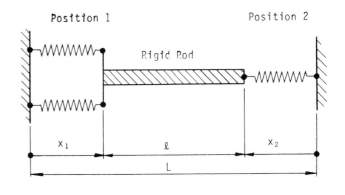

Fig. 5   One Dimensional Problem

In attempting to perform the above task it is beneficial to consider the one dimensional problem shown in Fig. 5. If springs of equal stiffness are placed at both positions 1 and 2, the equilibrium position is obviously

$$x_1 = x_2 = \frac{L-\ell}{2} \qquad (5)$$

By placing an additional spring of equal stiffness in position 1 the equilibrium position is attained at

179

$$x_1 = \frac{x_2}{2} = \frac{L-\ell}{3} \tag{6}$$

This is exactly the situation encountered in Figure 4. If, however, non-linear hardening springs are employed i.e., springs for which the force displacement relationship is

$$F = x^n \qquad n>1 \tag{7}$$

then equilibrium for the above nonsymmetric example is attained when

$$x_1 = 2^{-1/n} x_2 \tag{8}$$

Hence as n becomes large x_1 approaches x_2 which is the desired behavior. The above discussion suggests that the FRT condition is satisfied when the following measure is minimized.

$$E(\underset{\sim}{v_i}) = \frac{1}{2n} \sum_{i=1}^{M} k_i (\underset{\sim}{v_i} \cdot \underset{\sim}{v_i})^n \tag{9}$$

In the case of the CPT scheme, the PLT requirement is a trivial question requiring only an initial check that each hole center lies in its respective PLT zone. However, for the PPMT scheme life is not so simple, since the PLT criterion requires an equilibrium configuration in which the rigid body points must reside within the PLT zones. One approach to the problem is to proceed exactly as in the CPT scheme. This, however, may be overly conservative. An alternative approach is to create a new augmented PLT zone (the APLT zone) by increasing the size of the rectangular PLT zone by the radius of the FRT zone, a typical example of such a zone is shown in Fig. 2c. One would again follow the procedure outlined above, substituting the larger APLT zones for the PLT zones.

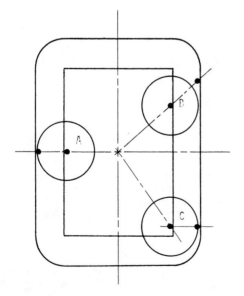

Fig. 6   Use of Secondary Springs

Unfortunately a class of sample problems can be devised for which this approach is nonconservative. A procedure found to be successful is to first test if the hole centers are in their respective APLT zones (rejecting those outside). If a hole center is found to lie in the APLT zone but outside of the PLT zone, then a properly proportioned secondary spring is attached between the PLT zone center and the corresponding rigid body point after which minimization proceeds as above. The difficulty with this procedure is illustrated in Fig. 6 wherein at position B the secondary spring is too stiff, and pulls the rigid body away from the hole center. This phenomena is due to the fact that the potential lines for standard springs are circular while the PLT zones are rectangular. A way of realizing nearly rectangular potential lines is observed in the equation

$$k_x x^n + k_y y^n = \text{constant} \qquad n > 2 \tag{10}$$

for which we obtain a family of closed curves which for n large approach a rectangle. Hence for the PPMT scheme the following augmented form of equation 9 is used.

$$E(\underset{\sim}{v_i}, \underset{\sim}{u_i}) = \frac{1}{2n} \sum_{i=1}^{M} k_i^{(1)} (\underset{\sim}{v_i} \cdot \underset{\sim}{v_i})^n$$
$$+ \frac{1}{2n} \sum_{i=1}^{M} (k_i^{(2x)} u_{ix}^{2n} + k_i^{(2y)} u_{iy}^{2n}) \tag{11}$$

where $k_i^{(1)}$, $k_i^{(2x)}$, $k_i^{(2y)}$ are properly tuned so as to achieve the conditions illustrated at position C in Figure 6. In addition $k_i^{(2x)}$ is zeroed if the i-th hole center is not in region 1 (Figure 2c) and $k_i^{(2y)}$ is zeroed if it is not in region 2. Obviously when implemented in a computer code such zeroing is never performed rather the unnecessary multiplication is eliminated by the logic of the program.

FORMULATION OF THE METHOD

We seek $\underset{\sim}{x} \epsilon \chi$ which for the CPT scheme minimizes $E \cdot v_i(\underset{\sim}{x})$ (equation 9) or for the PPMT scheme minimizes $E \cdot (\underset{\sim}{v_i}, \underset{\sim}{u_i})(\underset{\sim}{x})$ (equation 11). Since this latter function is a generalization of the former we will restrict our analysis to it. When n is an even integer $E \cdot (\underset{\sim}{v_i}, \underset{\sim}{u_i})(\underset{\sim}{x})$ is a convex real valued function defined on the convex finite dimensional set χ hence there exists a unique $\underset{\sim}{x}^* \epsilon \chi$ such that

$$E \cdot (\underset{\sim}{v_i}, \underset{\sim}{u_i})(\underset{\sim}{x}^*) = \inf \{E \cdot (\underset{\sim}{v_i}, \underset{\sim}{u_i})(\underset{\sim}{x}): \underset{\sim}{x} \epsilon \underset{\sim}{\chi}\} \tag{12}$$

Equivalently at $\underset{\sim}{x}^* \epsilon \chi$ the following are necessary and sufficient conditions for the existence of the global minimum

$$\underset{\sim}{D} E \cdot (\underset{\sim}{v_i}, \underset{\sim}{u_i})(\underset{\sim}{x}^*) = \underset{\sim}{\Theta} \tag{13}$$

$$\underset{\sim}{D}^2 E \cdot (\underset{\sim}{v_i}, \underset{\sim}{u_i})(\underset{\sim}{x}^*): \underset{\sim}{x} \otimes \underset{\sim}{x} > 0 \quad \forall \underset{\sim}{x} \epsilon \underset{\sim}{\chi} \tag{14}$$

where $DE \cdot (v_i, u_i)(x^\star)$, $D^2 E \cdot (v_i, u_i)(x^\star)$ are respectively the first and second derivatives of $E \cdot (v_i, u_i)(x)$ at $x^\star \epsilon \chi$. A technique found effective in minimizing equation 11 is the Newton method with a line search. The procedure involves defining a vector $f^{(k)} \epsilon \chi$ which for the k-th iteration is the solution of the linear system of equations

$$DE \cdot (v_i, u_i)(x^{(k)})$$
$$+ D^2 E(v_i, u_i)(x^{(k)}) f^{(k)} = 0 \qquad (15)$$

Defining a unit vector $e^{(k)} \epsilon \chi$ as

$$e^{(k)} = f^{(k)} / \| f^{(k)} \|_{\chi_2} \qquad (16)$$

and a real valued function

$$\phi(\lambda_k) = E \cdot (v_i, u_i)(x^{(k)} + \lambda_k e^{(k)}) \qquad (17)$$

A line search is performed to obtain $\lambda_k^\star$ such that

$$\phi(\lambda_k^\star) = \inf \{\phi(\lambda_k) : \lambda_k \epsilon R\} \qquad (18)$$

then

$$x^{(k+1)} = x^{(k)} + \lambda_k^\star e^{(k)} \qquad (19)$$

which is the new approximation for $\chi^\star$ in the k + 1 iteration. Convergence is determined when

$$\| DE(v_i, u_i)(x^{(k+1)})$$
$$- DE(v_i, u_i)(x^{(k)}) \|_{\chi_\infty} < \epsilon_1 \qquad (20)$$

$$|\lambda_{k+1}^\star| < \epsilon_2 \qquad (21)$$

where ϵ_1, ϵ_2 are two preselected convergence parameters. In order to achieve explicit expressions for the first and second derivatives variations are taken to equation 11 i.e.

In component form equation 3, 4 can be written as

$$v_{ix} = p_{ix} - \bar{x} - \ell_i \cos(\bar{\Theta} + \Theta_i)$$
$$v_{iy} = p_{iy} - \bar{y} - \ell_i \sin(\bar{\Theta} + \Theta_i) \qquad (24)$$
$$u_{ix} = q_{ix} - \bar{x} - \ell_i \cos(\bar{\Theta} + \Theta_i)$$
$$u_{iy} = q_{iy} - \bar{y} - \ell_i \sin(\bar{\Theta} + \Theta_i) \qquad (25)$$

Adopting the notation

$$S_i = \sin(\bar{\Theta} + \Theta_i)$$
$$C_i = \cos(\bar{\Theta} + \Theta_i) \qquad (26)$$
$$V_{ix} = v_{ix} + \ell_i C_i$$
$$V_{iy} = v_{iy} + \ell_i S_i \qquad (27)$$

then

$$\delta v_i = [-i \quad -j \quad \ell_i(S_i i - C_i j)] \begin{Bmatrix} \delta x \\ \delta y \\ \delta \Theta \end{Bmatrix} \qquad (28)$$

$$\delta^2 v_i = [\delta x \; \delta y \; \delta \Theta] \begin{bmatrix} 0 & 0 & 0 \\ & 0 & 0 \\ (*) & & \ell_i(C_i i + S_i j) \end{bmatrix} \begin{Bmatrix} \delta x \\ \delta y \\ \delta \Theta \end{Bmatrix} \qquad (29)$$

$$\delta u_{ix} = [-1 \; 0 \; \ell_i S_i] \begin{Bmatrix} \delta x \\ \delta y \\ \delta \Theta \end{Bmatrix} \qquad (30)$$

$$\delta E(v_i, u_i) = \sum_{i=1}^{M} k_i^{(1)}(v_i \cdot v_i)^{n-1}(v_i \cdot \delta v_i) + \sum_{i=1}^{M} (k_i^{(2x)} u_{ix}^{2n-1} \delta u_{ix} + k_i^{(2y)} u_{iy}^{2n-1} \delta u_{iy}) \qquad (22)$$

$$\delta^2 E(v_i, u_i) = 2(n-1) \sum_{i=1}^{M} k_i^{(1)}(v_i \cdot v_i)^{n-2}(v_i \cdot \delta v_i)^2 + \sum_{i=1}^{M} k_i^{(1)}(v_i \cdot v_i)^{n-1}(\delta v_i \cdot \delta v_i + v_i \cdot \delta^2 v_i)$$

$$+ (2n-1) \sum_{i=1}^{M} k_i^{(2x)} u_{ix}^{2n-2}(\delta u_{ix})^2 + \sum_{i=1} k_i^{(2x)} u_{ix}^{2n-1} \delta^2 u_{ix}$$

$$+ (2n-1) \sum_{i=1}^{M} k_i^{(2y)} u_{iy}^{2n-2}(\delta u_{iy})^2 + \sum_{i=1}^{M} k_i^{(2y)} u_{iy}^{2n-1} \delta^2 u_{iy} \qquad (23)$$

$$\delta^2 u_{ix} = [\delta x \; \delta y \; \delta\Theta] \begin{bmatrix} 0 & 0 & 0 \\ & 0 & 0 \\ (*) & & \ell_i C_i \end{bmatrix} \begin{Bmatrix} \delta x \\ \delta y \\ \delta\Theta \end{Bmatrix} \qquad (31)$$

$$\delta u_{iy} = [0 \; -1 \; \; -\ell_i C_i] \begin{Bmatrix} \delta x \\ \delta y \\ \delta\Theta \end{Bmatrix} \qquad (32)$$

$$\delta^2 u_{iy} = [\delta x \; \delta y \; \delta\Theta] \begin{bmatrix} 0 & 0 & 0 \\ & 0 & 0 \\ (*) & & \ell_i S_i \end{bmatrix} \begin{Bmatrix} \delta x \\ \delta y \\ \delta\Theta \end{Bmatrix} \qquad (33)$$

After some algebraic manipulation the above relations together with equations 22, 23 yield the following form for the first and second derivatives of $E(\underset{\sim}{v}_i, \underset{\sim}{u}_i)$.

Presented here are several examples of the application of an algorithm of this type. These sample problems are obviously not real world, they were selected to present results in a clear easy to plot form in addition to their being good checks of the programs numerical stability. In Fig. 7 is shown a typical CPT problem, as seen there, the procedure appears to have performed quite well. The hardening exponent n was given a value of ten (10), this value was found by experience to be sufficient. This problem bears a resemblance to the problem shown in Figure 4, in fact, if the problem is resolved using n equal to 2 (a linear spring) then the aforementioned difficulty arises (i.e. the component fails the FRT test). For this example, the diameters of the FRT zones were all

$$\underset{\sim}{DE} = \sum_{i=1}^{M} k_i^{(1)} (\underset{\sim}{v}_i \cdot \underset{\sim}{v}_i)^{n-1} \begin{Bmatrix} -v_{ix} \\ -v_{iy} \\ \ell_i(V_{ix}S_i - V_{iy}C_i) \end{Bmatrix}$$

$$+ \begin{Bmatrix} -k_i^{(2x)} u_{ix}^{2n-1} \\ -k_i^{(2y)} u_{iy}^{2n-1} \\ \ell_i(k_i^{(2x)} u_{ix}^{2n-1} S_i - k_i^{(2y)} u_{iy}^{2n-1} C_i) \end{Bmatrix} \qquad (34)$$

$$\underset{\sim}{D}^2 E = \sum_{i=1}^{M} 2(n-1) k_i^{(1)} (\underset{\sim}{v}_i \cdot \underset{\sim}{v}_i)^{n-2} \begin{bmatrix} v_{ix}^2 & v_{ix}v_{iy} & -v_{ix}\ell_i(V_{ix}S_i - V_{iy}C_i) \\ & v_{iy}^2 & -v_{iy}\ell_i(V_{ix}S_i - V_{iy}C_i) \\ (*) & & \ell_i^2(V_{ix}S_i - V_{iy}C_i)^2 \end{bmatrix}$$

$$+ k_i^{(1)} (\underset{\sim}{v}_i \underset{\sim}{v}_i)^{n-1} \begin{bmatrix} 1 & 0 & -\ell_i S_i \\ & 1 & \ell_i C_i \\ (*) & & \ell_i(V_{ix}C_i + V_{iy}S_i) \end{bmatrix}$$

$$+ (2n-1) \begin{bmatrix} k_i^{(2x)} u_{ix}^{2n-2} & 0 & k_i^{(2x)} u_{ix}^{2n-2} \ell_i S_i \\ & k_i^{(2y)} u_{iy}^{2n-2} & k_i^{(2y)} u_{iy}^{2n-2} \ell_i C_i \\ (*) & & \ell_i^2(k_i^{(2x)} u_{ix}^{2n-2} S_i^2 + k_i^{(2y)} u_{iy}^{2n-2} C_i^2) \end{bmatrix} \qquad (35)$$

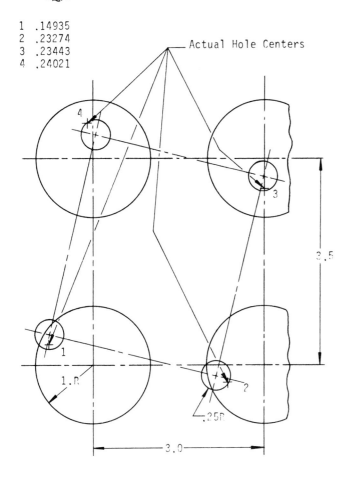

i	$\|\underline{v}_i\|$
1	.14935
2	.23274
3	.23443
4	.24021

Fig. 7   CPT Problem n=10

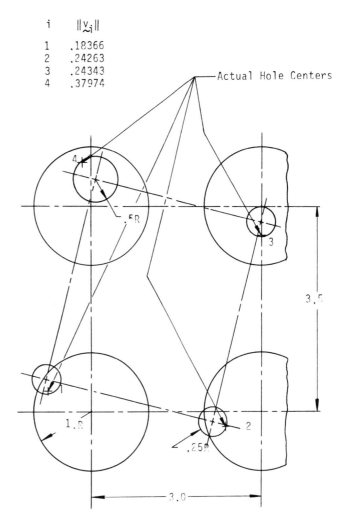

i	$\|\underline{v}_i\|$
1	.18366
2	.24263
3	.24343
4	.37974

Fig. 8   CPT Problem Different Size
FRT Zones

equal, however, in practice this may often not be the case (in fact when an MMC condition (1) is specified it never is). In the algorithm this phenomena can be modeled by properly proportioning the spring constants $k_i^{(1)}$ in equation 11. Fig. 8 shows a case where one of the FRT zones is larger than the others, again the results look quite good. Fig. 9 shows a PPMT type problem where the hole centers are at the inside edge of the APLT zones. The algorithm correctly found the pattern to be acceptable. It is interesting to note that had we used secondary springs with circular rather than rectangular potential lines this pattern would have been incorrectly rejected. For the above examples a CPT analysis required less than 3 CPU secs and a PPMT analysis less than 5 CPU secs for execution on a Univac 1108 computer. In verifying the computer code an assortment of additional patterns where tested and in every case the algorithm responded correctly.

CONCLUSION

Presented herein is an algorithm for the approximate simulation of a mechanical gauge for testing tolerances for patterns of features on mechanical components. These tolerances may be specified as either composite positional tolerances or as plus minus positional tolerances. The method has proved to be both efficient and accurate.

i	$\|\underline{v}_i\|$
1	.24801
2	.24801
3	.24801
4	.24801

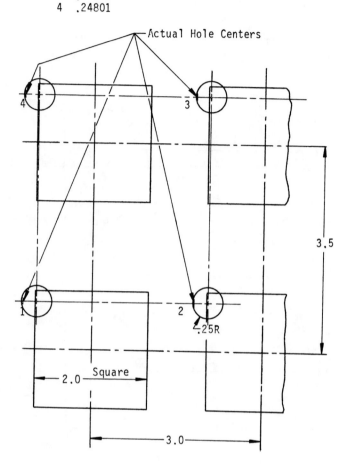

Fig. 9   PPMT Problem Secondary Springs
with Square Potential Lines

ACKNOWLEDGEMENT

The author would like to express his appreciation to Mr. Victor Dube, Project Engineer for Tracor Aerospace Co. for his invaluable suggestions during the performance of this project.

REFERENCES

1.   American National Standard Engineering - Drawing and Related Practices - Dimensioning and Tolerancing, ANSI Y14.5 - 1973, ASME, New York, 1973.

Reprinted from *Marine Technology*, July 1981

Shaft Alignment Using Strain Gages

Albert W. Forrest, Jr.,[1] and Richard F. Labasky[1]

A detailed description of the strain-gage shaft alignment procedure is presented, including a comparison between bearing reactions obtained using strain gages and hydraulic jacking. The various gage configurations available for measuring bending moments are discussed and estimates of the resulting error in bending moment are given. A simplified procedure is presented to calculate bearing reactions from the measured moments. Gage site requirements to produce a determinate system are established and a method is outlined to establish bearing reaction error bounds for a combination of gage configurations and sites.

Introduction

THERE ARE a number of steps which must be taken to achieve an acceptable shaft alignment. As discussed in T&R Report R-25 [1],[2] a variety of procedures exists to achieve these goals. All of them include the basic steps of (i) establishing the desired alignment, (ii) installing the shaft, and (iii) checking the alignment in some fashion. The strain-gage method of alignment applies only to the final phase of this operation and it can be used to replace, for example, the weighing of bearing reactions. The strain-gage procedure has been known and used for a number of years [1–4]. During this time, many practical suggestions about its proper application have been made and these will not be repeated here. Direct information verifying the accuracy of this procedure and describing its proper application has been lacking.

The strain-gage procedure has a number of advantages that make it an attractive alternative to the more commonly used alignment-check procedures. Once the gages are installed, both horizontal and vertical alignment can be checked simultaneously and a complete set of strain readings can be taken in less than an hour. The ease with which the readings may be taken (that is, no support structure as with the jacking technique) allows for measurements to be taken at sea under ballast and full-load conditions and periodically during the ship's life to insure that safe bearing and gear operating conditions are maintained. The procedure may be used to check the alignment after repairs or prior to the delivery of a new ship. Even reactions of bearings inaccessible for jacking can be determined accurately.

The purpose of this paper is to describe the details of the procedure, as understood by the authors, so that anyone with a good engineering background wishing to use the technique may be able to do so. A discussion is included to describe how strain gages are used to measure moments in shafts and to show the various methods available. A procedure is presented which allows an engineer to optimize the location of strain gages on a given shaft system and estimate maximum errors in the resulting bearing reactions. Results of application experience also are included where both strain gages and jacking were used to determine bearing reactions. It is hoped that this information will aid in both the use and acceptance of the strain-gage procedure.

Shaft moment measurement procedures

For shaft alignment, the primary strains are those induced by the moment in the shaft due to its own deadweight. Several strain-gage techniques are available to measure this moment, which subsequently is used to determine bearing reactions or elevations or both. These methods are shown in Figs. 1 through 3 along with the Wheatstone-bridge arrangement commonly used for each.

Figure 1 shows a single temperature-compensated gage oriented parallel to the axis of the shaft. To measure the moment, the shaft need only be rotated through 360 deg and the strains noted as a function of angle. Figure 4 shows the strain versus rotation angle that will result. This sinusoidal variation occurs since the strain is dependent on the bending moment and on the distance between the neutral axis of the shaft and the point where the strain is measured; it can be described using

$$\epsilon = \epsilon_a \cos(\theta + \lambda) + \bar{\epsilon} \tag{1}$$

where all variables are defined in the Nomenclature. The shaft moment is obtained by applying the basic beam relationship

$$\sigma = \frac{Mc}{I} = \epsilon_a E \tag{2}$$

which holds for the uniaxial stress condition on the shaft surface. It should be noted that the magnitude of the average strain ($\bar{\epsilon}$ in Fig. 4) is not important since only the amplitude is needed to determine the moment. From these relationships the vertical and horizontal moments can be written as

$$M_v = \frac{\epsilon_a EI}{c} \cos\lambda \tag{3}$$

$$M_h = \frac{\epsilon_a EI}{c} \sin\lambda \tag{4}$$

The reason for using this procedure is that a number of readings are incorporated in the determination of both the larger vertical and smaller horizontal moments. In this manner the error in the horizontal moment should be reduced. An alternative procedure would be to use the 12 and 6 o'clock readings to produce the strain amplitude (ϵ_a) and vertical moment. This would certainly be adequate when no significant horizontal moment (misalignment) exists.

The more complex strain-gage arrangements pictured in Figs. 2 and 3 employ exactly the same procedure as discussed for the single-gage method but lead to an improved level of accuracy. These arrangements produce strain indications two and four times larger, respectively, than the single gage strain reading and they are inherently temperature-compensated. Increasing the strain readings by these factors tends to improve the strain resolution available from standard instrumentation by a similar factor. To simplify the installation for the four-gage method, double gages are available from suppliers.

[1] Sun Ship, Inc., Chester, Pennsylvania. Mr. Forrest is now with E. I. du Pont de Nemours & Company, Inc., Circleville, Ohio.

[2] Numbers in brackets designate References at end of paper.

Presented at the March 21, 1980 meeting of the Philadelphia Section of THE SOCIETY OF NAVAL ARCHITECTS AND MARINE ENGINEERS.

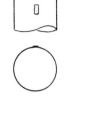

TOP

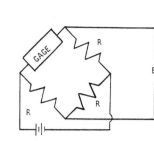

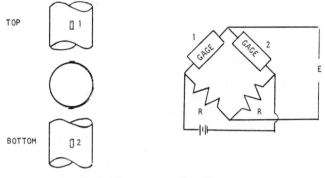

Fig. 1   Single-gage configurations

TOP

BOTTOM

Fig. 2   Two-gage configuration

TOP

BOTTOM

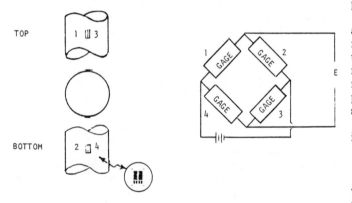

Fig. 3   Four-gage configuration

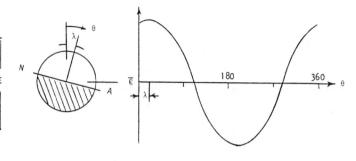

▨ COMPRESSION

Fig. 4   Typical shaft strain curve

2. *Gage factor variation*—The gage factor translates the output bridge voltage into strain and, therefore, directly affects the accuracy of the measured moment. Typically gage factors are given within ±0.5 percent. Transverse gage sensitivity is no problem in the uniaxial stress field that exists during alignment and has no effect on moment measurement.

3. *Instrumentation sensitivity*—A strain-gage indicator with a single temperature-compensated gage has been found to reproduce changes in strain to within ±1 microinch (μin.) (strain level errors may be higher). The two and four gage methods improve this to approximately 0.5 and 0.25 μin., respectively.

4. *Physical properties and dimensions*—The elastic modulus and shaft diameter enter into the moment calculation from strain measurements. Usually, the diameter is known accurately, but the elastic modulus of shaft steels typically varies from 29 to 30 million psi (203 000 to 210 000 MPa) and it is not normally measured. This represents a possible error of approximately 2 percent which could be eliminated or at least reduced by measuring E.

Consequently, the maximum error expected in the strain (η) is

$$\eta = (1 - \cos 2\alpha) \frac{\epsilon_a(1 - \nu)}{2} \pm 1/CF \times 10^{-6} \qquad (5)$$

where the circuit factor (CF) is either 1, 2, or 4, depending on the multiplication factor on strain introduced by the gage arrangement employed. The resulting error in moment can be written as

$$m = \frac{EI}{c}\left\{ (1 - \cos 2\alpha) \frac{\epsilon_a(1 - \nu)}{2} \pm 0.005\,\epsilon_a \pm \frac{1}{CF} \right. $$
$$\left. \times 10^{-6} \pm 0.02\epsilon_a \right\} \qquad (6)$$

where second-order effects have been dropped. Note that the effect of the strain introduced by the torque required to rotate the shaft can be eliminated by taking readings over several revolutions with pauses or changes in direction or both.

Accuracy of procedure

A number of parameters affect the accuracy of the moment determination. These include:

1. *Gage misalignment*—The misalignment error in a uniaxial stress field [5] depends on the amount which the gage is aligned off the axial direction, and it can be characterized as $(1 - \cos 2\alpha)$. If a good procedure is employed, this error can be reduced to a negligible amount (see Appendix 1).

Nomenclature

b = dimension in shaft free bodies
c = shaft radius
CF = bridge/gage circuit factor
E = elastic modulus
I = shaft moments of inertia
L = dimension in shaft free bodies
M = bending moment in shaft
m = error in shaft bending moment
R = bearing reaction
r = error in bearing reaction

V = shear in shaft
v = error in shaft shear
X = shaft axial dimension
α = gage misalignment angle
β = moment-influence coefficient
δ = bearing displacement from straight line
ϵ = strain
η = error in strain
θ = angle defining location of gages relative to vertical

λ = angle between normal to neutral axis and vertical
$\mu(x)$ = shaft weight per unit length

Subscripts
a = amplitude
h = horizontal
i = strain-gage location indicator
j = bearing location indicator
sl = straight-line condition
v = vertical

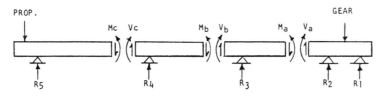

Fig. 5 Simple stern bearing system

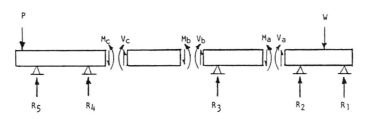

Fig. 6 Shafting arrangement requiring a free section

If the gage is positioned well (small α), the moment error equation can be reduced to

$$m = \pm 0.025\, M \pm \frac{EI}{cCF} \times 10^{-6} \qquad (7)$$

and m can be used to estimate the maximum error in moment which can be expected at a particular strain-gage installation.

Determining shaft alignment by moment-influence number approach

A typical shaft with N bearing supports is N-2 degrees statically indeterminate and N-2 added pieces of data are needed to determine the shaft alignment. Moment measurements or bearing reaction measurements at N-2 or more points can be used for this purpose. The sophisticated way of doing this using strain gages is to determine a system of influence coefficients for the moment in the shaft at the measured locations. Using the typical shafting system shown in Fig. 5 and assuming that the shaft reference line is fixed by setting the elevations at Bearings 1 and 2 to zero, the moment influence coefficient equations can be written as

$$M_i = M_{isl} + \beta_{i3}\delta_3 + \beta_{i4}\delta_4 + \beta_{i5}\delta_5 \qquad (8)$$

where N-2 equations are produced by varying i from 1 to N-2 or 3 in this case.

The moment influence coefficients (β_{ij}) can be obtained by perturbing each bearing individually using a shaft-alignment computer program. This would be described mathematically as

$$\beta_{ij} = \frac{\partial M_i}{\partial \delta_j} = \frac{\Delta M_i}{\Delta \delta_j} \qquad (9)$$

The measured moments (M_i) would then be inserted into equation (8) and the bearing elevations determined from a solution to N-2 simultaneous equations.

This procedure is direct and easily automated but offers little insight as to the proper location of strain-gage stations on a particular shaft system. A quick look might lead one to think that the engineer has a great freedom of choice in locating gage sites, even though intuitively spreading them out over the shaft system seems appropriate.

Determining shaft alignment by "free-body approach"

A less-sophisticated way of looking at the problem leads to more physical intuition and an error analysis. If the sites are

properly chosen, the shaft can be divided into a series of free bodies with measured moments and unknown shears at the intersections as shown in Figure 5. Note that the free bodies containing Bearings 1 through 4 are initially indeterminate with three unknowns, both shears and bearing reactions. However, the free body containing R_5 is completely determinant and R_5 and V_c can be calculated by summing forces and moments. Knowing V_c makes the adjacent free body determinate and R_4 and V_b can be determined. This process is continued up to the gear free body, which is also determinate after V_a has been calculated.

This procedure works well as long as a single stern-tube bearing is employed. When a double stern-tube bearing is used, the procedure must be modified as shown in Fig. 6. Here, a free section (one with no bearing) has been inserted and it has only two unknowns, the shears at each end. This section is inherently determinate and can be used between any two bearings to act as the starting point in either direction for the determination of the remaining shaft alignment conditions as just described. When inserted adjacent to a double stern-tube arrangement, for example, it provides a value for the unknown shear and leaves only the two bearing reactions as unknown. This allows for the determination of both bearing reactions and provides a known shear on the forward end for the determination of the remaining reactions.

Three bearings with inaccessible connecting shafting (that is, a double stern-tube and strut bearing arrangement) cannot be analyzed using the strain-gage method alone. The addition of a free section does not make the aft shaft free body statically determinate since the three bearing reactions remain as unknowns. The addition of a third strain-gage station as shown in Figure 7 does not help since the shear at Station d can be calculated using moments at either b and d or c and d. If in addition the moments at b and d are known, the value of the moment at c can be calculated using simple beam relationships. This fact shows that only two linearly independent pieces of data are obtained from this configuration. Consequently, this three-gage station arrangement cannot be employed to generate three linearly independent influence equations (8) since only two will be truly independent. For the double stern-tube/strut configuration, bearing reactions forward of free section could still be determined by inserting a free section to determine the shear at the forward end and proceeding in a manner similar to the cases considered previously. The three aft bearing reactions could be determined if one bearing is weighed using conventional procedures (that is, if the forward stern tube bearing could be jacked). This in conjunction with a free section would make the situation determinate with only the two aft bearing reactions, 4 and 5, unknown.

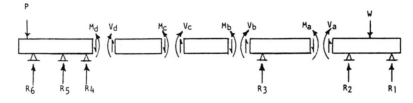

Fig. 7 Incorrect attempt to analyze a triple stern bearing arrangement

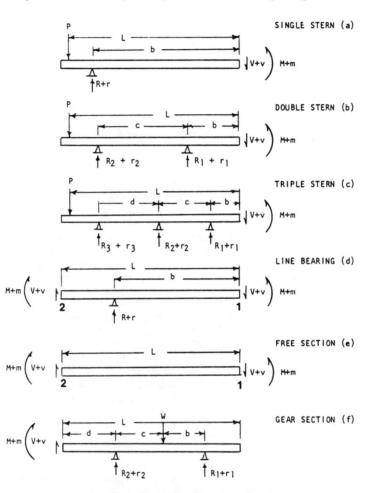

Fig. 8 Free-body sections for shaft alignment

Error analysis and siting recommendations

Both the free-body and influence coefficient approach are completely equivalent, since a properly arranged set of moment measurements uniquely defines the bearing reactions. Consequently, the method used to calculate them, if it is done accurately, is not important. The primary difference is that the influence coefficient method generates the bearing elevations directly and the free-body approach yields reactions. By using the normally produced reaction to elevation influence coefficients, reactions can be translated to elevations and vice versa, and this difference is not important. It is also noteworthy that any conclusions drawn about locating gage sites and estimating magnitudes of expected errors introduced by measurements also will be independent of the calculation technique used. Since the free-body technique is simpler and algebraically maintains the geometric character of the shafting arrangement, it will be used to evaluate the accuracy of the strain-gage alignment procedure and to establish gage siting recommendations.

Error analysis and siting recommendations

The procedures described to determine the shaft alignment

using strain gages are relatively simple, but to make them useful some method must be found to establish the bearing reaction or elevation error that will result for a particular system. To accomplish this task, the error in the calculated bearing reaction or shear or both can be estimated by assuming each moment (M) is in error by some small amount (m). Where appropriate, the shear (V) also is considered as having an error (v) so that the error transmitted to adjacent free bodies via the shear can be estimated. These shears and moments are inserted into the static relationships required to calculate the unknown bearing reactions and shears, and the resulting errors for each free-body type are algebraically determined. As previously discussed, the errors calculated using the free-body approach apply to the influence coefficient method as well as to any other procedure for determining shaft alignment from a series of measured moments. Consequently, error bounds predicted by this method hold for any calculational procedure used to determine shaft alignment from strain-gage measurements.

Figure 8 shows the six shaft free bodies needed to consider shaft alignments. The lineshaft-bearing free body [Fig. 8(d)] has measured moments M_1 and M_2 and requires one known shear

Table 1   Error bound formulation and recommendations [a]

Free Body Type	Reaction Error(s)	Shear Error(s)	Recommendations
Single stern (a)	$r = \dfrac{m}{b}$	$v = \dfrac{m}{b}$	maximize b
Double stern (b)	$r_1 = \dfrac{m + v(b + c)}{c}$	(requires known v)	minimize b
	$r_2 = \dfrac{m + vb}{c}$		
Triple stern (c)	$r_2 = \dfrac{m + r_1(c + d) + v(d + c + b)}{d}$	(requires known v)	minimize b
	$r_3 = \dfrac{r_1 c + m + v(b + c)}{d}$		
	(r_1 must be measured)		
Line shaft (d)	$r = \dfrac{m_1 + m_1 + v_2 L}{b}$	$v_1 = \dfrac{m_1 + m_2 + v_2(L - b)}{b}$	maximize b
Free section (e)	none	$v_1 = \dfrac{m_1 + m_2}{L} = v_2$	maximize L
Gear section (f)	$r_1 = \dfrac{m + vd}{c + b}$	none	minimize d
	$r_2 = \dfrac{m + v(b + c + d)}{b + c}$		

[a] Signs have been adjusted to reflect maximum possible values as positive.

(V_2). Summing moments about 1 yields

$$\sum M = 0 = M_1 \pm m_1 - M_2 \pm m_2 - (V_2 \pm v_2)L$$
$$- (R + r)b + \int_o^L x\mu(x)dx \quad (10)$$

or

$$(R + r) = \frac{M_1 - M_2 - V_2 L + \int_0^L x\mu(x)dx}{b} + \frac{m_1 + m_2 + v_2 L}{b} \quad (11)$$

where the signs on the various error components have been adjusted to indicate the worst case. Subtracting the exact value of R from equation (11) leads to a reaction error defined by

$$r = \frac{m_1 + m_2 + v_2 L}{b} \quad (12)$$

Summing forces in the vertical direction shows that

$$\sum F = 0 = V_2 + v_2 + R + r - (V_1 + v_1) - \int_0^L \mu(x)dx \quad (13)$$

or

$$V_1 + v_1 = V_2 + R - \int_0^L \mu(x)dx + v_2 + r \quad (14)$$

and the resulting shear error is

$$v_1 = v_2 + r = v_2\left(1 + \frac{L}{b}\right) + \frac{m_2 + m_1}{b} \quad (15)$$

This shows that the transmitted error (v_1) is at least twice the incoming error (v_2) and that maximizing b/L within the physical limits of the system tends to minimize this error. The importance of minimizing transmitted error on achieving good strain-gage alignment results subsequently will be emphasized by actual shipboard experience. It should be remembered that beam theory does not fully apply at the support points and strain-gage stations should be located approximately one shaft diameter from any support point [6] and at least one total flange width from the fillet of flanges [6,7].

The errors introduced in the reactions and shears for all of the six basic free bodies required to evaluate a shaft system are included in Table 1. Also included in this table are recommendations for locating strain gages based on minimizing the introduced errors; these were obtained from the error relationships by inspection.

After choosing a series of these free-body sections to approximate a particular shaft system, the magnitude of the error of each bearing reaction can be determined in terms of the errors in each measured moment. These errors can be quantified by inserting the expected error bound on moment given in equation (7).

Shipboard examples

To establish the credibility of the strain-gage alignment procedure, bearing reactions have been obtained for two ships using both the strain-gage alignment procedure and the commonly used jacking method. The first example represents the initial experience at Sun Ship and was done without the benefit of a preliminary error analysis to establish an acceptable gage siting plan. The ships considered include a tanker and a containership, but both of them have similar shafting arrangements as shown in Figs. 9 and 10.

Initial experience

The strain-gage locations chosen for the initial test of the strain-gage alignment procedure also are shown in Fig. 9. Ease of access was used as the primary consideration for choosing these locations since the error-bound procedure had not been developed. All gage installations employed the two-gage method (see Fig. 2) and 0.5-in.-long (12.7 mm) gages. Both gage readings and jacking data were taken shortly after the dock trial when the system was hot. Gage data were taken by rotating the shaft several times on the jacking gear and noting the several gage readings every 90 deg. The resulting strain-determined bearing reactions are shown compared with the measured bearing reactions in Table 2, where differences were generally less than 5 percent, with the exception of the aft gear bearing, which shows a difference of 25.3 percent.

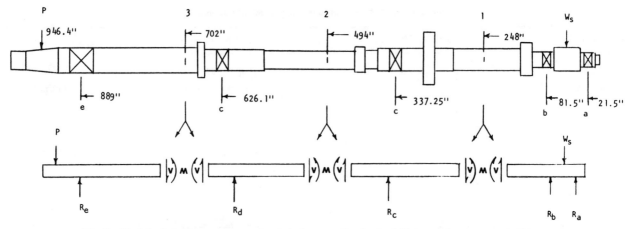

Fig. 9 Sketch of shafting arrangement and strain-gage sites for the initial experience case—tanker

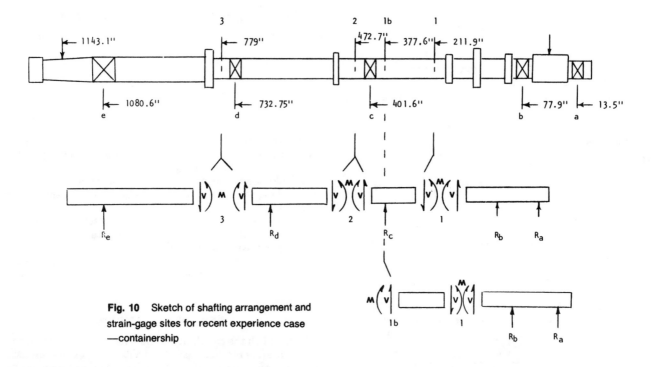

Fig. 10 Sketch of shafting arrangement and strain-gage sites for recent experience case —containership

It should be stated that the bearing jacking procedure is in no way exact. The bearing reaction is obtained by jacking the shaft adjacent to a bearing and obtaining the force-displacement curve. At some point there will be a discontinuity in the slope of this curve when the journal clears the bearing surface. The slope of the curve above this point is used to establish the zero-displacement force or the bearing reaction by extrapolation. This value is corrected to the bearing center to obtain the "measured" reaction. As can be seen, there are a number of places where errors can be introduced into this procedure. Those who perform this task routinely at Sun Ship estimate that errors on the order of 10 percent are inherent to the procedure.

In an attempt to understand the relatively large discrepancy between the reactions determined via strain gages and jacking, the error-bound procedure described earlier was developed. The application of the procedure to the configuration employed is summarized in Table 3. Notice how the error bounds are over 19 percent for the two gear bearings and relatively small for the stern-tube and lineshaft bearings. This is caused by the cumulative effect of the error transmitted forward by the shear since only the stern-tube free body is fully determinate and the remaining free bodies require shears determined from the section adjacent and aft.

Recent experience

In light of the difficulties encountered with the initial test of the strain-gage alignment system, recent applications have included a preliminary error-bound analysis to aid in the proper siting of the strain gages. Figure 10 shows a shafting arrangement for which this procedure was used. This system has five bearings and requires three gage installations, and Locations 1, 2, and 3 were selected based on recommendations in Table 1 and acces-

Table 2   Comparison of strain gage and jacking method bearing reactions for initial experience

Bearing	(1) Jacked Warm 6/20	(2) Jacked Warm 12/5	(3) Strain Gage 12/5	Difference,[a] %
a. Forward gear	41 300	39 400	40 642	3.2
b. Aft gear	30 200	34 500	25 767	25.3
c. Forward line	50 700	⋯	53 117	4.8
d. Aft line	18 600	⋯	17 744	4.6
e. Stern tube	⋯	⋯	160 261	⋯

[a] Difference is based on weighing (2) where applicable.

Table 3 Error-bound calculation for initial experience

Gage Location	Moment Magnitude, in.-kips	Diameter, in.	m, in.-kips	v, lb
1	280	23.875	26.7	2176
2	490	23.875	31.95	866
3	580	31.25	58.7	314

Bearing	Design Reaction (R), lb at Temperature	r, lb	Error Bound r/R, %
e. Stern tube	162 060	314	0.2
d. Aft line	13 010	1180	9.1
c. Forward line	55 363	3042	5.5
b. Aft gear	33 486	8732	26.1
a. Forward gear	33 502	6556	19.6

1 lb = 0.45 kg.
1 in. = 25.4 mm.

Table 4 Error-bound calculation for recent experience

Gage Location	Shaft Diameter, in.	Expected Moment, in.-kips	m, in.-kips	v, lb
1	23.25	2560	82.2	693 (699)
1b	23.5	600	33.8	(699)
2	23.25	370	27.5	308
3	23.25	1100	45.7	152

Bearing No.	Reaction, lb	r	r/R, %
e. Stern tube	125 091	152	0.1
d. Aft line	45 382	460	1.0
c. Forward line	45 338	1001	2.2
b. Aft gear	41 083	3409 (3427)	8.3 (8.3)
a. Forward gear	40 878	2716 (2729)	6.6 (6.7)

1 lb = 0.45.
1 in. = 25.4 mm.
Numbers in parenthesis represent values calculated using a free section defined by Gage Locations 1 and 1b.

sibility. An error-bound analysis was performed for this arrangement and the results are included as Table 4, where error limits at the gear are approximately 8 percent. Even though this was not considered excessive, it was decided to add a free section [see Figure 8(e)] aft of the thrust bearing, and the redundant gage location, 1b, was included. Error-bound calculations for the gear reactions determined using this added free section are given in Table 4 (enclosed in parentheses). Note that in this case no improvement in the error bound was achieved by adding the free section, but the alternative method of calculating the gear reactions is used to ensure no faulty gages are in the system.

The results of the jacking and strain-gage procedure for this case are given in Table 5. To determine these values, strain-gage and jacking data were collected at an ambient (cold) condition all within a period of several hours. The differences between the jacked and strain-gage-determined reactions are within the error bounds given in Table 4 with the exception of the aft line bearing, where a difference of 9.2 percent is noted. Because of the excellent agreement at the other bearings, this difference probably is attributable to errors in the jacking procedure.

Conclusions

The strain-gage method of determining shaft alignment is an acceptable technique and is capable of establishing bearing reactions about as closely as the jacking method. The most important advantage it has is the speed with which an alignment check may be made, especially after the gages have been installed. In addition to that, loads on inaccessible bearings can often be accurately determined.

The proper siting of gages on the shaft is crucial to both ensuring accuracy and to making the resulting system statically determinate. Excessive errors can result if gages are not sited so as to minimize the accumulation of error from one free body to the next via the discrepancies in calculated shear at the gage locations. Improper siting can also lead to a system of free bodies that are not statically determinate. A well-posed arrangement is ensured by dividing the shaft into free bodies at the gage locations, identifying fully determinate sections and determining that a sequence of calculations exists to determine unknown shears and bearing reactions. If all of the reactions can be found, the system has been rendered fully determinate.

The free-body approach of siting the strain gages, evaluating errors and calculating reactions provides both physical insight and more flexibility than more sophisticated procedures (that is, the moment-influence equation approach). Once the system is transformed into a series of moment-influence equations, all geometry affects are lost in a series of constants and the resulting

Table 5 Comparison of strain gage and jacking method bearing reactions for recent experience

Bearing	Jacked Cold 4/22/80, lb	Strain Gage 4/22/80			
		Full Model		Free and Gear Sections	
		lb	% difference	lb	% difference
a Forward gear	48 400	47 290	2.3	46 553	3.8
b. Aft gear	32 400	30 123	7.0	31 214	3.7
c. Forward line	56 400	56 708	0.5	...	...
d. Aft line	32 200	35 176	9.2	...	...
e. Stern tube	...	218 535	...	...	...

1 lb = 0.45 kg.

191

effect on accuracy of relocating a gage cannot be readily determined. Using the free-body approach, geometry effects can be determined simply by inspecting the various equations of static equilibrium employed. The free-body approach also provides a means of using redundant gage locations to improve accuracy as illustrated by the use of the free section in the recent experience example. Furthermore, reactions can be calculated as the readings are being taken by evaluating deadweight effects and reducing the results to several simple relationships for each free body. These may be evaluated using a hand-held calculator on the ship (see Appendix 2).

Recommendations

A strain-gage alignment plan should be conceived using the free-body approach as described. This allows for an evaluation of errors and insures that a fully determinate system is defined. Once the data are collected, any accurate calculation method may be used to transform the measured moments into bearing reactions or elevations.

Good instrumentation should be employed to determine bending moments in the shaft. Typical gage site plans require strain accuracies on the order of ± 1 μin. per inch or better. Even with the more complex bridge circuits (see Figs. 1–3), this is near the limit of currently available instrumentation.

Certainly, one of the potential advantages to the strain-gage procedure is that it may be capable of determining shaft alignment during operation. It should be emphasized that the static bending strains are relatively low when compared with the other operating strains (that is, torsion and thrust) and that these strains are unsteady because of the propeller blade-wake interaction. A reliable means must be established to separate these signals and the procedure must be checked, preferably by an alternate method.

References

1 SNAME Technical and Research Report R-25, 1978.
2 Grant, Robert B., "Shaft Alignment Methods with Strain Gages and Load Cells," MARINE TECHNOLOGY, Vol. 17, No. 1, Jan. 1980, pp. 8–15.
3 Kvamsdal, Rolf, "Shaft Alignment," *European Shipbuilding*, Nos. 1 and 2, 1969.
4 Kvamsdal, Rolf, "Shaft Alignment Control by Means of the Strain Gage (Bending Moments) Technique," Det norske Veritas, Research Department Report 68-19-M, Oslo, 1968.
5 "Strain Gage Misalignment Errors," Micro-Measurements Tech Note TN-138-4, Measurements Group, Romulus, Michigan.
6 Sechler, Ernest E., *Elasticity in Engineering*, Dover Press, New York, 1952, pp. 131–136 and pp. 157–162.
7 Peterson, R. E., *Stress Concentration Factors*, Wiley, New York, 1974, pp. 80–82.

Discussers

F. Everett Reid
George Laing

Appendix 1

Axial alignment of strain gages

The strain gages must be accurately aligned in the axial direction before the strain-gage procedure can be successfully employed. The reason error is introduced into the moment calculation by misalignment as shown in Fig. 11. Here, the error is the difference between the actual strain (ϵ) and the measured strain (ϵ_α) and it can be expressed for uniaxial stress in the principal direction [5] as

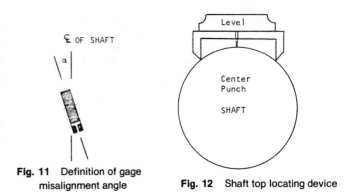

Fig. 11 Definition of gage misalignment angle

Fig. 12 Shaft top locating device

$$\epsilon - \alpha = \epsilon(1 - \cos 2\alpha) \frac{(1 - \nu)}{2} \qquad (16)$$

The significance of this error can be eliminated if the misalignment angle is kept below approximately 2 deg since the cosine of small angles is nearly one.

The gage alignment procedure used at Sun Ship involves finding several points on the top of the shaft a few feet apart. This is done using a machinist's level and the apparatus shown in Fig. 12. It is estimated that the top of the shaft can be located within $\pm \frac{1}{8}$ in. (3.175 mm) using this approach. By locating two points several feet apart, a reference line for gage installation can be constructed that is within 1 deg of the axial direction. The gages themselves are accurately made and come with points for alignment purposes. Using a large gage length [0.5 in. (12.7 mm)] and the procedure outlined in the preceding, an experienced technician has little difficulty in locating the gage with a misalignment angle well below the allowable 2 deg.

Appendix 2

Example Calculations from recent experience

For the recent experience case the shafting arrangement, gage locations and error-bound results are summarized on Fig. 9 and Table 4. A brief description is included here to illustrate how the bearing reactions were obtained from the measured strains. The first step was to modify equation (3) to account for the gage configuration employed and to allow for the input of the maximum and minimum strains directly. The preliminary alignment procedure employed at Sun usually eliminates appreciable horizontal misalignment and it is not considered unless the readings indicate a significant problem. The resulting vertical moment relationship becomes

$$M_v = \frac{(\epsilon_b - \epsilon_t) \, EI}{4c} \qquad (17)$$

where t and b refer to the condition where Gage 1 (see Fig. 2) is on the top and bottom of the shaft, respectively.

A series of relationships is derived prior to taking the measurements to allow for calculations of reactions as the data are taken. This is accomplished by algebraically solving for the unknown reactions and shears in the free bodies in terms of the measured moments and calculated shears. A typical example of this can be seen by examining the gear free body from Fig. 10. Summing moments about Point a yields

$$\sum M_1 = 0 = M_1 + V_1(198.4) + R_b(64.4)$$
$$+ \int_0^{211.9} (13.5 - x)\mu(x)dx \qquad (18)$$

Table 6   Reaction and shear equations for recent experience

Stern-tube section:

$$Re = \frac{38.616 \times 10^6 + m_3}{301.53}$$

$$V_3 = Re - 140\,372$$

Aft line shaft:

$$Rd = \frac{m_2 - m_3 - v_3\,(306.3) + 5.8175 \times 10^6}{206.05}$$

$$v_2 = v_3 + Rd - 40\,166$$

Forward line shaft:

$$Rc = \frac{m_1 - m_2 - v_2\,(260.8) + 4.1518 \times 10^6}{189.8}$$

$$v_1 = v_2 + Rc - 31\,772$$

Free section:

$$v_{1b} = \frac{m_1 - m_{1b} + 1.6808 \times 10}{165.8}$$

$$v_1 = v_{1b} - 20\,173 \qquad\qquad {}_6$$

Gear section:

$$Rb = \frac{-m_1 - v_1\,(198.4) + 5.6148 \times 10^6}{64.4}$$

$$Ra = -v_1 - Rb + 85\,522$$

and summing vertical forces leads to

$$\sum F = 0 = R_a + R_b + V_1 - \int_0^{211.9} \mu(x)dx \qquad (19)$$

where all gear weights are included in $\mu(x)$ and all the integrals can be evaluated by segmenting the shaft. Solving (18) and (19) for R_a and R_b results in

$$R_a = -R_b - V_1 + 85522 \qquad (20)$$

$$R_b = \frac{-M_1 - 198.4\,V_1 + 5.6148 \times 10^6}{64.4} \qquad (21)$$

This process is continued for each shaft free body and the results of these calculations are given on Table 6.

The equations for horizontal reactions and shears are obtained in a similar manner but no shaft weights are included. These relationships may be obtained by inspection from the equations in Table 6 simply by removing all constant terms. For the recent

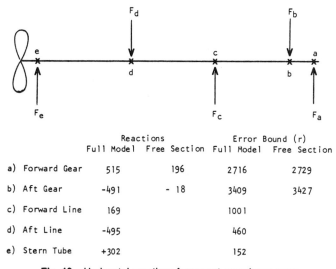

		Reactions		Error Bound (r)	
		Full Model	Free Section	Full Model	Free Section
a)	Forward Gear	515	196	2716	2729
b)	Aft Gear	-491	- 18	3409	3427
c)	Forward Line	169		1001	
d)	Aft Line	-495		460	
e)	Stern Tube	+302		152	

Fig. 13   Horizontal reactions for recent experience case

experience case, horizontal reactions were obtained for a warm-plant condition with all bearings well lubricated. These are included in Fig. 13 where reactions in the port direction are given as positive. These reactions are subject to the same error bounds (r's in Table 4) as the vertical reactions and only reactions larger than r can be considered as significant. Since horizontal reactions have not been obtained for a large number of ships over a significant time period, a method for evaluating the importance of horizontal forces in the shaft system is not available.

Obtaining horizontal reactions using strain gages is complicated slightly by the physical situations. No weight loads exist in the horizontal direction to seat the shaft in the bearings, and the shaft may move due to frictional forces induced during shaft rotation. To minimize this problem, the shaft should be rotated in both directions with well-lubricated bearings while taking strain readings and then the strains should be compared. If discrepancies exist, they should be corrected before any attempt is made to obtain horizontal reactions from these strains. A second complication arises from the way strain changes during the rotation of the shaft. As shown in Fig. 4, the strain is changing most rapidly near 90 to 270 deg or at the points where the horizontal moment is being obtained. Care must be taken, therefore, to get acceptable strain readings in the horizontal plane.

Inspection Setups That Simulate Functional Assembly Conditions

By Edward S. Roth
Productivity Services

INTRODUCTION

A significant percentage of the parts that do not pass surface plate inspection appear to function satisfactorily when assembled to their mating parts. This dilema may occur because the "rejected" part is held in a unique and functional orientation when assembled that was not simulated during the measurement process when the part was set up for inspection. Non-functional set ups can occur during inspection if product drawings do not reference datum features as the inspector must interpret the datums by implication. The use of non-functional part datum features as set ups can rotate and or translate measurement data so the part appears to be out of tolerance.

The datum referencing methodology defined in ANSI Y14.5 allows the designer to define the functional assembly orientation of any part. The methodology allows the designer to relate part datum features to a datum reference frame which can then in turn be simulated by the proper arrangement of inspection equipment.

This report describes the relationship between properly referenced datum features on drawings and inspection set ups that simulate the functional assembly orientation. It presents a format that could form the basis for standardizing the inspection operation and better controlling both the manufacturing and inspection processes. Also included are several illustrations of symmetrical parts that can be defined without the use of any datum references that are included as special cases at the end of this report. These cases are included to indicate that datum referencing is not always mandatory. In addition, two unique cases involving non-rigid part set ups conclude this report.

The material included is particularly useful to those Manufacturing Engineers who review preliminary designs for manufacturability.

INSPECTION PHILOSOPHY

The inspection set up, to include the clamping forces used to restrain the part, should simulate the functional assembly orientation of the part to the extent that the expected difference between the measured and functional state will not exceed a suitably small increment of the part tolerances being measured, typically ten percent (10%). It is recognized that this goal cannot always be economically or physically attainable.

The examples covered in this report (not in this order) fall into several measurement set up categories. In the first category, the part is set up stationary in relation to the datum reference frame. In the second, the part may rotate or translate in the inspection set up. In the third, a part/fixture assembly may rotate. The fourth category involves a series of independent measurements as no datums are referenced or required. The final category includes the use of special fixtures that restrain thin-walled parts when the fixture contains features that establish the part datum.

RUNOUT

Dial indicator readings obtained by rotating a part about a datum axis are runout measurements.

Entire Part Surfaces Used As Primary Datum Features

Fig. 1 shows the part drawing where the entire cylindrical surface of an external datum-feature establishes the datum axis for measurement. The following figures, which have no preferred order, show several setups that may be employed depending on the tolerances specified and the equipment available to the inspector. Each setup shown will influence the measurements obtained and no two will give identical results. In each case, the part will be rotated to ascertain if the specified runout is to specification.

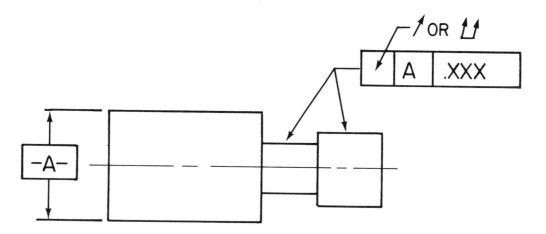

Fig. 1
Drawing Callout

Fig. 2 shows a precision chuck engaging the entire datum feature. The chuck axis is the datum axis. The part is rotated 360° with the indicators positioned at several locations on the part surfaces during measurement.

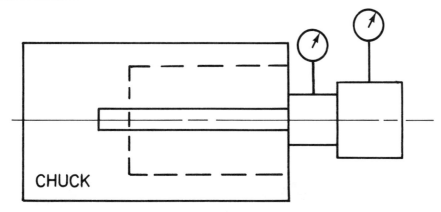

Fig. 2
Setup

Fig. 3 shows the datum feature in contact with a surface plate. The part must be repositioned at several regular angular orientations during measurement with the indicators positioned at several locations on the part surfaces during measurement.

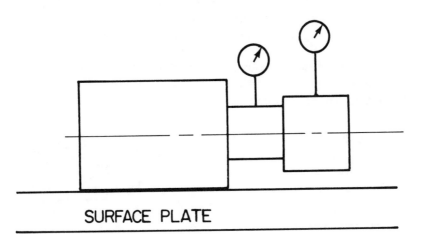

SURFACE PLATE

Fig. 3

Fig. 4 shows the non-functional end of the datum feature resting on the leveling plate of a precision spindle machine. The leveling plate must be so adjusted that the datum-feature axis is coaxial to the spindle axis before runout measurements are made. This part is rotated part 360° with the indicators positioned at several locations on the ons on part surfaces during measurement.

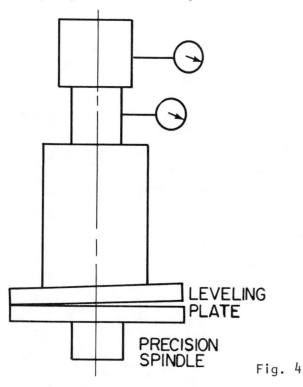

LEVELING
PLATE

PRECISION
SPINDLE

Fig. 4

Fig. 5 shows the datum-feature nested in a V block. The part is rotated 360° with the indicators positioned at several locations on the part surfaces during measurement.

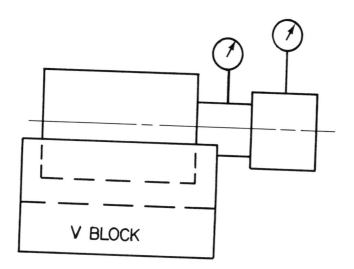

Fig. 5

Fig. 6 shows a part drawing where the entire length of the periphery of an internal datum-feature establishes the axis for measurement. Since only the cylindrical surface is designated as the datum-feature, the bottom shoulder is not contacted.

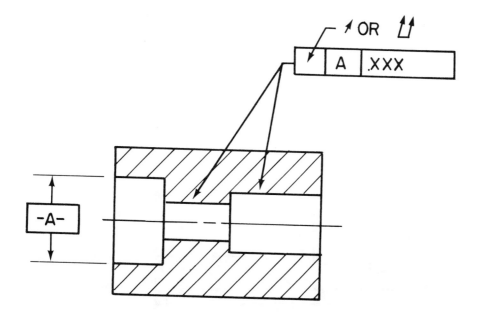

Fig. 6

Fig. 7 shows a master plug inserted into the datum-feature and the assembly resting on the leveling plate of a precision spindle machine. The leveling plate must be so adjusted that the master plug axis is coaxial to the spindle axis before runout measurements are made. The part is rotated 360° with the indicators positioned at several locations on the part surfaces during measurement.

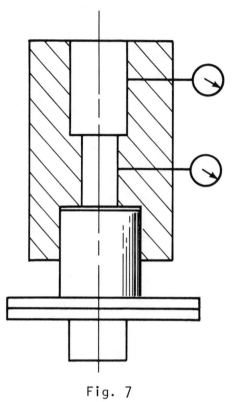

Fig. 7

Fig. 8 shows a master plug inserted into the datum-feature and the assembly inserted into a female precision chuck. The part is rotated 360° with the indicators positioned at several locations on the part surfaces during measurement.

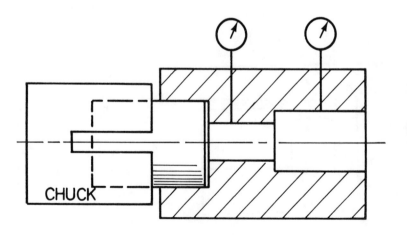

Fig. 8

Fig. 9 shows a master plug inserted into the datum-feature and the assembly used for measurement in conjunction with a V Block. The part is rotated 360° with the indicators positioned at several locations on the part surfaces during measurement.

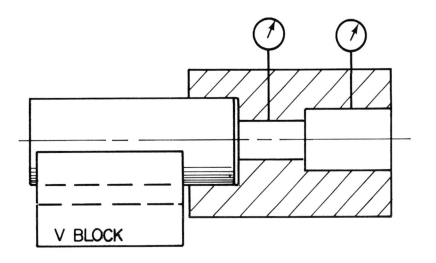

Fig. 9

Portions of Part Surfaces Used as Primary Datum-Features

Fig. 10 shows the part drawing where only a specifically designated portion of a feature, or datum target area, is the datum-feature. The following figures, which have no preferred order, show several setups that may be employed depending on the tolerances specified and the equipment available to the inspector. Again, each setup shown will influence the measurements obtained and no two will give identical results. In each case, the part will be rotated 360° to ascertain if the specified runout is to specification.

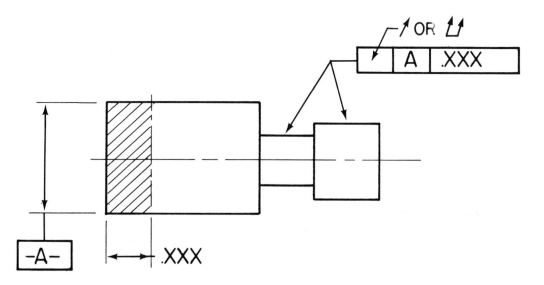

Fig. 10

Fig. 11 shows a female precision chuck engaging only that portion of the datum-feature surface designated as the datum target area. The part is rotated 360° with the indicators positioned in several locations on the part surfaces during measurement.

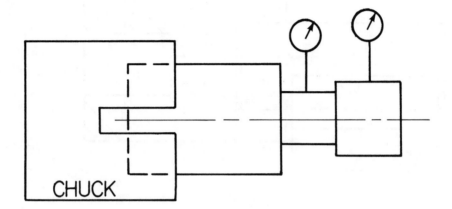

Fig. 11

Fig. 12 shows the part drawing where the extreme ends of two datum-features are designated as datums.

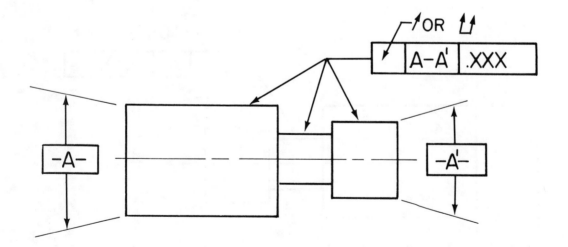

Fig. 12

Fig. 13 shows two female centers contacting the extreme ends of the datum-features. Part adjustment may be required to achieve the least runout near where the datum-features contact the female centers, before runout measurements are made on the remainder of the part. The of the part is rotated 360° with the indicators positioned in several n locations on the part surfaces during measurement.

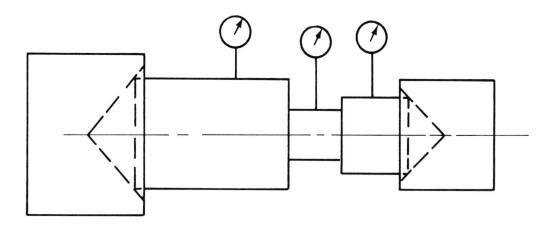

Fig. 13

Combination Primary and Secondary Datum Surfaces

Fig. 14 shows the datum axis perpendicular to the primary datum surface and coaxial to a datum target area at maximum material condition (MMC) on the secondary datum-feature.

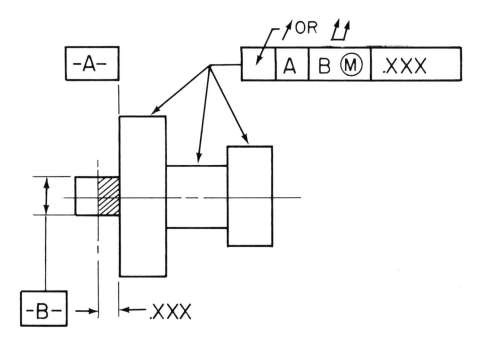

Fig. 14

Fig. 15 shows the fixture required to meet the setup indicated by part drawings in Fig. 14. The inside diameter of the fixture contacting the datum target area is the maximum material condition size M of Datum B.

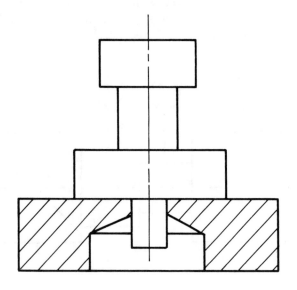

Fig. 15

Both part and fixture should be rotated as one assembly about the fixture axis. If the part does not meet specification, the part-fixture assembly axes may be translated by repositioning the part in the fixture. This can be repeated until the part either meets the specifications at one translated position or it is determined it cannot. The part is rotated 360° with the indicators positioned in several locations on the part surfaces during measurement. Note: The MMC callout to datum-feature B allows this deviation from exact coaxiality if diameter B is not at MMC.

Machining Centers Used As Datum-Features

Fig. 16 shows the part drawing where machining centers are designed as datum-features that establish the axis for measurement.

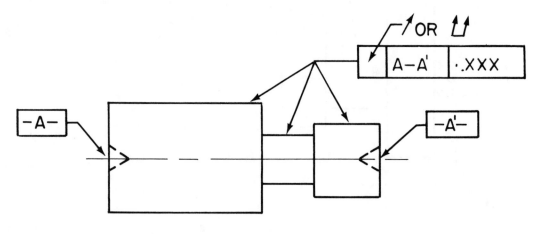

Fig. 16

Fig. 17 shows the setup utilizing male centers. The part is rotated 360° with the indicators positioned in several locations on the part surfaces during measurement.

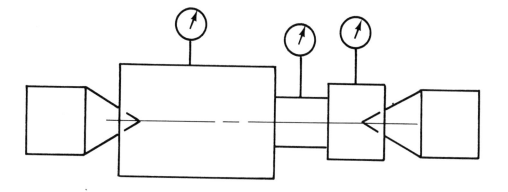

Fig. 17

PARALLELISM

Entire Part Surface Used as Primary Datum-Feature

Fig. 18 shows the part drawing where the entire surface of datum-feature A establishes the datum plane.

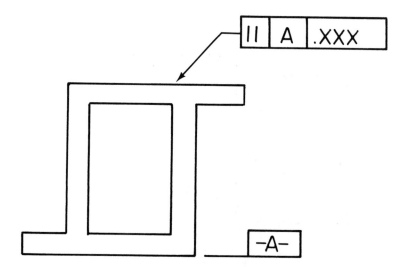

Fig. 18

Fig. 19 shows the setup where datum surface A is placed against a surface plate. The surface plate mounted indicator must be positioned at several locations on the part surface during measurement.

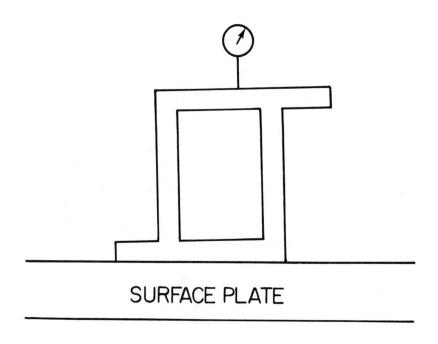

SURFACE PLATE

Fig. 19

Fig. 20 shows the part drawing where the entire inside diameter of datum-feature A establishes the datum axis for measurement.

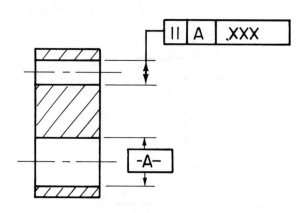

Fig. 20

Fig. 21 shows the setup where a master plug is inserted into datum
diameter A and the assembly placed into a V Block. A second master
plug is inserted into the hole to be measured for parallelism. (Indicator readings must be mathematically adjusted in some cases as only
the actual feature surface length must meet the tolerance.)

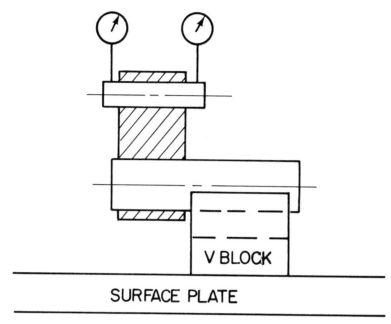

Fig. 21

Portions of Part Surfaces Used as Primary Datum-Features

Fig. 22 shows the part drawing where only two ends of datum-feature A
establish the datum axis for measurement.

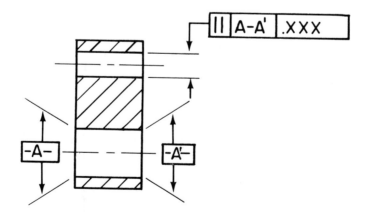

Fig. 22

Fig. 23 shows the setup utilizing male centers to pick up on the datum-feature and establish the datum axis for measurement and a master plug inserted into the hole to be measured for parallelism. Indicator readings must be mathematically adjusted in some cases as only the actual feature surface length must meet the tolerance.

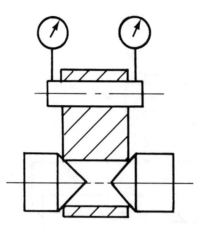

Fig. 23

PROFILE TOLERANCES

Machine Centers Used as Datum-Features

Fig. 24 shows a part drawing where Centers A and A^1 establish the datum axis.

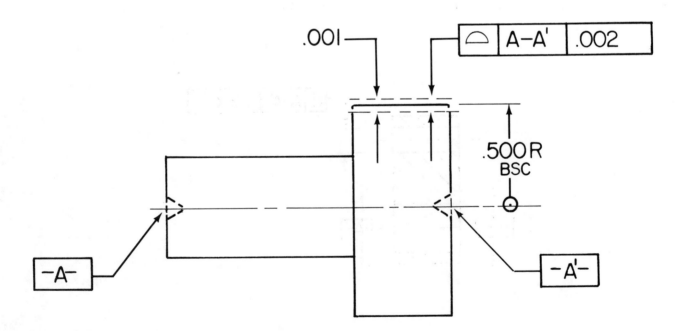

Fig. 24

Fig. 25 shows the setup. The indicator is mastered .500 inch from the datum axis in a measuring direction perpendicular to the datum axis. The indicator can only be moved parallel to the datum axis and no reading can exceed plus or minus .001 inch. The part is rotated 360° with the indicator at several locations during measurement.

Set and traverse the indicator perpendicular to the Datum Axis at .500 inch.

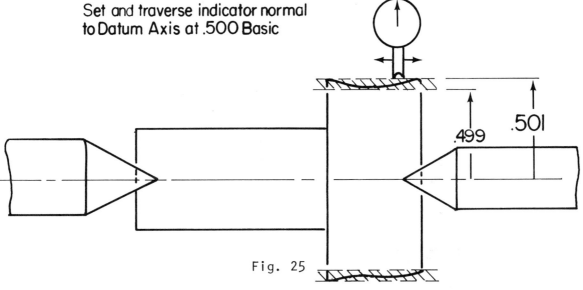

Fig. 25

Datum Surfaces Used to Establish the Datum Reference Frame

Fig. 26 shows a part drawing when surface A, B, and C establish the datum for profile.

Fig. 27 shows the setup utilizing these three surfaces as datums. Surface A is clamped to angle plate 2 when surface B is resting upon the surface plate, and surface C is contacting angle plate 1 which is perpendicular to angle plate 2. Indicators are "zeroed" at the basic coordinate locations. Each indicator must not vary more than plus or minus .005 inch when normal to the basic contour. Indicators may be repositioned to explore the surface in detail through the depth of the part.

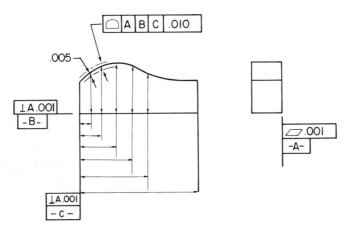

Fig. 26

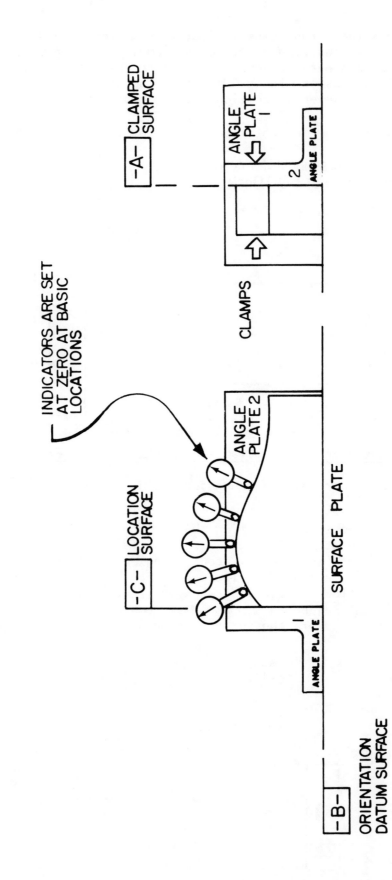

INDICATORS ARE SET AT ZERO AT BASIC LOCATIONS

-A- CLAMPED SURFACE

ANGLE PLATE 1

ANGLE PLATE

2

CLAMPS

-C- LOCATION SURFACE

ANGLE PLATE 2

SURFACE PLATE

ANGLE PLATE

1

-B- ORIENTATION DATUM SURFACE

Some Other Methods of Controlling and Inspecting Symmetry

The following examples do not require a set up based on a defined datum reference. All symmetry measurements involve the use of micrometers with suitable attachments to locate properly on the features shown.

Positional tolerance of symmetry can be adequately measured if the dimensioning is equal in value on both sides of the part axis. The following illustrations   this technique. Figures 28 and 29 show symmetrical dimensioning that controls this requirement.

Symmetry can also be specified by the use of notes that require each side of the part to be like the other within specific values. Fig. 30 & 31 show the control of symmetry in this manner.

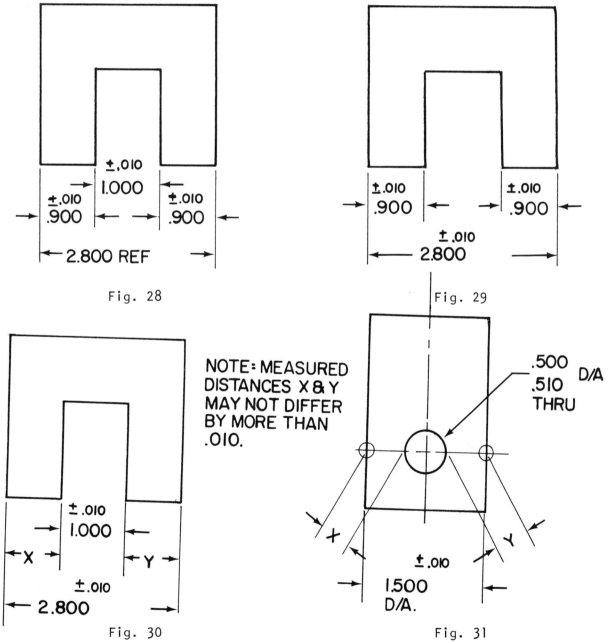

Fig. 28

Fig. 29

Fig. 30

Fig. 31

NOTE: MEASURED DISTANCES X & Y MAY NOT DIFFER BY MORE THAN .010.

Non-Rigid Part Control

All parts distort to some extent when removed from manufacturing fixtures or assembly constraints. Thin-walled parts that fall outside of drawing limits in their free state can be handled in two ways: (1) restraint of the part in a fixture that simulates the MMC assembly condition, or (2) the use of a method of dimensioning that enables the part to be measured in the free state with the assurance that it will function when restrained in a subsequent assembly.

Fig. 32 shows the constraint method utilizing a fixture that simulates the maximum material condition of assembly. All runout measurements would be measured from fixture axis A - A when the part is so constrained.

Fig. 33 utilizes a method of dimensioning that does not include a datum axis. This method is practical when the part is large in diameter as thin walled parts do not have a true axis. A minimum of four diameter measurements should be taken to determine the average diameter. The "constant within .005 inch" requirement controls the coaxiality as it limits the thickness variation of the .450 and .500 inch walls to .005 inch. This part can be inspected in its free state with little or no fixturing.

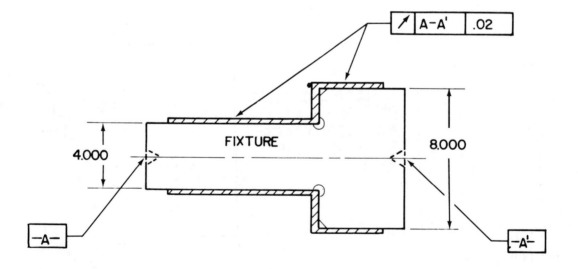

Fig. 32

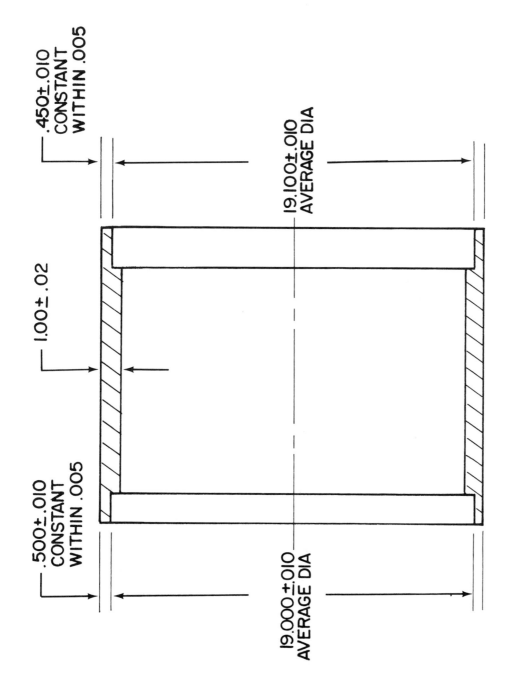

Fig. 33

SUMMARY

Inspectors should attempt to contact designers for datum reference information if they are forced to interpret setups because of incomplete drawings. If the designer is unavailable, the inspector should record the <u>exact</u> setup used by a sketch or photograph, attached to the inspection report. This setup clarification will be useful in any material review board actions when evaluating why accepted parts did not function or why rejected parts did function. It is not the current practice to include datum setup information in inspection reports, and the measurement data is quite useless without this information.

Reprinted with permission from *QUALITY*, (August, 1980) a Hitchcock Publication

Screw thread management

The recently-issued standard, ANSI B1.3, may bring order to dimensional screw thread technology.

Robert S. Chamerda, director of marketing,
The Johnson Gage Company, Bloomfield, CT

■ On June 21, 1979, the Board of Standards Review at the American National Standards Institute (ANSI) approved the standard, *Screw Thread Gaging Systems for Dimensional Acceptability,* ANSI BI.3, 1979. The standard has been printed and the date of issuance is October 15, 1979.

This document applies to all "UN", "UNR" and "UNJ" product screw threads. There is support to extend the document to cover metric threads having both the "M" and "MJ" thread profile. Additionally, adoption of this standard by the government for federal and military use in all or part of its content is anticipated. As such, federal standard H-28, MIL-S-7742 and MIL-S-8879 would all be affected.

Standardization of the screw thread on a formal national basis began with the creation of the National Screw Thread

> *. . . several generations of screw thread standards have been constructed and promulgated upon a tenuous and weak foundation of technical inconsistency, dimensional distortion and, sometimes, outright error.*

Commission, enacted by the Congress in 1918 under the administration of Woodrow Wilson. Since that time, several generations of screw thread standards have been constructed and promulgated upon a tenuous and weak foundation of technical inconsistency, dimensional distortion and, sometimes, outright error.

It is not the intent of this article to fix blame, motive or reason except to say that it should no longer continue. It remains to be seen whether or not ANSI B1.3, as a screw acceptability document, provides order and stability to a technology long overdue for clarification and reevaluation.

Before some of the specifics of ANSI B1.3 are discussed, some of the specific problems that distort or otherwise confuse the process of acceptance or rejection of product screw threads should be confronted. The best place to start would be to define a dimensional screw thread standard. It is a document that can be separated into three general components. The components are sequential, both in substance and engineering logic:

● First, there is an engineering drawing of the generic screw thread. This drawing is called the design form of

thread, and it is dimensional at its maximum material condition.

● Second, there are engineering tables that define the variations that are permitted to depart from the engineering drawing. The variations are in terms of tolerance and allowance applied to specific dimensions.

● Third, there is the narrative. This is explanatory material and deals with nomenclature, definitions, letter symbols, etc. It also explains the ways and means to achieve conformance to the engineering drawings and tables. It is here that the problems of screw thread acceptability were born.

It would be assumed by many that the product thread callout and the reference back to a dimensional standard, such as H-28 or ANSI B1, provide the dimensional information that would satisfy engineering, manufacturing and quality functions.

Is there any argument as to what constitutes a dimensionally acceptable or rejectable 1/2-20 UNF-2A thread? There is a good deal of argument. It makes no difference what document you look at, the tabulated limits for pitch diameter are .4662/.4619.

● Should this screw thread be acceptable to the dimensional standard when the actual pitch diameter is .4502?

● Should this screw thread be acceptable to the dimensional standard when the lead is .0438 instead of .050?

● Should this screw thread be acceptable to the dimensional standard when the pitch cylinder is tapered by .0097.

● Should this screw thread be acceptable to the dimensional standard when the included flank angle is 51°38' instead of 60°?

● Should this screw thread be acceptable to the dimensional standard when it is out-of-round by .0074?

Can the thread have all of the above errors and still be acceptable to the dimensional standard?

The answer to the above is yes; by virtue of application perhaps the thread can be used. But, should this thread still be allowed to be identified as an acceptable 1/2-20 UNF-2A thread (to the dimensional standard) when virtually each of the numbers specified, in terms of dimensions and tolerance as defined by drawings and tables, stand violated?

● Should the gaging used hide this fact? Even from manufacturing?

● Should "fudge" factors for E_n and E_s be used in calculations for strength?

● Should the ultimate user know that the numbers he thought meant something—don't?

● Should it be known how far out of tolerance the thread can be and still perform its intended function before we provide blanket acceptance for its use?

● Is there something wrong with all of this?

The thread described is not imaginary. It is very real. Others like it or worse can be accepted every day by unsuspecting users who have no idea that the thread is out of tolerance as defined by the numbers in the standard.

The before-mentioned thread will go into a "GO" thread ring gage and stay out of a "NOT GO" (LO) thread ring gage. A mating part with the same amount of error will accept the "GO" thread plug and refuse entry of the "NOT GO" (HI) thread plug. The case that a product screw thread out of tolerance should not be used is not being made here. The case that no standard mislead the user is also not intended.

For many years, both federal and commercial documents dealt with the acceptability of product screw threads in terms of gages rather than dimensions. This may sound innocent enough since, often, gages are thought of as dimensions in physical form. This oversimplification falls apart completely, however, when the gage specified cannot technically relate to the dimension for which it is intended. Such has been the case with screw threads.

The problem has become a monumental misunderstanding as to just what dimension or dimensions are actually inspected with the plugs and rings. Yes, they are marked with the maximum and minimum pitch diameters; and yes, the actual pitch diameters of the plugs and rings themselves may be precisely what the markings say they are. (They should be within "X" gage makers' tolerance.) From this point on, however, enter the clouds of confusion. Pitch diameter is not what is yielded from this gaging. The dimension that is actually being inspected is the functional diameter. This is the diameter of the screw thread with all of its error. The dangerous thing about this diameter is that it creates an illusion of material that does not exist. You have to be very careful here because most designs are predicated on material present—not absent.

The functional diameter of any given screw thread is an elusive dimension to visualize. Try to draw one on a piece of paper in terms of the thread profile and its error. The functional diameter is satisfied merely by the first piece of metal that arrives at a defined limit (even a burr). After that

The functional diameter of any given screw thread is an elusive dimension to visualize ... the word "functional" applies to its ability to assemble, not its ability to perform.

occurs, it ceases to care whether or not anything else is in place. It is like gaging for true position, having absolutely no knowledge of the sizes of the features involved. Remember, the word "functional" applies to its ability to assemble, not its ability to perform.

If ANSI B1.3 does nothing else, it will force recognition of the specific product screw thread dimension(s) with which you are dealing. The standard itself comprises only nine pages; however, an enormous amount of information has been compacted between its covers. The user of this standard will have to become familiar with four primary tables:
- Screw thread gages and measuring equipment for external product thread characteristics (Table 1).
- Screw thread gages and measuring equipment for internal product thread characteristics (Table 2).
- Gaging systems for external threads (Table 3).
- Gaging systems for internal threads (Table 4).

The only difference between Tables 1 and 2, and Tables 3 and 4 is that they separate external from internal threads. Tables 1 and 2 are constructed so that the horizontal columns contain all product screw thread dimensions. Also, where it is appropriate, the dimensions are further divided into limit and size. Limit refers to *attributes* inspection. Size refers to *variables* inspection.

The vertical columns contain all gages, gaging systems and general-purpose type measuring instruments that are employed to dimensionally gage or measure product screw threads. To the best knowledge of the committee, nothing has been excluded.

Table 1

Thread gages
and
measuring equipment

1.	Threaded rings (ANSI B47.1 split or solid):
	1.1 GO
	1.2 LO
2.	Thread snap gages:
	2.1 GO segments
	2.2 LO segments—two pitches approximately
	2.3 GO rolls
	2.4 LO rolls—two pitches approximately
	2.5 Minimum material—pitch diameter type—cone & vee
	2.6 Minimum material—thread groove diameter type—cone only
3.	Plain diameter gages
	3.1 Maximum plain cylindrical ring for major diameter
	3.2 Major diameter snap type
	3.3 Minor diameter snap type
	3.4 Maximum/minimum major diameter snap type
	3.5 Maximum/minimum minor diameter snap type
4.	Indicating thread gages: having either two contacts @ 180° or three contacts @ 120°
	4.1 GO segments
	4.2 LO segments—two pitches approximately
	4.3 GO rolls
	4.4 LO rolls—two pitches approximately
	4.5 Minimum material—pitch diameter type—cone & vee
	4.6 Minimum material—thread groove diameter type—cone only
	4.7 Major diameter/pitch diameter eccentricity gage
	4.8 Differential segment or roll: (GO profile for one pitch in length) used in combination with a GO indicating gage to yield a diameter equivalent for deviation in lead (including uniformity of helix); and a minimum material indicating gage to yield a diameter equivalent for deviation in flank angle.
5.	Indicating plain diameter gages:
	5.1 Major diameter type
	5.2 Minor diameter type
6.	Pitch micrometer w/standard contacts (approximately LO profile) cone & vee
7.	Pitch micrometer w/modified contacts (approximately P.D. contact) cone & vee
8.	Thread measuring wires with suitable measuring means
9.	Optical comparator/toolmakers microscope w/suitable fixturing
10.	Profile tracing equipment w/suitable fixturing
11.	Lead measuring machine w/suitable fixturing
12.	Helical path attachment used w/GO type indicating gages
13.	Helical path analyzer
14.	Plain micrometer/calipers—modified as required
15.	Surface measuring equipment
16.	Roundness equipment

Note 1. Maximum minor diameter limit is acceptable when product passes GO gage.

The black dots in the body of the table correlate the product screw thread dimension specified with the gages acceptable for use. As an example, refer to Table 1 for external threads. Select the dimension "minimum material" (pitch diameter). This dimension appears under the column "C_1" for limit and "C_2" for size. Scanning down these columns, we find the black dots appearing at 2.5, 4.5 and 7. At 2.5 (thread snap gage) limit only is controlled. At 4.5 (indicating thread gage) and at 7 (pitch micrometers having modified contact), both limit and size are controlled. By definition, control size and you automatically control the limits. With regard to any product screw thread dimension, the user is

214

For external product thread dimensions unified inch threads

Column key:
- A1, A2 = maximum material GO (func. limit, func. size)
- B1, B2 = LO functional diameter (func. limit, func. size)
- C1, C2 = minimum material, pitch dia. (limit, size)
- D1, D2 = minimum material, thd. groove dia. (limit, size)
- E1, E2 = roundness of pitch cylinder, oval 180° (limit, size)
- F1, F2 = roundness of pitch cylinder, multi lobe 120° (limit, size)
- G1, G2 = taper of pitch cylinder (limit, size)
- H1 = lead incl. helix deviation
- I = flank angle deviation
- J1, J2 = major diameter (limit, size)
- K1, K2 = minor diameter (limit, size)
- L = root rad.
- M = major to pitch
- N = surface texture

A1	A2	B1	B2	C1	C2	D1	D2	E1	E2	F1	F2	G1	G2	H1	I	J1	J2	K1	K2	L	M	N
●																		note 1				
		●																				
●								●										note 1				
		●						●				●										
●								●				●						note 1				
		●						●				●										
				●				●				●										
						●		●				●										
																●						
																●						
																		●				
																●						
																		●				
●	●	●	●					●	●	●	●							note 1				
	●		●					●	●	●	●	●	●									
●	●	●	●					●	●	●	●							note 1				
		●	●					●	●	●	●	●	●									
				●	●			●	●	●	●	●	●									
						●	●	●	●	●	●	●	●									
																					●	
								●	●	●	●	●	●	●	●							
																●	●					
																		●	●			
		●	●					●	●			●	●									
				●	●			●	●			●	●									
						●	●	●	●			●	●									
								●	●	●	●			●	●			●		●	●	
															●						●	
														●								
														●								
														●								
																●	●					
																						●
								●	●	●	●											

free to select any piece of equipment that satisfies the tables. It should also be noted that a specific piece of equipment may satisfy more than one dimension.

The reader may already be gnawing at the question, "What about differences in gages?". This was not overlooked by the committee. The standard states that for a specified dimension, if it's acceptable to one—it's acceptable to all. This does not mean that there is a free-for-all in selecting or finding some gage that accepts nonconforming product. Obviously, any gage is disqualified if it is worn or damaged out of tolerance, used incorrectly, or set to size incorrectly.

With that out of the way, some of the problems that do exist can be addressed. One of the more common arguments has nothing to do with differences in gages; it is again, a misunderstanding of dimensions.

A product thread is rejected because the pitch diameter is out of tolerance. The opposing party argues that the thread does not enter a "NOT GO" (LO) thread ring gage and, therefore, should not be rejected. This argument has nothing at all to do with gages. The two parties are not talking about the same dimension. The "NOT GO" (LO) thread ring does not control the pitch diameter. ANSI B1.3 will no longer allow that frequent error to be made.

There is another common problem that causes endless argument, and it does have to do with gages. It is the differ-

215

ence between two ring gages. This issue has never been dealt with satisfactorily in any standard. It has been of particular concern to the originator of the single unit adjusting and locking threaded ring gage, The Johnson Gage Company.

The word "adjustable" applies specifically to the manufacturing of the ring itself. The word "adjustable" had nothing to do with the user of the gage. The idea was to split the ring into sectors with an adjusting screw installed so that the last manufacturing operation was with a calibrated, cast-iron lap.

The gage was qualified to this lap, set to size, sealed and sent on its way. When this gage is fitted to a set master after wear occurs, there is play which is compensated for by adjusting the screw. The errors begin here. The gage is actually being distorted out-of-round. Worse, the helix is being misaligned (on 60 degree threads, the ratio is 1.7 to 1). The result is predictable. You argue a lot. If a solid ring is lapped oversize, it's lost forever. The original design intent—clearly stated in the patent—was that the gage could be saved by re-manufacturing after wear occurred; but, this could only be accomplished by qualifying the circle, realigning the helix and then resetting.

How was this confusion perpetuated? To take up the work of standardizing various types of common gages, including thread gages, the American Gage Design Committee was created in 1926 as a subcommittee of the National Screw Thread Commission. A number of large industrial concerns,

Control size and you automatically control the limits.

representatives of government departments and several of the leading gage manufacturers were already directing effort towards simplifying gaging practice through the adoption of standard designs for gage blanks and component parts. The committee was to consolidate these efforts for the benefit of industry at large.

The AGD Committee was charged to only standardize the construction details of the several kinds of then commonly used fixed-limit gages. These included the following:

- Plug and ring blanks for thread gages and plain cylindrical gages, machine taper gages and spline gages.
- Master disks and indicator-mounting dimensions
- Adjustable plain-limit snaps, built-up snaps, plug gages and length gages (all commercial standard CS-8 series).

The original AGD Committee, therefore, was not given either the charter or authority to standardize gages for acceptability. They were given charter and authority only to standardize blank sizes for gages already in use. How this committee, intentionally or unintentionally, switched gears and converted its activity from blank sizes to developing acceptability gages is an interesting chapter in the phenominal complexities of standards development.

Tables 3 and 4 are of profound significance because they define the systems that will be specified on the drawing or procurement package.

ANSI B1.3 identifies three levels of dimensional conformance for external and internal product screw threads. They are identified as System 21, System 22 and System 23. System 23 requires the most exacting dimensional control of product screw thread dimensions. System 22 requires less than System 23. System 21 requires the least in dimensional control and dimensional integrity of product screw threads.

Tables 3 and 4 are arranged so that the individual systems appear in the vertical columns. The horizontal columns contain the dimensions to be inspected for each system along with the gage or gaging systems that apply to that dimension,

again divided into attributes and variables. Correlating the two are the gage numbers and dimension column locators that refer back to Tables 1 and 2.

System 21

One common expression used to describe a System 21 thread was "a glob on the end of a stick." Obviously for its mating part, we would have a "glob inside a hole." System 21 should not be used if there is concern or confidence needed with regard to the integrity or conformity of those product screw thread dimensions contained in the engineering drawings and tables.

System 21 specifies and provides for nothing more than control of the functional diameter along with external major and internal minor diameters. There is no control specified

Table 2
Thread gages and measuring equipment

1. Threaded plugs (ANSI B47.1):
    1.1 GO
    1.2 HI
    1.3 Full form gage GO plug (UNJ only)
2. Thread snap gages:
    2.1 GO segments
    2.2 HI segments—two pitches approximately
    2.3 GO rolls
    2.4 HI rolls—two pitches approximately
    2.5 Minimum material—pitch diameter type—cone & vee
    2.6 Minimum material—thread groove diameter type—cone only
3. Plain diameter gages
    3.1 Minimum plain cylindrical plug for minor diameter
    3.2 Major diameter snap type
    3.3 Minor diameter snap type
    3.4 Maximum/minimum major diameter snap type
    3.5 Maximum/minimum minor diameter snap type
4. Indicating thread gages:
    having either two contacts @ 180° or the contacts @ 120°
    4.1 GO segments
    4.2 HI segments—two pitches approximately
    4.3 GO rolls
    4.4 HI rolls—two pitches approximately
    4.5 Minimum material—pitch diameter type—cone & vee
    4.6 Minimum material—thread groove diameter type—cone only
    4.7 Minor diameter/pitch diameter runout gage
    4.8 Differential segment or roll:
        (GO profile for one pitch in length) used in combination with a GO indicating gage to yield a diameter equivalent for deviation in lead (including uniformity of helix); and a minimum material indicating gage to yield a diameter equivalent for deviation in flank angle.
5. Indicating plain diameter gages:
    5.1 Major diameter type
    5.2 Minor diameter type
6. Pitch micrometer w/standard contacts (approximately HI profile) cone & vee
7. Pitch micrometer w/modified contacts (approximately P.D. contact) cone & vee
8. Thread measuring balls with suitable measuring means
9. Optical comparator/toolmakers microscope w/suitable fixturing & cast replica
10. Profile tracing equipment w/suitable fixturing
11. Lead measuring machine w/suitable fixturing
12. Helical path analyzer
13. Plain micrometer/calipers—modified as required
14. Surface measuring equipment
15. Roundness equipment

Note 1. Minimum major diameter limit is acceptable when product passes GO gage.

or required for minimum material. Putting it another way, the external thread has no floor; the internal thread has no ceiling. As such, any amount of lead error, helical variation, flank angle error, as well as taper and out-of-roundness are permitted to go undetected. These errors, by their very nature, drastically reduce the percentage of flank-to-flank engagement in the assembly.

It is important for the reader to recognize the strange way that this approach to System 21 differs from normal design engineering practice. Normally, when it is deemed that a tolerance is too tight, proper engineering cognizance may choose to loosen the requirement. That is to say, tolerance may be opened up or expanded. In System 21, that is not the case. Instead, the entire dimension is eliminated. In this case, several dimensions are eliminated. Admittedly, this flies in the face of the original reason for having different classes. If class "3A" or "3B" were deemed too tight, class "2A" and "2B" provided more tolerance. The dimensions still existed—not so in System 21.

In summary, System 21 is intended to provide dimensional assurance that parts assemble; dimensionally, nothing more is promised.*

---

* When System 21 is used, it should be clearly understood by everyone concerned that the product thread is under no obligation or restriction to be manufactured or inspected for any dimensions other than what is specified for System 21. It would be prudent for each company to consult with their legal counsel to determine if there could be product liability risks inherent in this selection.

For internal product thread characteristics unified inch threads

Column groups:
- maximum material, GO — func. limit A_1, func. size A_2
- HI functional diameter — func. limit B_1, func. size B_2
- minimum material, pitch dia. — limit C_1, size C_2; thd. groove dia. — limit D_1, size D_2
- roundness of pitch cylinder, oval 180° — limit E_1, size E_2; multi lobe 120° — limit F_1, size F_2
- taper of pitch cylinder — limit G_1, size G_2
- lead incl. helix variation — H_1
- flank angle variation — I
- major diameter — limit J_1, size J_2
- minor diameter — limit K_1, size K_2
- root rad. — L
- dia. runout minor to pitch — M
- surface texture — N

A_1	A_2	B_1	B_2	C_1	C_2	D_1	D_2	E_1	E_2	F_1	F_2	G_1	G_2	H_1	I	J_1	J_2	K_1	K_2	L	M	N
•															note 1							
		•																				
•															note 1						•	
•								•							note 1							
		•						•				•										
•								•							note 1							
		•						•				•										
				•				•				•										
						•		•				•										
																		•				
																•						
																•						
																		•				
•	•							•	•	•	•				note 1							
		•	•					•	•	•	•	•	•									
•	•							•	•	•	•				note 1							
		•	•					•	•	•	•	•	•									
				•	•			•	•	•	•	•	•									
						•	•	•	•	•	•	•	•									
																					•	
								•		•	•	•	•	•	•							
																•	•					
																		•	•			
		•	•					•	•			•	•									
				•	•			•	•			•	•									
						•	•	•	•			•	•									
				•	•									•		•	•			•		
															•		•			•		•
														•								
														•								
																•	•	•	•			
																						•
								•	•	•	•											

System 22

The floor that was missing from the external thread and the ceiling that was missing from the internal thread in System 21 are constructed in System 22. The thread now has an envelope: maximum material to assure assembly and minimum material to provide integrity.

This system does not provide specific control over lead, helix, flank angle, taper or roundness. They are, however, controlled by proxy since the combined variation of all these dimensions is confined within the material envelope. The principle is an established one: when no tolerance of form is specified, the tolerance for size dictates the tolerance for form.

There is a proviso in System 22 that allows substituting direct control of all thread elements in place of minimum material. Frankly, this was a bad compromise. Limits and/or methods for thread elements of lead, helix, angle, taper and roundness must be established. The problem with this remains to be seen because the material limits—both maximum and minimum—are different manifestations of a datum. The pitch diameter establishes the datum while the functional diameter portrays it fictitiously, after all errors are added in.

The concern with arriving at either limit by accumulating data for error is one in which you must be positive you have "captured" the worst condition. While academically this is certainly possible; practically, it creates problems.

Several years ago, the Bureau of Standards, under the direction of the Inter-Departmental Screw Thread Committee chairman, undertook a study with numerous actual threaded product and tried to establish material limits by this exact method. The study proved unsatisfactory because of the haphazard results.

The concept of tooling or process control should not be misconstrued. Out-of-tolerance parts refute the control.

System 23

System 23 is System 22 with specific control added for lead (including helix), flank angle, as well as taper and roundness. System 23 produces a precision thread and should be used where a magnitude of precision is required by the dictates of the design.

It should be clear that the existence of three different levels of dimensional conformance does not require three different gaging systems.

It will be up to someone to specify what system is to be used. In some cases, other product standards will define the level of dimensional quality. In this case, serious deliberation is warranted by these standards committees. This is particularly true if the items covered by these standards end up in any type of general distribution system where the end use of application is not always known, anticipated or predictable. In many cases however, it will be an item-by-item deter-

Table 3—External

System	Dimensions inspection	Applicable thread gages and measuring equipment			
		Attributes/fixed limit		Variables/indicating	
	For dimension/gage combinations to be used — (refer to Table 1)	control	Col.	control	Col.
21	GO maximum material	1.1, 2.1, 2.3, 4.1, 4.3	A1	4.1, 4.3	A2
	LO functional diameter	1.2, 2.2, 2.4, 4.1, 4.2 4.3, 4.4, 6	B1	4.1, 4.2, 4.3, 4.4, 6	B2
	Major diameter	3.1, 3.2, 3.4, 5.1, 14	J1	5.1, 14	J2
22	GO maximum material	1.1, 2.1, 2.3, 4.1, 4.3	A1	4.1, 4.3	A2
	Minimum material —Pitch diameter	2.5, 4.5, 7	C1	4.5, 7	C2
	or—Thread groove diameter	2.6, 4.6, 8	D1	4.6, 8	D2
	LO functional diameter combined with mandatory examination of: or	1.2, 2.2, 2.4, 4.1, 4.2 4.3, 4.4, 6	B1	4.1, 4.2, 4.3, 4.4, 6	B2
	Lead (including helix) and			4.8, 9, 11, 12, 13	H
	Flank angle over the length of full thread.			4.8, 9, 10	I
	Major diameter	3.1, 3.2, 3.4, 5.1, 14	J1	5.1, 14	J2
	Minor diameter (UNJ only)	3.3, 3.5, 5.2, 9	K1	5.2, 9	K2
	Root profile (UNJ only)			9, 10	L
23	GO maximum material	1.1, 2.1, 2.3, 4.1, 4.3	A1	4.1, 4.3	A2
	Minimum material: —Pitch diameter	2.5, 4.5, 7	C1	4.5, 7	C2
	or—Thread groove diameter	2.6, 4.6, 8	D1	4.6, 8	D2
	Major diameter	3.1, 3.2, 3.4, 5.1, 14	J1	5.1, 14	J2
	Minor diameter (UNJ only)	3.3, 3.5, 5.2, 9	K1	5.2, 9	K2
	Root profile (UNJ only)			9, 10	L
	Roundness of pitch cylinder: —Oval 180°	2.1, 2.2, 2.3, 2.4, 2.5, 2.6, 4.1, 4.2, 4.3, 4.4, 4.5, 4.6, 4.8, 6, 7, 8, 9, 16	E1	4.1, 4.2, 4.3, 4.4, 4.5, 4.6, 4.8, 6, 7, 8, 9, 16	E2
	—Multi lobe 120°	4.1, 4.2, 4.3, 4.4, 4.5, 4.6, 4.8, 9, 16	F1	4.1, 4.2, 4.3, 4.4, 4.5, 4.6, 4.8, 9, 16	F2
	Taper of pitch cylinder	2.2, 2.4, 2.5, 2.6, 4.2, 4.4, 4.5, 4.6, 4.8, 6, 7, 8	G1	4.2, 4.4, 4.5, 4.6, 4.8, 6, 7, 8	G2
	Lead including helix deviation			4.8, 9, 11, 12, 13	H
	Flank angle deviation			4.8, 9, 10	I
	Major/pitch diameters			4.7	M
	Surface texture			10, 15	N

Table 4—Internal

System	Dimensions inspected	Attributes/fixed limit		Variables/indicating	
	For dimension/gage combinations to be used — (refer to Table 2)	control	Col.	control	Col
21	GO maximum material	1.1, 1.3, 2.1, 2.3, 4.1, 4.3	A1	4.1, 4.3	A2
	HI functional diameter	1.2, 2.2, 2.4, 4.2, 4.4, 6	B1	4.2, 4.4, 6	B2
	Minor diameter	1.3, 3.1, 3.3, 3.5, 5.2, 13	K1	5.2, 13	K2
	GO maximum material	1.1, 1.3, 2.1, 2.3, 4.1, 4.3	A1	4.1, 4.3	A2
	Minimum material Direct method				
	— Pitch diameter	2.5, 4.5, 7	C1	4.5, 7	C2
	or — Thread groove diameter	2.6, 4.6, 8	D1	4.6, 8	D2
	Indirect method — HI functional diameter combined with control of:	1.2, 2.2, 2.4, 4.2, 4.4, 6	B1	4.2, 4.4, 6	B2
22	— Lead (including helix)			4, 8, 9, 11, 12	H
	— Flank angle			4.8, 9, 10	I
		1.3, 3.1, 3.3, 3.5, 5.2, 13	K1	5.2, 13	K2
	— Roundness of pitch cylinder: — Oval 180°	2.1, 2.2, 2.3, 2.4, 2.5, 2.6, 4.1, 4.2, 4.3, 4.4, 4.5, 4.6, 4.8, 6, 7, 8, 15	E1	4.1, 4.2, 4.3, 4.4, 4.5, 4.6, 4.8, 6, 7, 8, 15	E2
	— Multi lobe 120°	4.1, 4.2, 4.3, 4.4, 4.5, 4.6, 4.8, 15	F1	4.1, 4.2, 4.3, 4.4, 4.5, 4.6, 4.8, 15	F2
	— Taper of pitch cylinder	2.2, 2.4, 2.5, 2.6, 4.2, 4.4 4.5, 4.6, 4.8, 6, 7, 8	G1	4.2, 4.4, 4.5, 4.6, 4.8, 6 7, 8	G2
	GO maximum material	1.1, 1.3, 2.1, 2.3, 4.1, 4.3	A1	4.1, 4.3	
	Minimum material: — Pitch diameter	2.5, 4.5, 7	C1	4.5, 7	C2
	or — Thread groove diameter	2.6, 4.6, 8	D1	4.6, 8	D2
	Minor diameter	1.3, 3.1, 3.3, 3.5, 5.2, 13	K1	5.2, 13	K2
	Roundness of pitch cylinder: — Oval 180°	2.1, 2.2, 2.3, 2.4, 2.5, 2.6, 4.1, 4.2, 4.3, 4.4, 4.5, 4.6, 4.8, 6, 7, 8, 15	E1	4.1, 4.2, 4.3, 4.4, 4.5, 4.6, 4.8, 6, 7, 8, 15	E2
23	— Multi lobe 120°	4.1, 4.2, 4.3, 4.4, 4.5, 4.6, 4.8, 15	F1	4.1, 4.2, 4.3, 4.4, 4.5, 4.6, 4.8, 15	F2
	Taper of pitch cylinder	2.2, 2.4, 2.5, 2.6, 4.2, 4.4 4.5, 4.6, 4.8, 6, 7, 8	G1	4.2, 4.4, 4.5, 4.6, 4.8, 6 7,8	G2
	Lead including helix deviation			4.8, 9, 11, 12	H
	Flank angle deviation			4.8, 9, 10	I
	Major/pitch diameters runout			4.7	M
	Surface texture			10, 14	N

mination. This is a major procedural change that differs from past practice.

As a package, the ANSI B1 series of documents provides the essential information for a generic screw thread. As pure dimensional standards, they do not deal with application. The thread can be manufactured on a bolt, nut, valve, fitting, connector, shaft, tubing, casting, forging, etc. The manner in which the thread is produced can be by single pointing, tapping, chasing, rolling, hobbing, milling, grinding, etc.

It would be impossible to describe all or even most applications involving the use of all threaded components. That is not the function of this standards group. However, regardless of the application, ANSI B1 has provided the following for each of them:
- An engineering drawing of the thread
- Engineering tables for dimensions, tolerance, and allowance for numerous diameter/pitch combinations, as well as provision for specials.
- Standards narrative and direction for conformance.

A new ANSI B1.3 plays no small part in the overall scheme of things. Is it a good standard? It can be—provided it stays technical. One thing is certain: more people are going to become concerned. There will be less complacency, more awareness, more study. And, the concept of dimensional screw thread management will have begun.

As a general approach to selection of a system, the following guidelines are offered. The specifier may wrestle with the particulars of his product.

- System 21

Not recommended where calculations of mechanical strength factors (tensile, stress, shear, fatigue, etc.) are based on existing pitch diameter limits of size.

Not recommended where the material used has been chosen for its strength.

Not recommended when the item is of such a general nature that marketing, distribution, installation and service are not directly under your control.

Not recommended where failure of the threaded connection would have as a possibility destruction of property, injury or death.

- System 22

Recommended for most applications that can be satisfied with the simple dimensional statement "Not too big and not too small." In other words, the marriage of assembleability and dependability.

- System 23

Not recommended when System 22 is adequate.

Presented at the SME Southeastern Conference, December 1976

Thread Gaging with the Ball Contact Method

By Richard Browning
Southern Gage Company

ABSTRACT

The measurement of threads has, for many years, been a source of controversy throughout the industrial spectrum. The principles, methods, types and designs of gages and gaging of threads have been governed by the National Bureau of Standards through their publication of "H-28 HANDBOOK-SCREW THREAD STANDARDS For FEDERAL SERVICES". The theory of using spheres (balls) has been defined in this standard but here-to-fore had not been developed. The design and procedure has now been introduced for measurement of the three elements of a thread (pitch diameter, lead, and flank angles) using balls based upon the same principle as using wires (pins) as is now accepted for accurate measurement of external threads.

INTRODUCTION

By simple definition, the pitch diameter is an imaginary cylinder along which the width of every thread ridge and every groove is one-half pitch (p/2). The term "Pitch Diameter" is used interchangeably with "pitch cylinder" and "simple effective diameter".

In the measurement of threads, this is the datum or gage point. Since the measurement across the groove is so much easier and more accurate, the method specified for determining the pitch diameter is stated (from H28 Handbook) as follows: "It is the diameter yielded by measuring over or under cylinders (wires) or spheres (balls) inserted in the thread groove on opposite sides of the axis and computing the thread groove diameter as thus defined".

Obviously, if you are "measuring over" you are measuring an external thread, and "measuring under" you are measuring an internal thread. Also notice there is no mention of measurement of the Ridge width to define this cylinder (diameter).

DETAILS OF THE BALL GAGING SYSTEM

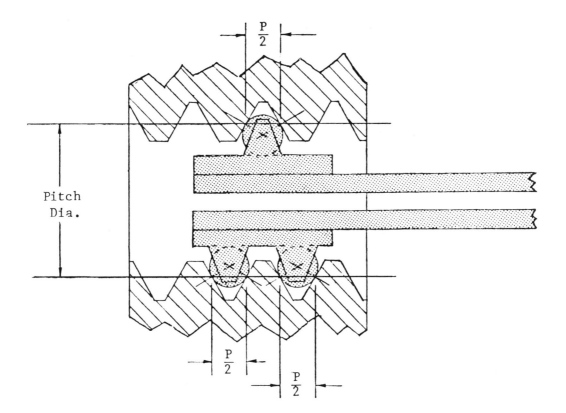

Unique Floating Feature

As shown above the pitch diameter fingers utilize
three balls of best wire size. The two balls in the
lower finger are free to float axially. It is important to note here that this linear freedom also
permits the rest of the gage to float until the single
ball in the upper finger seats to maximum available
depth in the thread groove. In so doing, no tilt is
created. The upper finger remains normal to the thread
axis as it aligns itself with the cut.

Because the fingers of the gage are interchangeable,
the lower P.D. finger can be used with a variety of
upper fingers to check many thread elements. In every
case, the upper contact will automatically align itself
with the cut, yet remain normal to the thread axis.

Proponents of the floating ball system are convinced that errors in lead and in included angle must be excluded from the pitch diameter gaging procedure, or the reading obtained will be meaningless.

How It Excludes Lead Error

In the floating ball system, one ball is positioned in a groove and two balls are positioned in grooves diametrically opposite. The balls are made to best wire size, and measure P/2 at their line of contact when seated in thread grooves of correct angles. the two lower balls are free to float along the thread axis, so all balls readily find seats in their respective grooves regardless of the width of ridges.

The balls measure the distance between grooves 180° apart at their P/2 widths to establish the thread groove diameter (effective pitch diameter). This measurement is compared to the nominal P.D. by a mechanical indicator or electronic comparator attached to the gage.

In this system the influence of lead error is completely nullified because the ridges, in which lead error evidences itself, are not used as contact points.

How It Identifies Error In Included Angle

In the floating ball system, the single (upper) p.d. ball may be replaced by other gaging elements to give the inspector procedures for verifying the included angle and its squareness to the thread axis. The equivalent pitch diameter tolerance consumed by such angular errors is easy to compute, and can also be obtained from prepared tables in Handbook H-28.

1. INCLUDED ANGLE

Verified by measuring width of angle at two depths. First use contacts A and J,

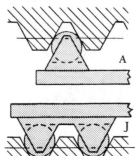

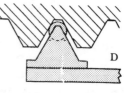

then D and J.

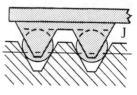

222

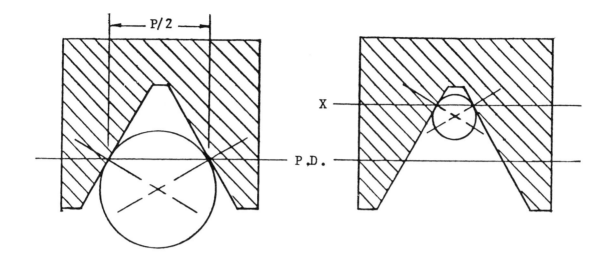

If the included angle of the groove is correct, contact A will seat at data point P/2 and contact D will seat at predetermined data point X. If angular error exists, the balls will not seat the proper distance apart. Indicated error can be transposed into degrees and minutes of angular error by using a formula published in Handbook H-28.

2. FLANK ANGLES

Unequal flank angles cause a cut thread to stand at a tilt to the thread axis. This condition can be checked by a two-part procedure, using contacts E and J.

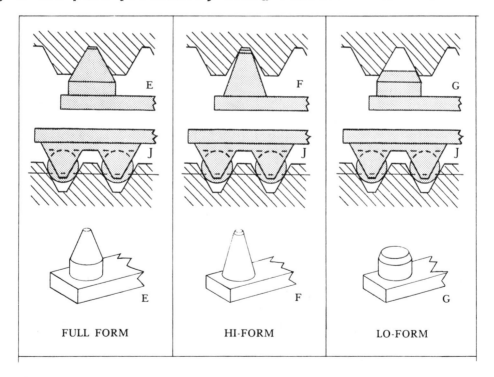

FULL FORM   HI-FORM   LO-FORM

When paired with lower P.D. contact J, each of the upper contacts has the same unvarying "underpinnings": two balls of Best Wire size firmly seated at the Effective pitch diameter. Readings obtained, therefore, pertain solely to the specific thread form being checked by the upper finger. In effect, measurements are taken from the center line. Working from the center line eliminates possibilities of double error caused by inaccuracies of lead, angle, etc. in the lower thread form.

3. FUNCTIONAL DIAMETER (VIRTUAL DIAMETER)

Variations in lead and flank angles from perfect will, in effect, add to the actual measured pitch diameter of an external thread, or subtract from the measured pitch diameter of an internal thread. To measure the functional diameter a set of thread segments is used to simulate the mating fit. The total effect of all errors will be read with these fingers.

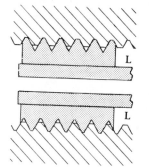

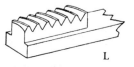

FUNCTIONAL
Reads net effect
of all errors on
allowable thread
envelope

4. LEAD ERROR

Lead error is established by calculation. Starting with the Pitch Diameter reading, subtract the reading obtained by the Functional 'L' Fingers. The result is the combined error of all thread elements except Pitch diameter. From this combined error substract the diametral effect of errors found in 1 and 2. The remaining error is the diametral equivalent of the lead error since all other elements have been eliminated.

5. MINOR DIAMETER

In effect, the minor diameter is a bored hole and can be checked for size, roundness and taper by various hole gages. If the gage is used to make these checks, use two C contacts (rather than C and J) so that the Pitch

Diameter is not contacted. This will eliminate the influence of any eccentricity between minor and P.D..

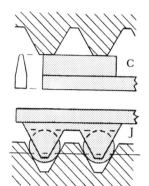

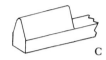

MINOR DIAMETER
Checks Minor Size
and concentricity

Setting the Gage

The gage is set "over the balls" or "over the crests" with gage block, Supermicrometer or plain cylindrical ring gages. No threaded masters are needed.

The pitch diameter setting equals the nominal P.D. plus the ball radius. For a 3/4-16 thread this would be .7094 P.D. + .0180 ball radius .7274 setting size.

All contacts (except Minor Diameter) that are used opposite contact J are marked with a constant dimension thus: + CD .0214; or - CD .0261. Setting size is equal to the P.D. setting plus or minus the constant dimension.

Functional fingers are marked with a specific setting size thus: .7504 O.D.

Paired minor diameter fingers are set exactly to Minor Diameter size. When a single Minor Diameter finger is used opposite a J finger to check concentricity, no precise setting is required. The fingers are brought into contact with the workpiece and the indicator is read for runout.

CONCLUSION

The developement of a Hand-Held comparator using spheres (balls) for gaging contacts offers accuracy never before obtainable in measurement of threads. The ability to set to a known standard that can be directly calibrated gives a precise comparison to an assigned numeric value. This eliminates the transfer of error using threaded masters.

Under proposed revisions to the standards governing screw threads, the measurement of all elements of the thread may be required. This system offers a portable, precise and economical method of satisfying that requirement.

Reprinted from *Technology of Machine Tools*, October 1981

Postprocess Gaging With Feedback

Daniel C. Thompson, Engineer
Machine Tool Development Group, Lawrence Livermore National Laboratory

INTRODUCTION

Postprocess gaging with feedback (PPGWF) is a technique to improve part accuracy by using the results of part inspection to compensate for repeatable errors in the machine tool path. The process is normally applied to NC machines by using inspection data to modify the part program, and on tracer machines by using the same data to modify the part template. PPGWF is normally best applied to machines lacking other error compensation capabilities and which are more repeatable than accurate.

Because the improvements to part or machine accuracy that can be attained with any precalibration technique are limited by the repeatability of the machine tool, it is useful to define what constitutes a repeatable error. In our terminology, a repeatable error of a machine component is one that recurs as the component goes through its mechanical motions and, by extension, the repeatable errors of the entire system are those that recur in the machining cycles of a series of identical workpieces. Nonrepeatable errors, on the other hand, are those caused by error sources within the machine or its environment that are not directly related to the use of the machine, even though source and error follow a cause-effect relationship. Thus, for example, slide straightness errors are generally repeatable, whereas the effects of variations in room temperature constitute nonrepeatable errors.

To improve part accuracy with PPGWF, the user must be able to separate the repeatable from the nonrepeatable machine errors; corrections for nonrepeatable errors, especially if they are the dominant error type for the machine, can actually result in a net decrease in part accuracy. Unfortunately, it is difficult to separate repeatable and nonrepeatable errors in a machine tool by the use of PPGWF. Thus the best candidate machines for error compensation by precalibration are those that are highly repeatable.

POSTPROCESS GAGING VS/CONVENTIONAL PRECALIBRATION

The primary differences between postprocess gaging with feedback and other precalibration techniques involve the nature of the master used for the calibration process, and the universality of the improvement to accuracy obtained. More conventional applications of precalibration, such as the use of a corrector cam for lead screw error correction, require the use of a master, such as a step bar or laser interferometer, to map positioning error over the full travel of each axis. The error data acquired is used to cut a corrector cam for each axis, which will result in improved positioning and part-cutting accuracy over the full range of travel of all axes. In principle the same approach can be used to correct for straightness errors by applying the correction to the position of the orthogonal axis, although no commercial examples are known. Another approach is the use of stored error data in a numerical control, which can compensate for positioning, straightness, squareness, and angular motion errors.

Rather than relying on a master to map repeatable errors over full slide travels, postprocess gaging with feedback requires the use of a master to evaluate the accuracy of the machine tool over a specific tool path. The master is the part itself, which, after an initial cut, is evaluated with suitable inspection equipment. The measured part error is then used to modify the commanded motions of the machine tool, and the part is recut under the same conditions as the original cut. The expected end result is a part, or a series of identical parts, more accurate than the machine tool is normally capable of producing.

BENEFITS OF POSTPROCESS GAGING WITH FEEDBACK

There are a number of benefits that can result from the application of postprocess gaging with feedback. Since any machine that is a good candidate for PPGWF is also suitable for the application of other precalibration techniques, one of the advantages is the absence of capital investment required to improve part accuracies, providing that inspection equipment of adequate accuracy is already available. Assuming the existence of the necessary inspection equipment, a machine can be programmed to produce a part or series of parts to higher accuracies than are normally expected of it without modification of the machine.

In addition, there are benefits that are unique to postprocess gaging with feedback. Some errors, while repeatable, are more dependent upon the part geometry, material, and machining process than upon the machine itself. An example of such an error is tool wear, which for identical cuts may show an acceptably repeatable wear pattern from tool to tool. The part error resulting from tool wear is not compensated for with other precalibration techniques, but may be reduced with PPGWF. Other examples would include repeatable chucking distortion errors, spindle growth, or other errors more closely related to the machining process than to the machine itself. A final advantage, related to the one just mentioned, is that postprocess gaging includes the effects of all repeatable errors that effect the part with a single measurement, eliminating the need for multiple accuracy evaluations, or the need to determine cause-and-effect relationships between source and error.

DISADVANTAGES

The obvious disadvantage associated with postprocess gaging with feedback is that a set of compensation data is only applicable to a single part or set of identical parts; thus whereas the capital investment required may be low, the process is labor intensive and time-consuming. The technique is therefore generally limited to a situation where higher accuracy parts than a machine is normally capable of producing are not required on a regular basis, or where the number of parts in a series justifies the effort necessary to meet the accuracy requirements. Experience has shown that, depending upon the accuracy required, several iterations of the machining/inspection/correction process may be needed, thus compounding the labor-intensive nature of the process.

ACCURACY LIMITATIONS

The part accuracy attainable using postprocess gaging with feedback is limited by the following factors:

- Machine repeatability. As discussed previously, PPGWF cannot be used to compensate for nonrepeatable errors. Thus the maximum improvement in tool path accuracy that can be attained is limited by the repeatability of the machine. In addition, because PPGWF is used to correct for all measured errors, the machine accuracy must be dominated by repeatable errors.

- Machining process repeatability. The repeatability of machining process errors, such as tool wear, spindle growth, and chucking distortion also limit the improvement to part accuracy.

- Machine resolution. The machine resolution, or smallest move the machine slides can make, acts as another limit to the maximum part accuracy improvement. A 1-μm resolution NC machine, for example, cannot compensate for a 0.1-μm error.

- Inspection accuracy. The accuracy of the final part can obviously be no better than the accuracy of the inspection process.

- Data handling/human error. Depending upon the process and equipment used to implement postprocess gaging with feedback, part accuracy can be influenced to some degree by data interpretation and human reading error.

- Correction frequency. The total number of corrections that may be implemented for a given cut with an NC machine is limited by such factors as memory space for a stored program control, tape length, or the limits of the humans involved. Part errors occurring at a frequency beyond this limit cannot be corrected.

IMPLEMENTATION AND EXAMPLE

Various methods can be used to implement postprocess gaging with feedback. Most differ only in the equipment and procedures used to translate inspection data into a form compatible with the machine control; the most rudimentary systems use analog part inspection data, which are used to generate a new NC tape or template by hand, whereas a more sophisticated system might use digitized part interferograms to automatically generate a corrected part program. The magnitude of the element of human error, capital costs, and labor costs are obviously all related to the data handling scheme selected. Rather than discussing the various systems and their relative merits, the manufacturing history will be presented of a specific part requiring the use of postprocess gaging with feedback to meet required accuracies.

1. Part description:
   Figure 9.9-1 describes the working surface of a grazing-incidence x-ray microscope manufactured at the Lawrence Livermore National Laboratory (LLNL) for laser fusion diagnostics. Figure 9.9-2 is a drawing of the complete part, and Figure 9.9-3 is a photograph of the finished product. The substrate material is stainless steel, and the material to be machined is electroless nickel.

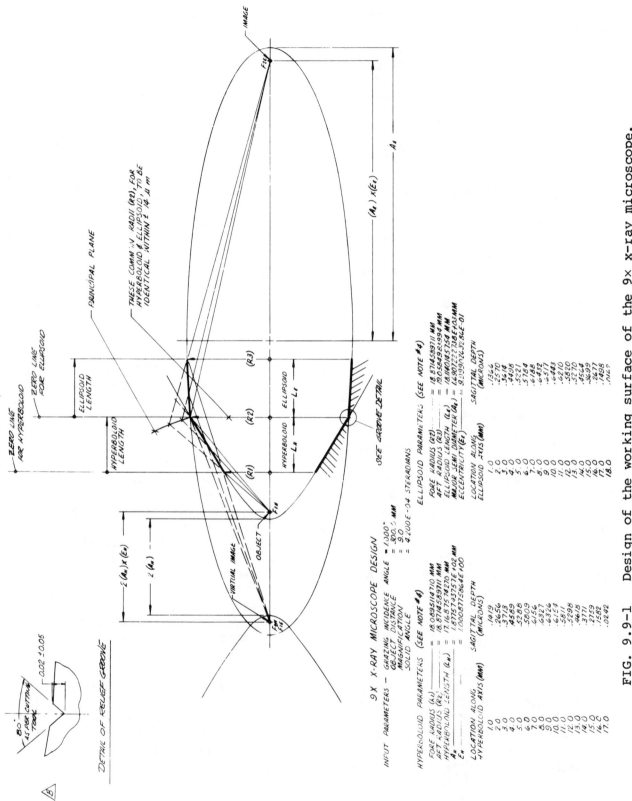

FIG. 9.9-1   Design of the working surface of the 9× x-ray microscope.

230

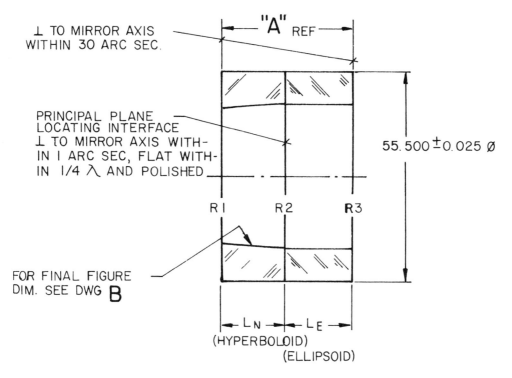

⊥ TO MIRROR AXIS
WITHIN 30 ARC SEC.

"A" REF

PRINCIPAL PLANE
LOCATING INTERFACE
⊥ TO MIRROR AXIS WITH-
IN I ARC SEC, FLAT WITH-
IN I/4 λ AND POLISHED

55.500 ±0.025 ∅

R I   R 2   R 3

FOR FINAL FIGURE
DIM. SEE DWG B

Lₙ — Lₑ
(HYPERBOLOID)
(ELLIPSOID)

FIG. 9.9-2   Drawing of completed 9× x-ray microscope.

FIG. 9.9-3   Photograph of completed 9× x-ray microscope.

Of particular interest are the sagittal depth dimensions tabulated in Figure 9.9-1, which are specified to 0.0001 μm (1 Å). While such accuracies are clearly unattainable with even the most accurate diamond turning machines, the job was accepted on a "best effort" basis.

2. Machine description:
The machine selected to manufacture the part was LLNL's Diamond Turning Machine No. 2 (DTM No. 2). The machine, which is numerically controlled with a 0.025 μm resolution, would normally be capable of turning the optical surfaces to an accuracy of 0.25 μm, a figure unacceptable to the designer. To improve upon this figure, the decision was made to use postprocess gaging with feedback; the 0.025 μm repeatability of the machine was expected to permit a substantial improvement in part accuracy.

3. Inspection equipment:
The unique shape of the part profile, with a sagittal depth of less than 0.5 μm for both the ellipsoidal and hyperbolic sections, permitted the use of a Clevite surface analyzer to measure part contour. The Clevite is an instrument having a high accuracy straight-line datum normally used for evaluating surface roughness of flat specimens; its application here allowed evaluation of the part to an accuracy of approximately 0.025 μm.

4. Procedure:
The part was machined on DTM No. 2 and removed to the inspection area for evaluation of the Clevite. The resulting strip chart record was compared to a template of the desired contour. The difference between the actual and desired sagittal depths was compared and digitized for 180 points over the 18-mm length of each section. A new NC tape that included corrections for each point was generated, and the part was recut under the same conditions as the original cut. It should be noted here that the original relationship between the slides and machine base was preserved by using a manual compound slide to advance the tool for the corrected cut.

5. Results:
Figure 9.9-4 shows the desired contour, a trace of the initial cut, and a trace of the corrected cut. The improvements in part accuracy using the correction technique are dramatic, with the final part accuracy of 0.05 μm representing a fivefold improvement. This accuracy is consistent with assuming independent errors due to machine repeatability, machine resolution, inspection repeatability, and variations in fixture distortion, all with magnitudes of 0.025 μm, and combining these in a root-sum-square manner.

FUTURE DIRECTIONS AND POTENTIAL

In examining the advantages associated with the use of postprocess gaging with feedback, it is apparent that the procedure is only superior to other precalibration techniques in the areas of capital investment, and in the ability to correct for nonmachine errors. With the advent of correction value tables in commercially available numerical controls, it is likely that the capital cost benefit will eventually disappear. Thus, future applications will most likely only be justifiable when otherwise unattainable accuracies are required.

The technology required to implement PPGWF is already mature, as was shown by the example given and other well-known cases. Thus there is no need for further research or development in this area.

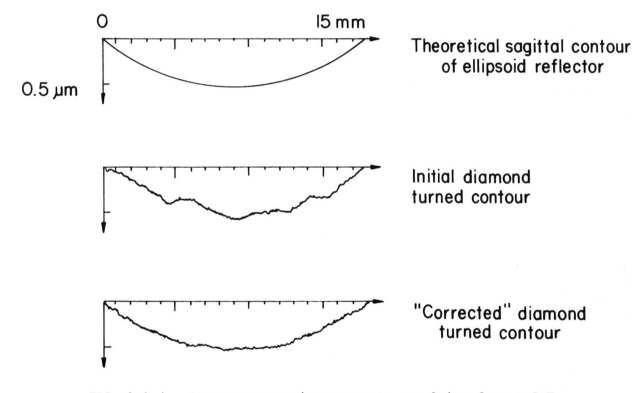

FIG. 9.9-4   Part accuracy improvements resulting from PPGWF.

Reprinted from *Modern Machine Shop*, June 1980

Microprocessors Revolutionize Gaging

Electronic gaging has been a fact of manufacturing life for over thirty years. But the low-cost microprocessor is bringing advanced quality control concepts into everyday use at the bench and machine level.

KEN GETTELMAN, Editor, interviews
RON LAVOIE, Marketing Manager
Federal Products Corporation
Providence, Rhode Island

Mr. Lavoie, we have seen the price of computers and calculators drop to the point that they are becoming everyday items—even for the home. You indicate this is also having a powerful impact upon gaging and quality control. In what areas?

There are five, and all of them are very significant. They include (1) low-cost precision comparative units, such as bore gages, that will outperform air gages, (2) complete low-cost statistical analysis, (3) low-cost special applications, (4) quick and inexpensive geometry definition and measurement, and (5) a much better surface finish evaluation.

Is the key to these developments either the minicomputer or the microprocessor?

Both. To give you some idea of the magnitude of what has happened in the past twenty years, look at Figure 1. In Today's $60 microprocessor we can get the same memory and computational capability that was in the $1 million computer offered just a score of years ago.

But hasn't electronic gaging been around for more than thirty years?

Yes, but early electronic gaging systems dealt primarily with analog signals. In the '60s, digital electronics became more popular, allowing very high resolution-to-range ratios. Where an analog meter value would be broken down into 80 graduations over the total measuring range, digital electronics allowed us to break down the total measuring range into 4000 parts, providing much greater precision for the same measuring range. With the advent of the microcomputer, it became possible to routinely combine analog signals in a complex fashion to achieve results that were heretofore achieved only by specially engineered analog computing systems.

The analog computing systems were expensive to engineer and to manufacture. Also, they were a one-of-a-kind device. Microcomputer systems perform the same function, but are developed via software only. This means that hardware remains the same; therefore, engineering and fabrication costs are significantly reduced. Finally, microcomputers combine these signals in a far more accurate fashion. Analog computers typically experienced severe fall-off in accuracy as the complexity of the signal combinations became greater. Today's microcomputers can handle very complex signal combinations with no reduction in accuracy.

Then the low-cost computational capability of today is the key to taking the analog gaging data input and quickly and inexpensively converting it to the digital data that is needed for complete statistical quality control analysis.

Yes, this is true. Not only are the inputs converted from analog to digital but they can be formatted for the best output by simply programming either the minicomputer or the microprocessor.

Is the programming done by the electronic gage user?

For the most part the answer is no. We build the system that is needed for the user. These usually make use of PROMS (programmable read-only memory units) that are programmed at the time the system is being built and are not changed. The user does not need to worry about programming.

You mentioned five areas of impact. Let's talk about the first

one—the simple gaging unit that compares a workpiece bore diameter to a master setting.

Fig. 1—The microprocessor has plunged in cost, increased in capacity, and shrunk in size by geometric progression within the past decade.

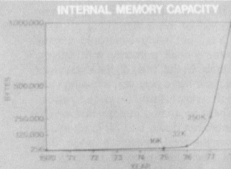

We have traditionally used three approaches for this measurement: (1) a mechanical setup utilizing a dial indicator, (2) air gaging, or (3) electronic gaging. For shop floor or at-the-machine use, both mechanical and air systems were widely used. The mechanical did well in the thousandths area, but for tenths or millionths, the air gage was often employed.

Isn't the air gage very accurate and relatively inexpensive?

It is both of these. However, the air gage does require a constant supply of clean, dry air whether or not gaging is actually taking place at the moment. In terms of energy consumption and operating costs, the air gage is high in comparison with the new electronic gages.

Are there other advantages of electronic over air?

In addition to simplicity, low initial cost and low operating cost, the new electronic plug and similar type gages are accurate across the full range of their scale. Air, on the other hand, tends to lose precision near the ends of the scale.

Are there other considerations?

Yes, it is quite easy to design amplification for different scale ranges and to transfer the gaging output signal to some other more sophisticated statistical or record gathering and reporting system.

But is it not true that the air bore gage uses a noncontact probe whereas the electronic plug does have contact points?

This is true. However, in very few situations is this a factor. With the electronic plug gage system, the contact points are tungsten carbide for long wear and the pressures are very light. The fact that there is contact is a very small disadvantage compared to the advantages of full accuracy over the entire range, ease of reading, no need to provide a constant supply of clean dry air, low initial cost, the ability to handle both inch and metric, and the ability to tie the readings into a full computer control quality reporting and analysis system.

The inexpensive microprocessor should have an impact far beyond a simple plug gage should it not?

Very definitely, and one of the new developments, shown in Figure 2, is the complete statistical analyzer with a price tag in the four-thousand-dollar area; thus, making it a working shop tool.

"Many more applications will be devised utilizing micro-computer technology and we are just seeing the start of a whole new productivity trend in metrology and quality control."—Ron Lavoie, Marketing Manager, Federal Products Corporation, Providence, Rhode Island.

Just what does it do?

It provides a complete analysis of a production lot from a sample of up to 255 pieces. The unit does this without the need of a statistician or experienced quality control expert to work out complex formulas from the gathered data. It is all computed and printed out by the microprocessor.

When you speak of a complete analysis, just what do you mean?

A complete analysis involves an extensive list that includes the following:

- The capability to handle either inch or metric
- A measuring mode of either the actual, maximum, minimum, total indicator reading or nominal
- Both minimum and maximum tolerances values as dialed in on the thumb wheels
- The total number of workpieces measured
- Both Minimum and maximum measured values within the sample

Fig. 2—Low-cost computing capability is the key to a stand-alone statistical analysis system for shop-floor use.

- The sequence number and measured value of each workpiece with the out-of-tolerance values indicated by an asterisk
- The difference between the minimum and maximum measured values
- A complete pictorial histogram of the measured sample with optimum display arrangements
- The average of all measured values
- The standard deviation of the measured values
- The range of three standard deviations which will include 99.7 percent of all pieces within the lot
- A true statistical estimate of the percent under and over tolerance in the lot being sampled
- A total percent out-of-tolerance based on the normal distribution computations
- The fit factor of the sample data, which is a measure of normality and
- A discretionary analysis with or without any stray pieces in the sample.

How does the unit present all of this information to either the operator or the QC inspector or manager?

It is all printed out as shown in Figure 3 within less than one minute after the sample has been gaged by the operator.

Is this all done without the use of separate computer programming or routines?

Yes, the statistical analyzer is a preprogrammed unit and the operator does nothing more than key-in certain facts such as the beginning and ending of the sample. He then proceeds in a logical order.

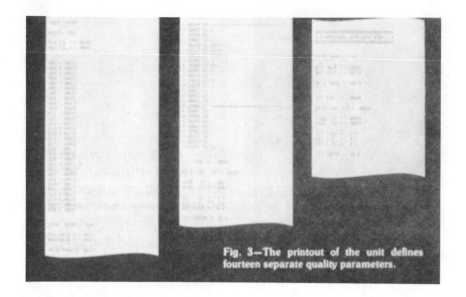

Is the printout of the fourteen different data items the only output?

The final output consists of the fourteen data items, with additional descriptions that we did not discuss in detail. However, all the data that appears on the printout can also be transferred from the statistical analyzer to an external recording device or mainframe computer system via an ASCII connector. There is a third connector, which allows the user to perform statistical analysis of data that can be keyed into the unit's minicomputer from a separate, optional alphanumeric keyboard.

In other words, it would be possible to take many other obtained measurements and enter them via keyboard and get a complete statistical analysis.

Yes, this is a capability of the unit.

But haven't computers been doing this type of thing for years?

Yes, they have. But the ever lower cost and greater capability of the minicomputer (in this instance) have allowed us to make a very economical stand-alone unit that functions in the hands of an inspector and produces a complete analysis on a real-time basis without any external programming. The programming is done at the time the analyzer is built. It provides a very powerful tool to allow the shop to quickly act upon the analyzed data. It is right on the shop floor—not lodged in a central computer facility.

You indicated the new age of computer and microprocessor capability is having an impact upon special gaging functions. Can you provide a good example?

An excellent example is the measurement of connecting rods. In the past, gages that measured connecting rods involved air or electronic devices of considerable mechanical complexity as well as a very elaborate cabinet of analog signal combination circuits. These gages were expensive to engineer, manufacture and maintain. Today, a relatively simple electronic gage with only four sensors per bore can explore both bores of a connecting rod, feeding data to a microcomputer while exploration takes place. The microcomputer then retrieves this data, combining it mathematically to determine such characteristics as hole-to-hole distance, bend, twist, taper, out-of-round as well as hole diameter. The net result is a simpler, more reliable and accurate gaging system.

Does it require a metrology expert to utilize this approach?

No, and that is one advantage of the microcomputer system. It is all preprogrammed at the time of unit fabrication. For example, in the past the measurement of roundness on a part required that the part be located precisely on the center of a precision rotary table. With a microcomputer, it need not be perfectly centered.

The computer will make provisions for any centering error as shown in Figure 4. Data from the sensing gage heads is processed with computer speed and accuracy to very quickly yield the roundness information. As a result, workpiece setup is much faster and data interpretation is more simplified.

You also indicated an advanced area of geometry measurement. Could you please amplify this?

It is often necessary to get a true geometric description of a workpiece. A good example is the need for manufacturing fuel injection systems to a very high precision level to meet stringent emission requirements. This means that plungers and housings must be true cylinders. We are now able to develop gaging systems that will probe several points along the shaft and the interior of the housing. From the data, the computer can construct a very good picture of true cylindricity and render a final "acceptable" or "not acceptable" verdict about the workpiece. The computer can even distinguish whether any geometric descriptions are the result of true part geometry or just simply a slight misalignment within the gaging fixture as shown in Figure 5.

The computer has no inherent capability of knowing these conditions, does it?

No, it does not. That is our function as metrology experts. We program the minicomputer with instructions so that the data picked up is properly processed. The computer offers only the tremendous speed and accuracy of calculation. It has absolutely no ability to think. But once programmed, it will produce very reliable and fast results.

In such systems, do you do the programming as part of unit construction?

Yes, these are preprogrammed units so that the user is not faced with the problem of writing extensive metrology programs. Again, the low

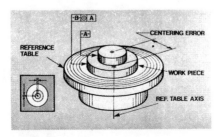

Fig. 4—In roundness measurement, the processor will make allowances for any centering error.

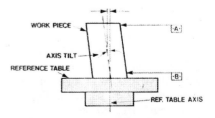

Fig. 5—In geometric analysis, any misalignment will automatically be calculated and allowed for in the final description.

cost of today's computing capability has allowed us to dedicate whole microprocessors or minicomputers to specific tasks. There is no need to share them with other areas and all of the associated programming requirements.

The last area you mentioned was surface texture. How can the low-cost computer element help here?

ANSI (American National Standards Institute) standard B46.1—1978 provides a clear definition of surface roughness, waviness and lay. It also covers such items as irreg-ularities. The traditional method of determining surface roughness involved passing a profilometer probe over the surface to be measured and the operator obtained an average reading from the indicator needle. It is now possible to have the data accepted by a computer, which will then sort out the various conditions as defined in the ANSI standard and determine if the surface is acceptable. Thus, we can economically get a much better description of work-piece surfaces. Here again, the speed and accuracy of low-cost com-puter elements provide a much better surface description with greater accuracy and at lower cost.

We have discussed the areas in which the low-cost computer is bringing greater metrology capability to the machine and bench level. Have we seen the end of such developments?

Absolutely not. We have just touched the tip of the iceberg. Many more applications will be devised utilizing microcomputer technology and we are just seeing the start of a whole new productivity trend in metrology and quality control. **MMS**

Graphical Inspection Analysis
By George O. Pruitt

FOREWORD

This paper is a presentation of an inspection technique that can be used to obtain the benefits of functional gaging without the expense and time required to design and manufacture functional gages. With the increased implementation of geometric dimensioning and tolerancing concepts where parts are dimensioned with functional relationships in mind, Graphical Inspection Analysis becomes advantageous to functionally verify part conformance to ensure full use of allowable manufacturing tolerances.

It is hoped that the entire product team (engineering, design, drafting, manufacturing, and quality control) will become familiar with the concept of Graphic Inspection Analysis, and use this inspection process to increase producibility.

Any comments, suggestions, and corrections from readers regarding information contained in this paper are welcomed by the author. There is a section for reader input at the end of the paper.

ACKNOWLEDGMENT

The author is grateful to his daughter, Karen, for her assistance on the original manuscript for this paper and to Mr. Edward Roth of Productivity Services for his valuable suggestions and encouragement.

INTRODUCTION

Graphical Inspection Analysis (GIA), also referred to as Layout Gaging, is an inspection verification technique employed to ensure functional part conformance to engineering drawing requirements without high-cost metal functional gages. With increased use of dimensioning and tolerancing methods, where we are concerned with concepts such as maximum material condition, cylindrical shaped tolerance zones, and "BONUS" tolerancies, we must use inspection methods that will evaluate parts conformance on a functional basis. Graphical Inspection Analysis provides the benefit of functional gaging without the expense and time required to design and manufacture a close toleranced, hardened metal functional gage. Graphical Inspection Analysis is well suited when small lots of parts are being manufactured; although, the inspection concept applies equally well to large quantity production. Also, inspection can be performed on parts too large and parts too small for conventional, functional receiver gages.

Inspection with functional receiver gages is the most desirable method to verify part conformance for interchangeability. Functional gages are really three-dimensional, worst case condition, mating parts. If the gage assembles to the part being inspected, we are assured assembly with all conforming mating parts. The drawbacks of functional gages are the expense and time required for design and manufacturing. Graphical Inspection Analysis provides a technique which will provide immediate, accurate inspection results without high cost and time delays.

GEOMETRIC TOLERANCING PRINCIPLES

Graphical Inspection Analysis is very dependent on the methods and principles used in geometric dimensioning and tolerancing. It is assumed that readers of this paper are knowledgeable in the concepts and rules of the Geometric Dimensioning and Tolerancing Standard as specified in American National Standards Institute, standard ANSI Y14.5. The ANSI principles will be reviewed to ensure common interpretation for an effective understanding and use of GIA. Applicable ANSI Y14.5 revision year and paragraph number of the information source for the following principles is shown in the margin for the readers' reference.

PRINCIPLE 1 - AUTHORITY STANDARD

It is imperative that the parts inspector knows which revision of the ANSI Y14.5 standard is used as the authority document in the preparation of the drawing. This paper will point out various differences existing between the three revisions of the standard (USASI Y14.5-1966, ANSI Y14.5-1973, and ANSI Y14.5M-1982).

66/1.4
73/5-1.4
82/1.1.2

PRINCIPLE 2 - BASIC DIMENSIONS

66/1.5-2
73/5-1.6.2
82/1.3.2
66/4.4.2
73/5-
   5.4.2.1
82/3.3.4
66/4.4.3
73/5-5.4.3
82/5.2.2

A basic dimension is a theoretical, exact dimension without tolerance. When used in conjunction with position tolerance specification, the basic dimension locates the exact center of the position tolerance zone. The center plane or axis of an acceptable feature is allowed to vary within the basic located position tolerance zone as shown in Figure 1. Basic dimensions on drawings are identified by the word BASIC or the abbreviation BSC near the dimension. Symbolically, the dimensions can be enclosed in a box to signify the basic notation. When features are located by basic chain dimensioning on the drawing, there is no accumulation of tolerance between features (Figure 2). Basic dimensions are absolute values and, when added together, they equal an absolute value. The basic dimensioning locates the position of the tolerance zone and not the manufactured feature locations.

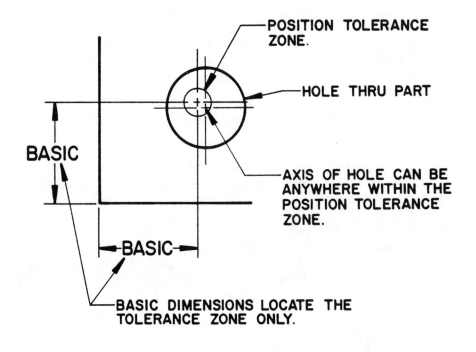

Figure 1. Positional Tolerance Zone.

240

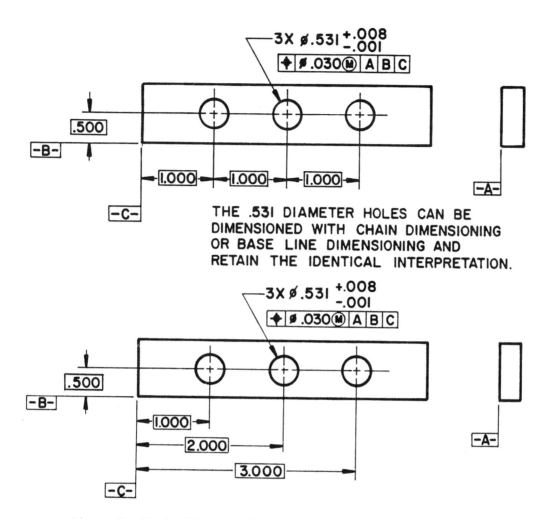

Figure 2. Chain Dimensioning Versus Base Line Dimensioning.

PRINCIPLE 3 – CONDITION MODIFIERS

There are three material condition modifiers used with position tolerancing: maximum material condition (MMC), regardless of feature size (RFS), and least material condition (LMC).

The symbol for maximum material condition is (M) . When the MMC symbol is associated with the tolerance or a datum reference letter in the feature control frame, the specified tolerance <u>only applies</u> to the feature if the feature is manufactured at its maximum material condition size.

66/4.5.1
73/5-5.5.1
82/2.8.2

Let's review maximum material condition as it applies to position tolerance. Maximum material conditon is the condition of a feature when it contains the maximum amount of material or weight within its allowable size limits. For example, a .500 ± .010 diameter pin will be at maximum material condition only if the machinist manufactures the pin at .510 diameter – a .509 diameter pin would not be at maximum material condition. Another example is a .525 ± .010 diameter hole in a part would be at maximum material condition, or its greater weight, if drilled at .515 diameter. In other words, maximum material condition is maximum allowable shaft size or minimum allowable hole size.

66/1.5.10
73/5-
1.6.11
82/5.3.2

As shown in the feature control frame (Figure 3), the .028 tolerance value only applies if the feature which the frame is associated is produced at MMC. What happens if the associated feature or datum departs from MMC? Principle 5 will discuss additional tolerance allowances.

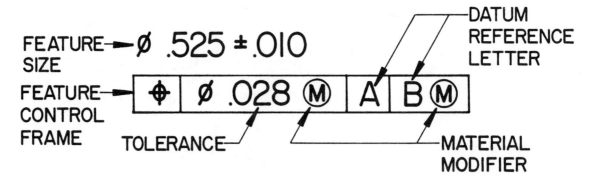

Figure 3. MMC Feature Control Frame.

66/1.5.11
66/4.5.2
73/5-
  1.6.13
73/5-5.5.3
82/5.3.4

The symbol for regardless of feature size is Ⓢ . When the RFS symbol is selected to modify the tolerance or datum reference in the feature control frame, the specified tolerance applies to the location of the feature regardless of the feature's size. In Figure 4, the .028 tolerance applies if the hole is manufactured at .515, .535, or any other value inbetween. The .028 is the total location tolerance for the feature regardless of its size.

.525 ±.010 DIA HOLE

| ⊕ | ⌀ .028 Ⓢ | A | B Ⓢ |

Figure 4. RFS Feature Control Frame.

73/5-
  1.6.12
82/1.3.13
82/5.3.5

The symbol for least material condition is Ⓛ . The least material condition concept and symbol were not introduced until the approval of ANSI Y14.5M-1982. Least material condition is the condition of a feature when it contains the least amount of material or weight; for example, smallest shaft size and largest hole size - just opposite of the MMC concept. When a feature control frame references LMC to the tolerance or datum reference letter, the specified tolerance only applies when the feature is produced at the LMC size. Additional discussion of what happens when the feature or datum size departs from LMC will be discussed in Principle 5.

PRINCIPLE 4 – FEATURE CONTROL FRAME

The feature control frame, formally called feature control symbol, tells the drawing user many things. The shape and size of the tolerance zone is specified, the surfaces and order of precedence for set-up of the part is dictated, and the material condition modifiers are assigned to the tolerance and datum reference letters.

66/3.4
73/5-3.4

As shown in Figure 5, the inspector is aware that the .014 tolerance value only applies when the feature being verified for locations is produced at its maximum material condition size. Also, in Figure 5, the diameter symbol (∅) preceding the tolerance indicates to the inspector that the .014 tolerance zone will have a cylindrical shape. Absence of the diameter symbol would indicate that the tolerance zone shape would be the area between two parallel planes or two parallel lines .014 apart. No symbol is specified to indicate the latter tolerance zone shape. The 1966 standard required the abbreviation DIA for diameter tolerance zone and R for radius tolerance zone to be included in the feature control frame. The 1966 standard also required the specifications of TOTAL for total wide tolerance zone and R for one-half of total wide tolerance zone.

66/3.6
73/5-3.6
82/3.6

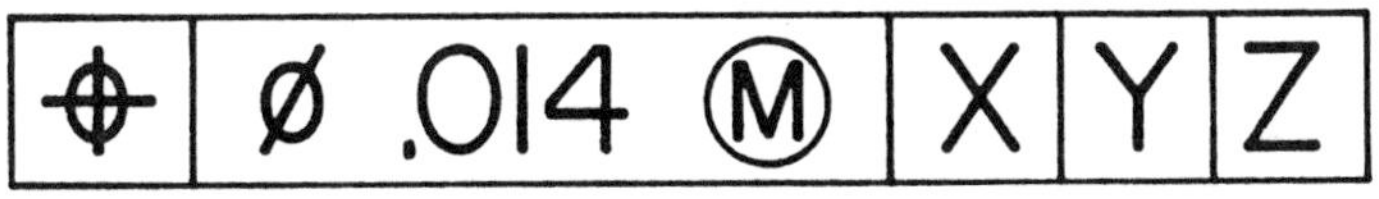

Figure 5. Feature Control Frame.

The order of specified datum references in the feature control frame is very important. The primary datum is shown at the left; the least important datum is shown at the right. This datum precedence allows manufacturing and inspection to determine part orientation for their respective functions. The part is fixtured to allow a minimum of three points of contact on the primary datum surface, minimum two points of contact for the secondary datum feature and minimum one point of contact for the tertiary datum feature.

66/3.4.2
73/5-
3.4.2.1
82/4.3

The 1966 revision of the ANSI Y14.5 standard required the datum reference letters to precede the tolerance compartment of the feature control frame – while the 1982 revision requires the datum reference letters to follow the tolerance compartment. Either method was accepted in the 1973 revision of the standard. The left to right datum procedure concept is identical in both methods (Figure 6).

66/3.4.1.1
73/5-3.4.2
82/3.4.2

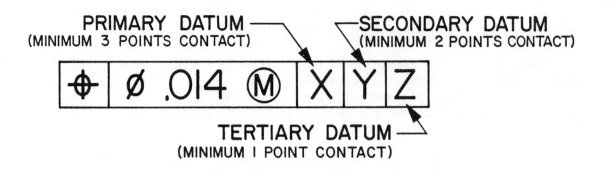

PRIMARY DATUM
(MINIMUM 3 POINTS CONTACT)

SECONDARY DATUM
(MINIMUM 2 POINTS CONTACT)

TERTIARY DATUM
(MINIMUM I POINT CONTACT)

Figure 6. Datum Precedence.

PRINCIPLE 5 – BONUS TOLERANCE AND ADDITIONAL TOLERANCE

66/4.5.1.2
73/5-
  2.12.2
82/5.3.2.1

As discussed in Principle 3, when a tolerance or datum reference letter is modified with the MMC symbol, the specified tolerance in the feature control frame only applies to the feature location when the feature is manufactured at its MMC size. As the feature departs from MMC size, the position tolerance is increased. The amount that the feature deviates from MMC size is added to the position tolerance specified in the feature control frame. This extra tolerance is called a "BONUS" tolerance. As shown in Figure 7, the .014 diameter position tolerance applies when the hole is drilled at .515 diameter, the MMC size. If hole number 2 was drilled at .525, a departure of .010 from the MMC size of the hole, we would gain a "BONUS" tolerance of .010. The .010 "BONUS" tolerance is added directly to the original .014 position tolerance to give an allowable positional tolerance of .024 (see table in Figure 7). The axis of hole 2 must be within the .024 diameter tolerance. The allowable positional tolerance zone size for each hole must be determined in conjunction with the actual manufactured hole size.

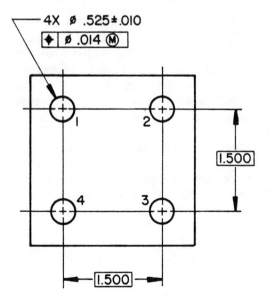

FEATURE SIZE	BONUS TOLERANCE	TOTAL POSITIONAL TOLERANCE
.515 (MMC)	.000	.014
.516	.001	.015
.517	.002	.016
.525	.010	.024
.533	.018	.032
.534	.019	.033
.535 (LMC)	.020	.034

Figure 7. "BONUS" Tolerances.

We can also gain extra locational tolerance as a datum feature of size, referenced in the feature control frame at MMC, departs from MMC size. This added tolerance is not called a "BONUS" tolerance and is treated somewhat differently than a "BONUS" tolerance.

66/4.6
73/5-
    5.5.1.3
82/5.3.2.2

In Figure 8, we could gain up to .005 "BONUS" tolerance as the .260 diameter hole departs from MMC. Additional tolerance could also be gained as the datum B hole departs from MMC. The added tolerance as datum B departs from MMC does not add directly to the original position tolerance as a "BONUS" tolerance would. The datum additional tolerance must be applied to the hole pattern as a group. For example, the added tolerance could allow the four hole pattern to shift to the right as a group. No additional hole-to-hole tolerance gain is realized within the four hole pattern.

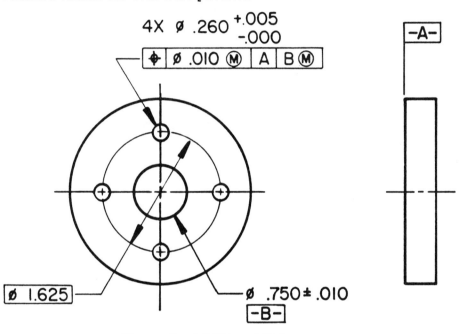

Figure 8. Additional Tolerance.

The rules and principles of MMC "BONUS" and additional tolerance also apply to tolerances and datum references that are modified at LMC. The "BONUS" and additional tolerances are determined from the feature's departure from LMC size.

82/5.3.5

The 1966 and 1973 revisions of ANSI Y14.5 allowed the MMC symbol to be implied by the drawing user when the material condition modifier had been omitted from the position feature control frame. It should be noted that this rule is contrary to international practice, which implies RFS in the same situations. The 1982 revision of ANSI Y14.5 requires the feature control symbol to be completed with the appropriate modifier (no implied modifier).

66/2.11.2
73/5-2.12
82/2.8

PRINCIPLE 6 - DATUMS

Datum specification in the feature control frame is of great importance to the drawing user. Proper functional datum selection allows the part to be fixtured for manufacturing and inspection as it will be assembled as a finished product. Consistency with quality control will be assured when manufacturing uses the specified datum features in the proper order of precedence for machining operations.

66/4.4.2.3
73/5-
  5.4.2.3
82/5.2.1.3

The 1966 and 1973 revisions of the ANSI Y14.5 standard allow the use of implied datums by not specifying datum reference letters in the feature control frame. This required the print users to make certain assumptions in the course of manufacturing and inspection; such as, which features to fixture upon and in what order of precedence. Use of implied datums can lead to problems if everyone concerned does not make the same assumptions. Designers and drafters should be encouraged to specify all necessary datum to provide uniform drawing interpretation. The 1982 revision of the standard has discontinued the use of implied datums.

66/APP A3
73/5-4.3.1
82/4.1.2
66/APP A5
73/5-4.5
82/4.3.2

The three datum plane reference frame is ideal for insurance of common drawing interpretation. For non-cylindrical parts the manufacturing/inspection fixture shown in Figure 9 can be constructed to assure uniformity during manufacturing and inspection operations. All related measurements of the part originate from the fixture datum planes. For cylindrical parts the three plane reference frame is more difficult to visualize. The primary datum is often described as a flat surface perpendicular to the axis of the cylindrical datum feature as shown in Figure 10. This axis can be defined as the intersection of two planes, 90 degrees to each other, at the midpoint of the cylindrical feature. The tertiary datum is used if rotational orientation of the cylindrical feature is required due to interrelationship of radially located features. The tertiary datum is often a locating hole, slot, or pin. If no angular located features are involved in the hardware requirements, the teritary datum is omitted.

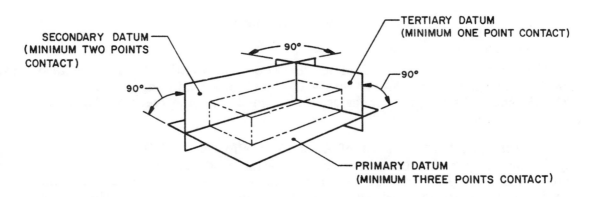

Figure 9. Three Plane Reference Frame for Non-Cylindrical Features.

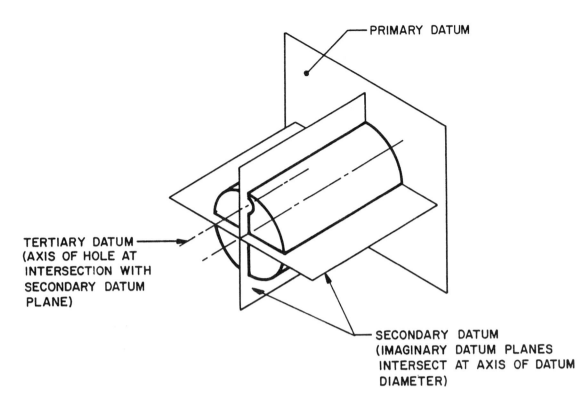

PRIMARY DATUM

TERTIARY DATUM
(AXIS OF HOLE AT
INTERSECTION WITH
SECONDARY DATUM
PLANE)

SECONDARY DATUM
(IMAGINARY DATUM PLANES
INTERSECT AT AXIS OF DATUM
DIAMETER)

Figure 10. Three Plane Reference Frame for Cylindrical Features.

There are two types of datum features: datums of size and non-size datums. Non-size datums are established from surfaces. A datum surface has no size tolerance since it is a plane from which dimensions or relationships originate. Datums of size are established from features that have size tolerance, such as, holes, outside diameter, and slot widths. The center plane or axis of datum feature of size is the actual datum. For example, a .250 + .005 hole specified as a datum would be a datum feature of size (+ .005 tolerance on feature size). The centerline or axis of the manufactured hole is the actual datum.

66/APP A6
73/5-4.6
82/4.4.1
82/4.4.2

PRINCIPLE 7 - POSITIONAL TOLERANCE

A position tolerance is the total permissible variation in the location of a feature about its exact position. For cylindrical features such as holes and outside diameters, the position tolerance is the diameter of the tolerance zone within which the axis of the feature must lie. The center of the tolerance zone is located at the exact position. For other than round features, such as slots and tabs, the position tolerance is the total width of the tolerance zone within which the center plane of the feature must lie. The center of the tolerance zone is located at the exact location. Position tolerance zones are three-dimensional and apply to the thickness of the part. As shown in Figure 11, a diameter position tolerance zone is really a cylindrical tolerance zone through the part within which the feature axis must lie. This cylindrical tolerance zone is 90 degrees basic from the specified datum reference surface. Note the tolerance zone will also control perpendicularity of the feature within the position tolerance requirement.

66/4.1
73/5-5.2
82/5.2

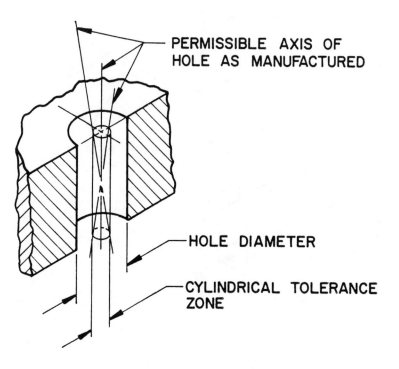

PERMISSIBLE AXIS OF
HOLE AS MANUFACTURED

HOLE DIAMETER

CYLINDRICAL TOLERANCE
ZONE

Figure 11. Position Cylindrical Tolerance Zone.

66/NA
73/5-5.6.1
82/5.4.1

Composite position tolerancing provides a method of allowing a more liberal locational tolerance for the pattern of feature as a group and then controlling the feature-to-feature interrelationship to a finer requirement. Each horizontal entry of the feature control frame shown in Figure 12 constitutes a separate inspection operation. The upper entry specifies the positional requirements for the pattern as a group, and the lower entry specifies the position tolerance within the pattern.

Figure 12. Composite Feature Control Frame.

248

PRINCIPLE 8 – SCREW THREAD SPECIFICATION

Where geometric tolerancing is expressed for the control of a screw thread or where a screw thread is specified as a datum reference, the application shall be applied to the pitch diameter. If design requirements necessitate an exception to this rule, the notation MINOR DIA or MAJOR DIA shall be shown beneath the feature control frame or datum reference as applicable.

66/2.11.4
73/5-2.13
82/2.9

PRINCIPLE 9 – VIRTUAL CONDITION AND SIZE

Virtual condition is the worst possible assembly condition of mating parts resulting from the collective effects of size and the geometric tolerancing specified to control the feature (see Figure 13). Virtual condition is primarily a tool used by product and tool/gage designers to calculate basic gage element size or to perform tolerance analysis to ensure assembly of mating parts. The following formulas can be used to determine virtual conditions:

66/NA
73/5-2.15
82/2.11

EXTERNAL FEATURES = MMC SIZE + TOLERANCE OF FORM ATTITUDE OR LOCATION.

INTERNAL FEATURES = MMC SIZE - TOLERANCE OF FORM, ATTITUDE OR LOCATION.

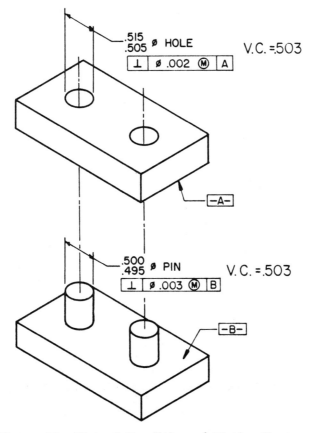

Figure 13. Virtual Condition of Mating Parts.

Virtual condition is calculated from information specified on the engineering drawing (such as size and geometric tolerancing), while virtual size is determined by the parts inspector from the actual measured size and location or attitude configuration of each feature. Virtual size of features must be considered for accurate GIA verification. See page 65 for additional discussions.

PRINCIPLE 10 – SEPARATE REQUIREMENTS

66/4.6.3
73/5–5.6.3
82/5.3.6.2

When more than one pattern of features (such as holes and slots) are located by basic dimensions from common datum features of size, and the feature control frame for each of the patterns contains the same datums in the same order of precedence and at the same material condition, all the features are considered as one single pattern. In Figure 14, the two .221 diameter holes appear as one pattern and the two .391 diameter holes are a separate pattern. Since the location feature control frame for both hole patterns contain the same datums, in the same order of precedence and at the same material condition, the patterns are considered one pattern of four holes. The parts inspector must verify the dimensional conformance of both hole patterns simultaneously. If the designer had felt this interrelationship of the four holes was not required between the two patterns of features in Figure 14, a notation such as SEPARATE REQUIREMENTS would have been placed beneath the feature control frames. This would allow each pattern of features to shift independently in relationship to the common datum system. The parts inspector would verify each pattern of holes separately.

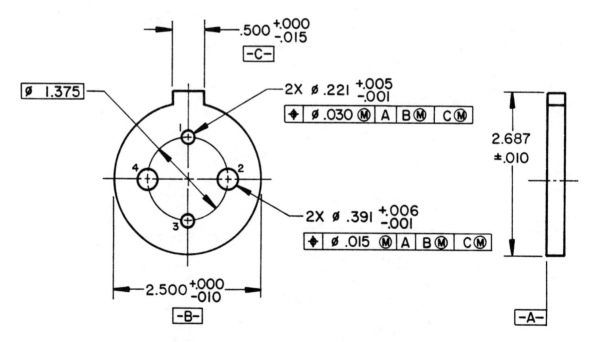

Figure 14. Composite Feature Patterns.

GRAPHICAL INSPECTION ANALYSIS PROCEDURE

The performance of Graphical Inspection Analysis is a relatively simple four-step procedure. The first step of the process requires plotting the basic location of the feature to be inspected in relationship to the datum features as described by the engineering drawing. This plot is called the DATA GRAPH and is drawn on 10 x 10 graph paper at any desirable scale; 8½ x 11 paper can be used for simple parts, whereas larger sheets will be required for complex parts. We will use the part shown in Figure 15 as an introductory explanation of a GIA procedure.

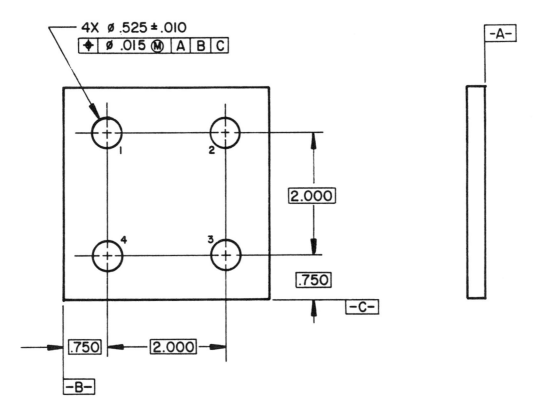

Figure 15. Example 1a Part.

Figure 16 shows the DATA GRAPH for Example 1a. The basic feature locations are plotted on the graph in relationship to the datum features at any convenient scale. We shall refer to this scale as the "CONFIGURATION SCALE."

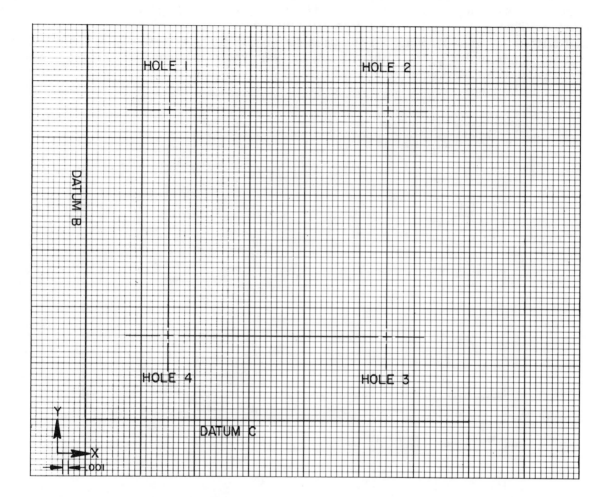

Figure 16. Example 1a DATA GRAPH.

Step 2 of the analysis requires plotting the locations of the actual inspected features on the DATA GRAPH in relationship to the basic plot. This data is established by the parts inspector through surface plate inspection techniques as shown in Figure 17, or with the coordinate measurement machine as shown in Figure 18. The data is recorded on the appropriate inspection report form (Table 1) to become part of the inspection report.

Figure 17. Surface Plate Inspection Process.

Figure 18. Coordinate Measurement Machine Inspection Process.

Table 1. Example 1a Inspection Data.

INSPECTION DATA

FEATURE NUMBER	FEATURE LOCATION						FEATURE SIZE			POSITION TOLERANCE			
	X AXIS			Y AXIS									
	SPECIFIED	ACTUAL	DEVIATION	SPECIFIED	ACTUAL	DEVIATION	SPECIFIED	ACTUAL	DEVIATION	MATERIAL CONDITION	SPECIFIED	BONUS	TOTAL
1	.750	.754	.004	2.750	2.751	.001	.515	.531	.016	MMC	.015	.016	.031
2	2.750	2.760	.010	2.750	2.750	.000	.515	.533	.018	MMC	.015	.018	.033
3	2.750	2.762	.012	.750	.748	.002	.515	.535	.020	MMC	.015	.020	.035
4	.750	.753	.003	.750	.749	.001	.515	.530	.015	MMC	.015	.015	.030

To establish accurate data, the part shown in Figure 15 must be fixtured for surface plate inspection as specified in the feature control frame. When set-up for inspection, the part must contact the angle plate at a minimum of three points for the primary datum (datum A) and a minimum of two points of contact on the surface plate for the secondary datum (datum B) as shown in Figure 19.

Figure 19. Surface Plate Inspection Set-up (Primary and Secondary Datums).

The tertiary datum surface (datum C) is established by placing the second angle plate against the original angle plate as shown in Figure 20. Since the part being inspected has a symmetrical configuration, the machinist must mark the surfaces that he used to establish datums A, B and C for the placement of the holes. This will ensure consistency between manufacturing and quality control. Notice the orientation of the two angle plates and the surface plate has created the three plane reference frame shown in Figure 9. The part will be clamped to the angle plate to restrict its movement when performing inspection measurements. Snug fitting gage pins are placed in each hole and a measurement is established from the surface plate to the top of the pin as shown in Figure 21. One-half of the gage pin diameter is subtracted from the measured value to establish the dimension from datum surface B (surface plate) to the center of the hole. Measurements are recorded for each hole in this set-up.

Figure 20. Surface Plate Inspection Set-up.

Figure 21. Surface Plate Inspection From Datum B.

The angle plate is rotated 90 degrees (without loosening clamps) to establish similar measurements from datum C as shown in Figure 17.

A three plane reference frame must also be employed when using the coordinate measurement machine to verify hole locations of the part shown in Figure 15. The table top establishes the primary datum (Figure 22a), and the two parallels provide contact surfaces for the secondary (Figure 22b) and tertiary datums (Figure 22c).

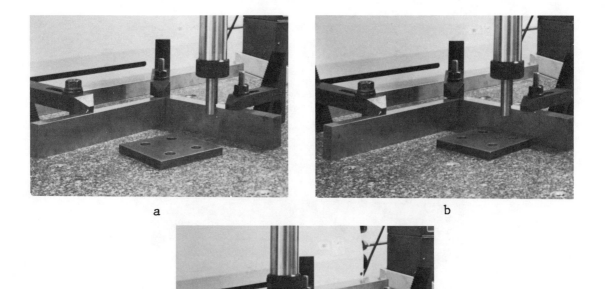

a   b

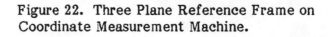

c

Figure 22. Three Plane Reference Frame on
Coordinate Measurement Machine.

The appropriate measurement probe is placed in the coordinate measurement machine and calibrated to a zero reading at the intersection of the parallels that establish datum B and C. The probe is placed in each hole of the part to verify the actual measurements from datums B and C as shown in Figure 18. This data is recorded on the inspection report form.

Step 2 of the GIA process, for example part 1a, can be completed with the data tabulated in Table 1. The feature location deviations from the basic specified location will be in the order of magnitude of thousandths of an inch. We will let each grid square of the graph equal .001 inch to allow sufficient accuracy for the gaging operations. We will refer to this scale as the "DEVIATION SCALE." The location deviations are plotted for each feature as shown in Figure 23.

In Step 3, a transparent TOLERANCE ZONE OVERLAY GAGE will be generated (see Figure 24) by placing a transparent material over the DATA GRAPH. With a compass, the allowable position tolerance zone will be drawn at each of the basic locations for the specified tolerance zone plus any "BONUS" tolerance. The "BONUS" tolerance data has been recorded in the feature size deviation block of the inspection report shown in Table 1. The tolerance zones will be generated at the same scale (DEVIATION SCALE) used to plot hole locational deviations in Step 2.

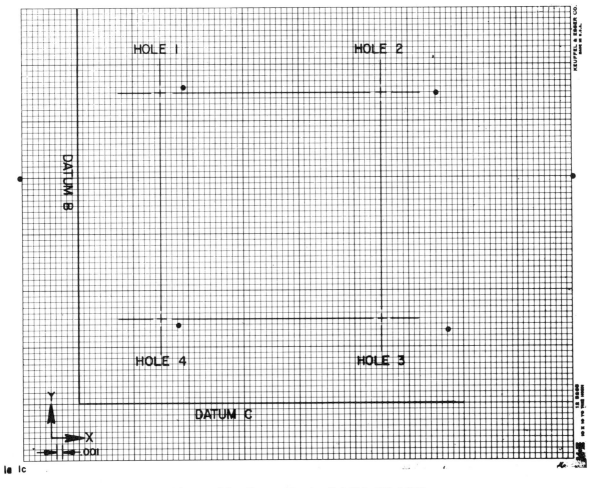

Figure 23. Example 1a DATA GRAPH.

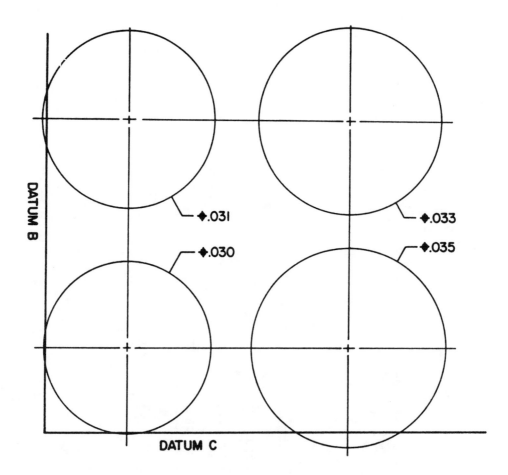

Figure 24. Example 1a TOLERANCE ZONE OVERLAY GAGE.

In Step 4, the transparent TOLERANCE ZONE OVERLAY GAGE is superimposed over the DATA GRAPH and rotated and/or translated to provide alignment with the datum feature. We have an acceptable part if the actual feature axis of the four holes falls within the allowable tolerance zones at proper setting of the overlay (shown in Figure 25).

The part drawing for example 1a is well done, in that datum surfaces have been specified and the feature control frame has indicated a datum set-up precedence. Both manufacturing and inspection will fixture the part with minimum three points contact of surface A, minimum two points contact of surface B, and minimum one point contact of surface C. The manufacturer marked the surface he used for datums A, B, and C in his drilling operations to ensure consistency with inspection. The MMC modifier associated with the .015 position tolerance zone in the feature control frame has alerted manufacturing and inspection of the "BONUS" tolerance possibilities as the hole size departs from .515 diameter. In this example the manufacturer used the largest standard drill available to take advantage of this additional production tolerance.

Inspection data for a number of parts can be plotted on the same DATA GRAPH as shown in Figure 26. Notice the TOLERANCE ZONE OVERLAY GAGE includes numerous tolerance zones to allow evaluation of the features as they depart from MMC. Each feature would be checked for conformance at its allowable position tolerance.

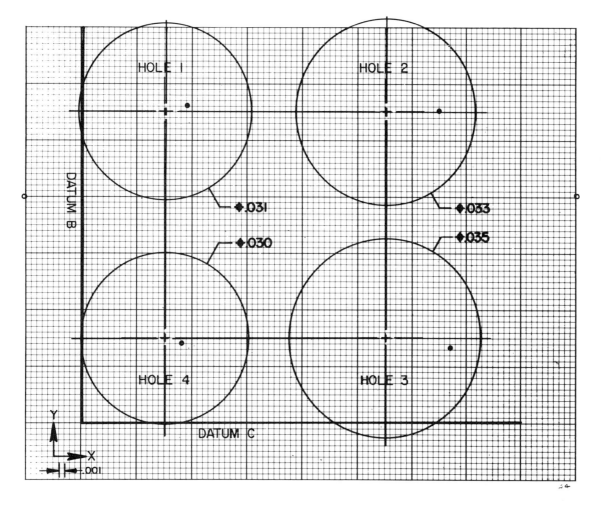

Figure 25. GIA of Example 1a.

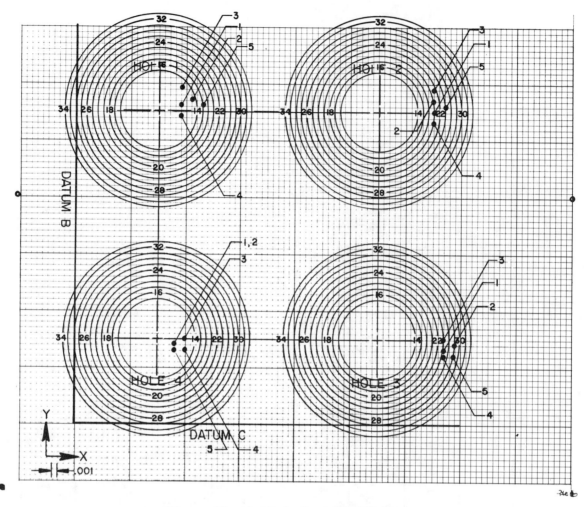

Figure 26. Multiple Part Verification.

Another technique of graphical inspection which employs common axis data plotting and a tolerance zone overlay of prepared concentric circles can be used for feature location verification of simple parts as shown in Example 1a. However, this technique does not lend itself to the evaluation of parts with secondary and tertiary datums of size (Example 1e) or where angular orientation is concerned. Therefore, this paper will only demonstrate the graphical inspection technique where data plotting resembles the geometry of the part.

EXAMPLE 1b

The zero position tolerancing concept of controlling feature locations is being used more often. Figure 27 shows the part from example 1a dimensioned with the zero position concept. Notice the maximum allowable hole size of both examples 1a and 1b is .535 diameter. Also, the position tolerance at the least material condition size (.535 diameter) is identical (.035 position tolerance), when the allowable "BONUS" tolerance is added to the positional tolerance specified in each of the feature control frames. The only difference between the two methods of dimensioning is the minimum allowable hole size and the specified position tolerance at MMC. The manufacturer of the part shown in Figure 27 could drill the holes at .500 diameter which would require a perfect location (zero position tolerance at MMC), or he could produce .515 diameter holes (.015 departure from MMC) to allow a "BONUS" tolerance of .015 for the location requirement of the holes. This simple example shows that economical manufacturing can be a reality with proper specification and interpretation of geometric dimensioning and tolerancing. If the manufacturer does not understand the zero position tolerance concept, he may find himself in trouble.

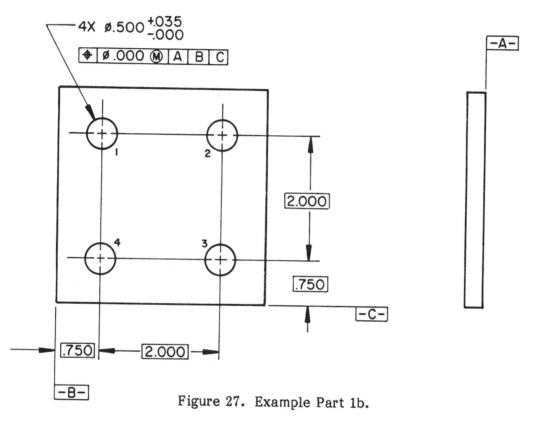

Figure 27. Example Part 1b.

The DATA GRAPH (Figure 28) for part 1b has been completed with the inspection data from Table 2.

Table 2. Example 1b Inspection Data.

FEATURE NUMBER	FEATURE LOCATION						FEATURE SIZE			POSITION TOLERANCE			
	X AXIS			Y AXIS									
	SPECIFIED	ACTUAL	DEVIATION	SPECIFIED	ACTUAL	DEVIATION	SPECIFIED	ACTUAL	DEVIATION	MATERIAL CONDITION	SPECIFIED	BONUS	TOTAL
1	.750	.754	.004	2.750	2.751	.001	.500	.531	.031	MMC	.000	.031	.031
2	2.750	2.760	.010	2.750	2.750	.000	.500	.533	.033	MMC	.000	.033	.033
3	2.750	2.762	.012	.750	.748	.002	.500	.535	.035	MMC	.000	.035	.035
4	.750	.753	.003	.750	.749	.001	.500	.530	.030	MMC	.000	.030	.030

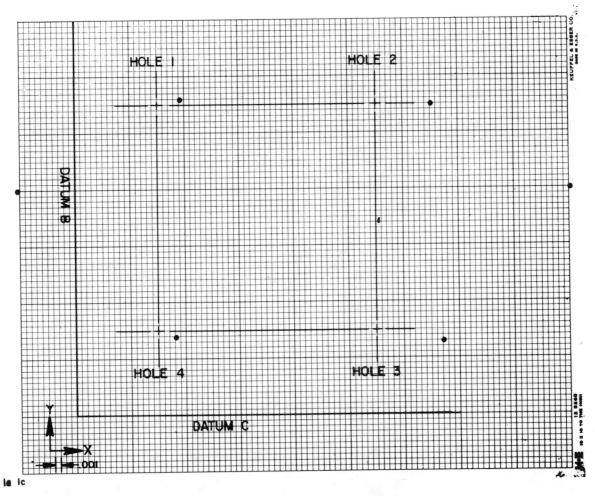

Figure 28. Example 1b DATA GRAPH.

Figure 29 shows the TOLERANCE ZONE OVERLAY GAGE superimposed over the DATA GRAPH. If the holes in part 1b had been produced at .500 diameter, the MMC size, the TOLERANCE ZONE OVERLAY GAGE circle diameters would be reduced to zeros. The circular tolerance zone diameter is completely dependent on "BONUS" tolerance allowance in this example.

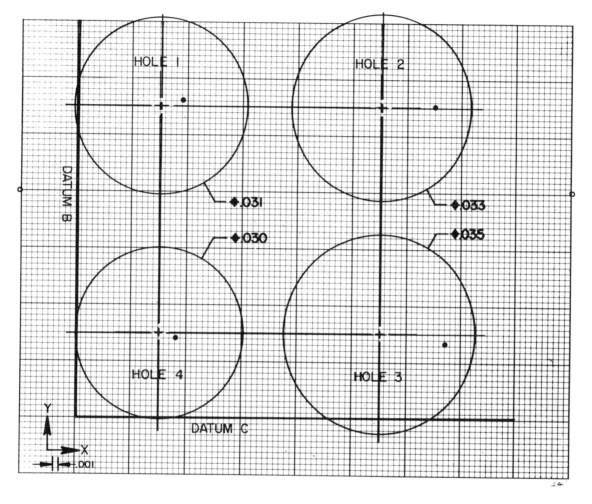

Figure 29. Example 1b DATA GRAPH/TOLERANCE ZONE OVERLAY GAGE.

EXAMPLE 1c

The drawing shown in Figure 30 will be used for Example 1c. Notice the differences between Examples 1a, 1b, and 1c. Example 1c does not specify datum features to be used in the inspection operation. The surfaces from which the .750 + .010 dimensions originate are implied datum surfaces. When the part is inspected we do not know which datum feature is primary, secondary, or tertiary. Also, the .750 dimension is no longer specified as a basic dimension. The designer has attempted to allow manufacturing to have a more liberal location tolerance (+ .010 = .020) for the group of holes while maintaining the more restrictive .015 position tolerance for hole-to-hole interrelationship. It should be pointed out that the + .010 tolerance locates the centers of the positional tolerance zone and not the centers of the actual holes.

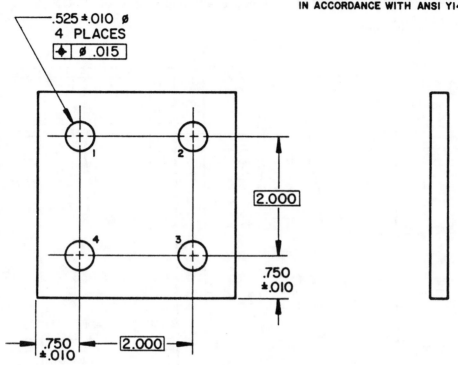

Figure 30. Example Part 1c.

From the inspection data shown in Table 3, we can plot the locations of the actual holes on the DATA GRAPH shown in Figure 31. Notice three of the hole centers are outside of the square tolerance zones. As previously stated, the ± .010 tolerance locates the position tolerance zone, not the actual hole.

Table 3. Example 1c Inspection Data.

INSPECTION DATA

FEATURE NUMBER	FEATURE LOCATION						FEATURE SIZE			POSITION TOLERANCE			
	X AXIS			Y AXIS						MATERIAL CONDITION	SPECIFIED	BONUS	TOTAL
	SPECIFIED	ACTUAL	DEVIATION	SPECIFIED	ACTUAL	DEVIATION	SPECIFIED	ACTUAL	DEVIATION				
1	.750	.739	.011	2.750	2.743	.007	.515	.516	.001	MMC	.015	.001	.016
2	2.750	2.737	.013	2.750	2.759	.009	.515	.517	.002	MMC	.015	.002	.017
3	2.750	2.751	.001	.750	.762	.012	.515	.516	.001	MMC	.015	.001	.016
4	.750	.753	.003	.750	.745	.005	.515	.515	.000	MMC	.015	.000	.015

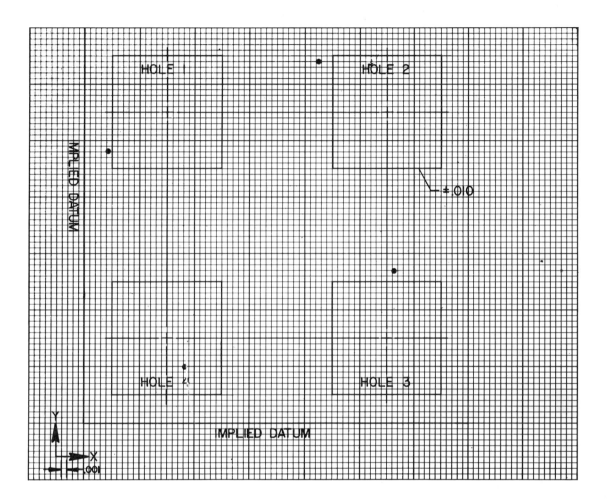

Figure 31. Example 1c DATA GRAPH.

The TOLERANCE ZONE OVERLAY GAGE is transposed over the DATA GRAPH. The centers of the position tolerance zones can float anywhere within the ± .010 square tolerance zones. The omission of datum reference letters in the feature control frame and the failure to provide basic dimensions to locate the position tolerance zones allow this tolerance zone to float. Figure 32 indicates that the part meets the drawing requirements.

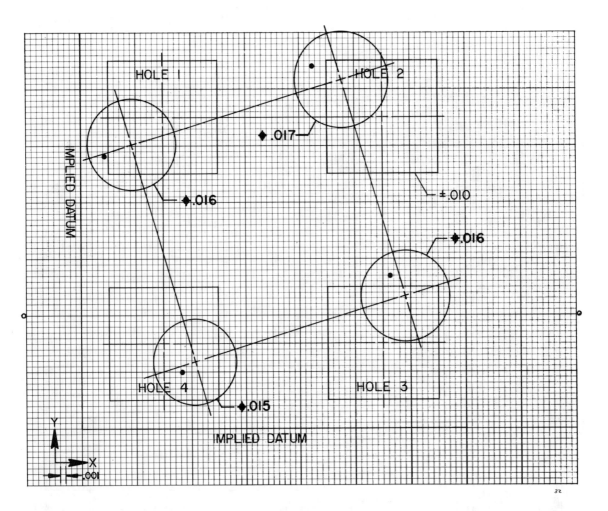

Figure 32. Example 1c DATA GRAPH/TOLERANCE ZONE OVERLAY GAGE.

EXAMPLE 1d

In example 1d, Figure 33, the composite position tolerancing method of dimensioning has been used to provide the same design intent as specified in Example 1c. Example 1d no longer combines the coordinate dimensioning system with the position tolerancing system. The .750 dimensions are now specified basic.

Each entry of the composite feature control frame must be verified separately. This is an expensive gaging operation when using functional receiver gages since separate gages must be designed and manufactured for each entry. The upper entry has specified selected datum features and order of precedence for common interpretation. The lower entry is independent of any datum location requirement. The four-step GIA verification process can be performed with the inspection data shown in Table 4.

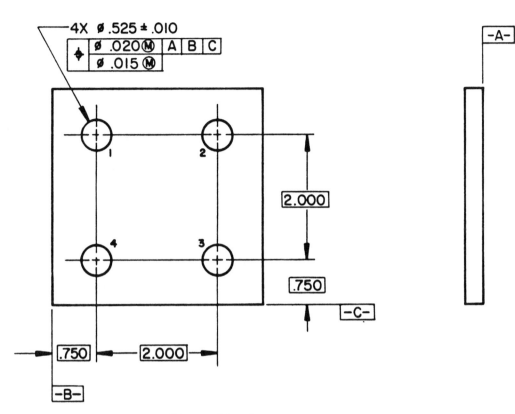

Figure 33. Example 1d Part.

Table 4. Example 1d Inspection Data.

INSPECTION DATA

FEATURE NUMBER	FEATURE LOCATION						FEATURE SIZE			POSITION TOLERANCE			
	X AXIS			Y AXIS						MATERIAL CONDITION	SPECIFIED	BONUS	TOTAL
	SPECIFIED	ACTUAL	DEVIATION	SPECIFIED	ACTUAL	DEVIATION	SPECIFIED	ACTUAL	DEVIATION				
1	.750	.754	.004	2.750	2.757	.007	.515	.515	.000	MMC	.020	.000	.020
1										MMC	.015	.000	.015
2	2.750	2.755	.005	2.750	2.743	.007	.515	.516	.001	MMC	.020	.001	.021
2										MMC	.015	.001	.016
3	2.750	2.745	.005	.750	.743	.007	.515	.516	.001	MMC	.020	.001	.021
3										MMC	.015	.001	.016
4	.750	.746	.004	.750	.757	.007	.515	.519	.004	MMC	.020	.004	.024
4										MMC	.015	.004	.019

Steps 1 and 2 require plotting the actual locations of the feature in relationship to their basic locations (Figure 34).

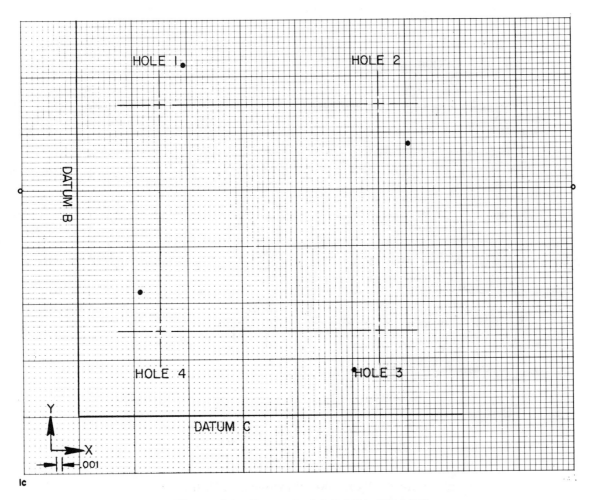

Figure 34. Example 1d DATA GRAPH.

In Step 3, TOLERANCE ZONE OVERLAY GAGES are prepared for both the upper and lower entry of the composite position feature control frame.

Notice the presence of datum feature lines in the upper TOLERANCE ZONE OVERLAY GAGE (Figure 35) to allow overlay alignment with the specified datum surfaces on the DATA GRAPH as specified by the feature control frame. The lower TOLERANCE ZONE OVERLAY GAGE (Figure 36) is not datum related and is, therefore, allowed to float at will to establish hole-to-hole locational requirements. Step 4, Figure 37, requires the plotted hole located to be within the tolerance zones of both the upper and lower TOLERANCE ZONE OVERLAY GAGES in order to meet drawing requirements. This part is acceptable.

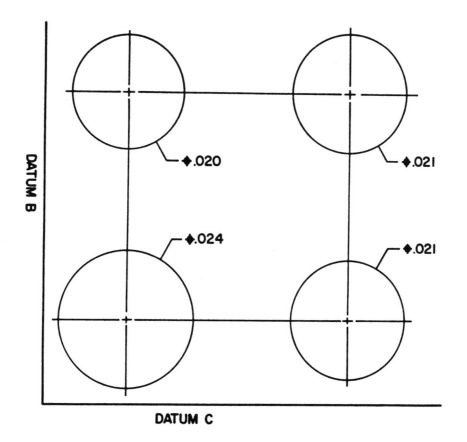

Figure 35. Upper TOLERANCE ZONE OVERLAY GAGE.

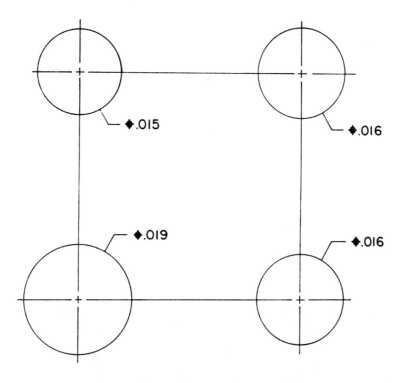

Figure 36. Lower TOLERANCE ZONE OVERLAY GAGE.

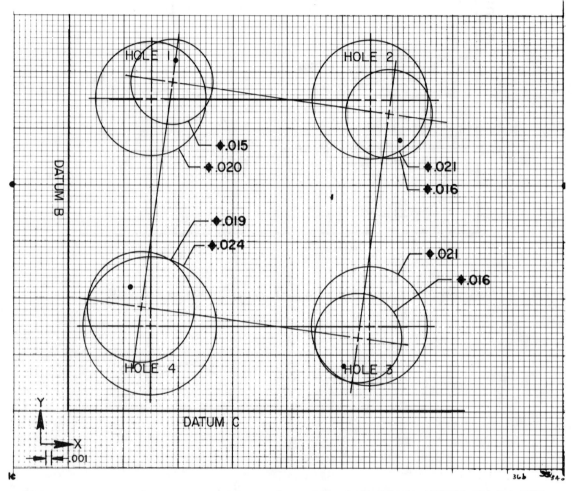

Figure 37. Example 1d DATA GRAPH/TOLERANCE ZONE OVERLAY GAGES.

EXAMPLE 1e

Example 1e (Figure 38) has specified datums B and C as datums of size. When this occurs, the axis or center plane of the feature is the real datum. Notice the inspection data in Table 5 has been measured from the center lines of the part rather than from the sides. As discussed in Principle 5, when datums of size are specified at MMC, it is possible to gain extra manufacturing tolerance as the datums depart from MMC size. This extra tolerance allows additional locational tolerance of the pattern of feature as a group. In this example, additional tolerance can be utilized as the widths of the part depart from their MMC size of 3.510.

Table 5. Example 1e Inspection Data.

INSPECTION DATA

| FEATURE NUMBER | FEATURE LOCATION | | | | | | FEATURE SIZE | | | POSITION TOLERANCE | | | |
| | X AXIS | | | Y AXIS | | | | | | | | | |
	SPECIFIED	ACTUAL	DEVIATION	SPECIFIED	ACTUAL	DEVIATION	SPECIFIED	ACTUAL	DEVIATION	MATERIAL CONDITION	SPECIFIED	BONUS	TOTAL
1	1.000	1.008	.008	1.000	1.002	.002	.515	.516	.001	MMC	.015	.001	.016
2	1.000	.991	.009	1.000	1.004	.004	.515	.517	.002	MMC	.015	.002	.017
3	1.000	1.004	.004	1.000	1.002	.002	.515	.516	.001	MMC	.015	.001	.016
4	1.000	.996	.004	1.000	1.006	.006	.515	.516	.001	MMC	.015	.001	.016
DATUM B	—	—	—	—	—	—	3.510	3.504	.006	MMC	—	—	—
DATUM C	—	—	—	—	—	—	3.510	3.500	.010	MMC	—	—	—

Steps 1 and 2 require preparation of the DATA GRAPH (Figure 39) and the plotting of the feature locations from the inspection data (Table 5). Notice datums B and C are the center planes of their respective widths. The inspection data has been generated from these datum planes rather than the outside surfaces as done for previous examples.

In Step 3, the TOLERANCE ZONE OVERLAY GAGE (Figure 40) is generated using the inspection data. Datum feature lines are included on the overlay to represent datums B and C. The length of the datum features are also marked in the TOLERANCE ZONE OVERLAY GAGE.

When the TOLERANCE ZONE OVERLAY GAGE is transposed on the DATA GRAPH in Step 4 (Figure 41), we see that holes 1 and 2 do not meet the drawing requirements.

In Principle 5, we discussed additional tolerances that could be gained as datum features of size depart from MMC. From the inspection data (Table 5) we see that datum B has departed from MMC size by .006 and datum C has departed from MMC size by .010. These additional tolerance allowances must be included on the DATA GRAPH to complete the tolerance analysis. The TOLERANCE ZONE OVERLAY GAGE used for feature-to-feature verification can be rotated and translated within these additional tolerance zones in an attempt to verify part conformance as shown in Figure 42. The entire length of each datum feature must remain within this additional tolerance zone. This part meets the drawing specification.

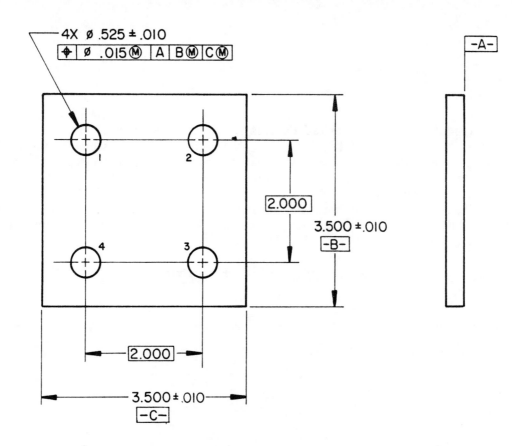

Figure 38. Example 1e Part.

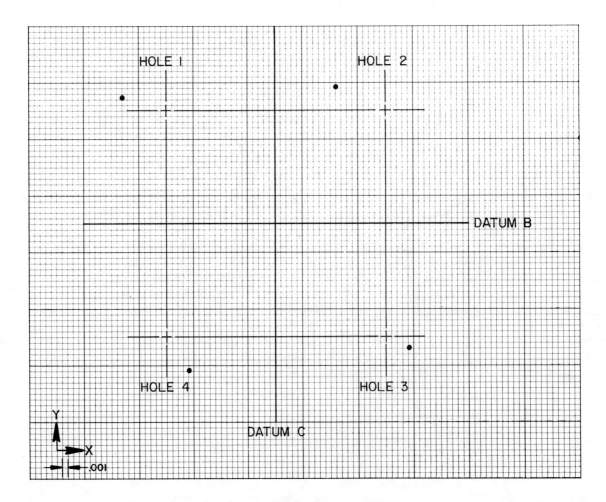

Figure 39. Example 1e DATA GRAPH.

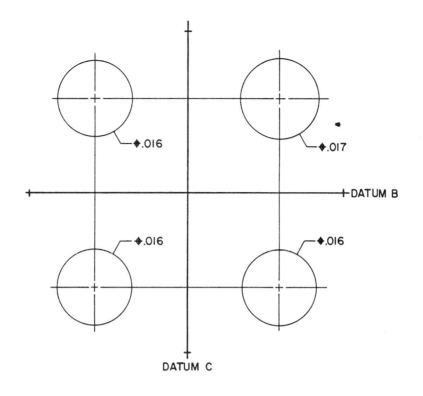

Figure 40. Example 1e TOLERANCE ZONE OVERLAY GAGE.

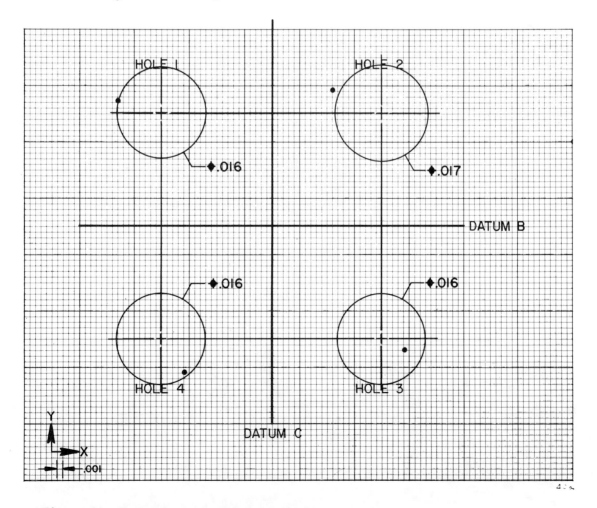

Figure 41. Example 1e DATA GRAPH/TOLERANCE ZONE OVERLAY GAGE.

273

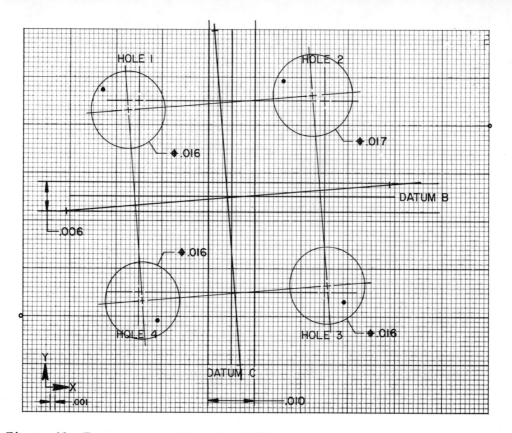

Figure 42. Example 1e DATA GRAPH/TOLERANCE ZONE OVERLAY GAGE.

WORK PROBLEMS

Examples 2 through 6 are provided for the reader to gain experience in the concepts of Graphical Inspection Analysis. The reader can refer to the part drawing, inspection data, and pictures of the inspection process for the information required to generate DATA GRAPHS and TOLERANCE ZONE OVERLAY GAGES (completed examples are provided for those who do not care for the hands-on experience).

EXAMPLE 2

Example 2 provides a part with radially located holes related to a datum of size as shown in Figure 43. The inspection procedure is shown in Figure 44 that was used to generate the inspection data shown in Table 6.

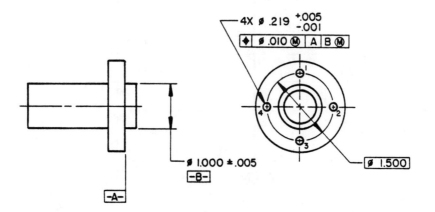

Figure 43. Example 2 Part.

Figure 44. Example 2 Inspection Procedure.

Table 6. Example 2 Inspection Data.

INSPECTION DATA

FEATURE NUMBER	FEATURE LOCATION						FEATURE SIZE			POSITION TOLERANCE			
	X AXIS			Y AXIS						MATERIAL CONDITION	SPECIFIED	BONUS	TOTAL
	SPECIFIED	ACTUAL	DEVIATION	SPECIFIED	ACTUAL	DEVIATION	SPECIFIED	ACTUAL	DEVIATION				
1	.000	.000	.000	.750	.754	.004	.218	.219	.001	MMC	.010	.001	.011
2	.750	.743	.007	.000	+.007	.007	.218	.220	.002	MMC	.010	.002	.012
3	.000	-.006	.006	.750	.745	.005	.218	.219	.001	MMC	.010	.001	.011
4	.750	.757	.007	.000	+.007	.007	.218	.219	.001	MMC	.010	.001	.011
DATUM B	—	—	—	—	—	1.005	.995	.010	MMC	—	—	—	

Steps 1 and 2 of the GIA procedure require the generation of the DATA GRAPH and the incorporation of the inspection data (see Figure 45). Notice from the inspection data that hole 1 has been used by the inspector to establish a starting point.

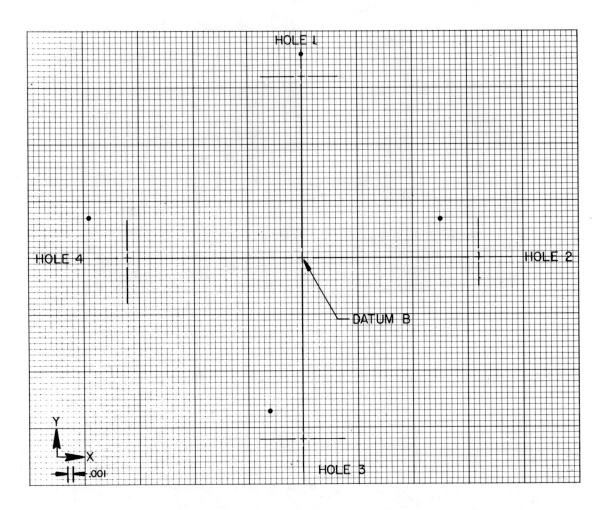

Figure 45. Example 2 DATA GRAPH.

Step 3 requires generation of the TOLERANCE ZONE OVERLAY GAGE for the holes (Figure 46) and a TOLERANCE ZONE OVERLAY GAGE (not shown) for any tolerance gained as the datum feature of size departs from MMC size.

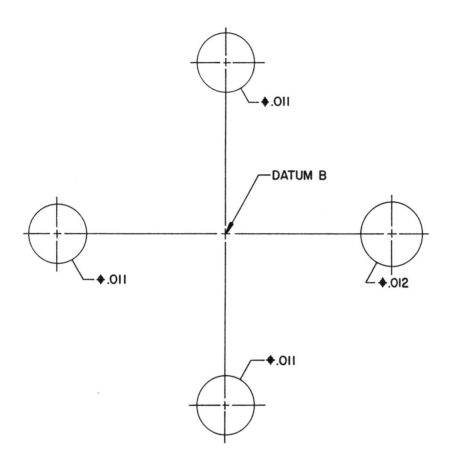

Figure 46. Example 2 TOLERANCE ZONE OVERLAY GAGE.

Step 4 shows the hole TOLERANCE ZONE OVERLAY GAGE rotated and translated within the added tolerance resulting from datum B departure from MMC in an attempt to accept the hole locations (Figure 47). This part would not be accepted if verified by a technique that was incapable of evaluating the additional tolerance allowed as datum B departs from MMC.

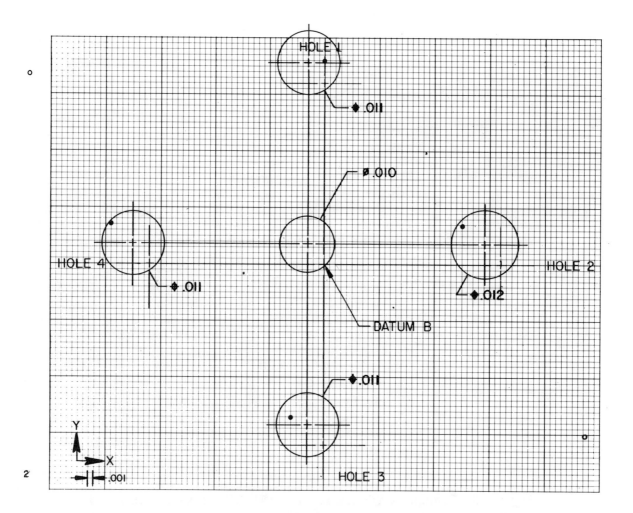

Figure 47. Example 2 DATA GRAPH/TOLERANCE ZONE OVERLAY GAGE

EXAMPLE 3

Example 3 is similar to Example 2 except the angular orientation of the holes are related to a tertiary datum feature as shown in Figure 48.

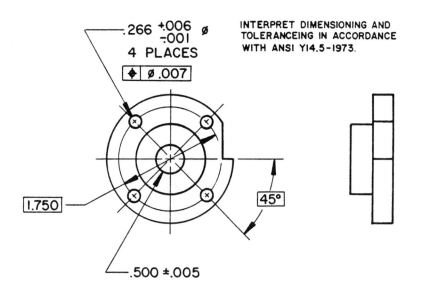

Figure 48. Example 3 Part With Implied Datums.

Notice the part drawing shown in Figure 48 does not include any datum reference. When the inspector fixtures the part for inspection, which flat surface will the part be set upon? Which diameter will be used to establish the axis of the part? What datum precedence will be used for the inspection set-up? How did manufacturing fixture the part? This drawing is not complete and its use will require assumptions to be made. Let's try a new drawing, as shown in Figure 49.

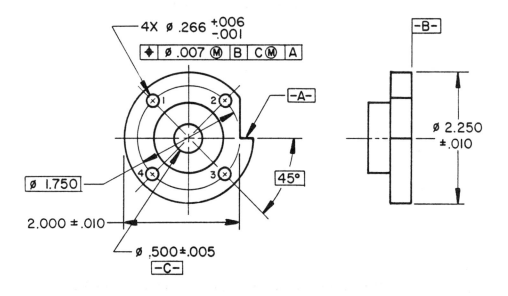

Figure 49. Example 3 Part With Specified Datums.

The part shown in Figure 49 will allow common interpretation by the manufacturer and inspector. The surface plate inspection process is shown in Figure 50, and the inspection data is shown in Table 7.

Figure 50. Example 3 Inspection Procedure.

Table 7. Example 3 Inspection Data.

INSPECTION DATA

FEATURE NUMBER	FEATURE LOCATION						FEATURE SIZE			POSITION TOLERANCE			
	X AXIS			Y AXIS						MATERIAL CONDITION	SPECIFIED	BONUS	TOTAL
	SPECIFIED	ACTUAL	DEVIATION	SPECIFIED	ACTUAL	DEVIATION	SPECIFIED	ACTUAL	DEVIATION				
1	.619	.625	.006	.619	.615	.004	.265	.266	.001	MMC	.007	.001	.008
2	.619	.617	.002	.619	.623	.004	.265	.268	.003	MMC	.007	.003	.010
3	.619	.624	.005	.619	.622	.003	.265	.268	.003	MMC	.007	.003	.010
4	.619	.618	.001	.619	.6265	.0075	.265	.267	.002	MMC	.007	.002	.009
DATUM C	—	—	—	—	—	—	.495	.501	.006	MMC	—	—	—
—	—	—	—	—	—	—	2.260	2.250	.010	—	—	—	—
—	—	—	—	—	—	—	2.010	2.000	.010	—	—	—	—

The DATA GRAPH with data incorporated is performed in Steps 1 and 2 for Example 3 (Figure 51). Notice the coordinate dimensions for the hole locations had to be mathematically calculated by the inspector from the drawing angular specifications and bolt circle.

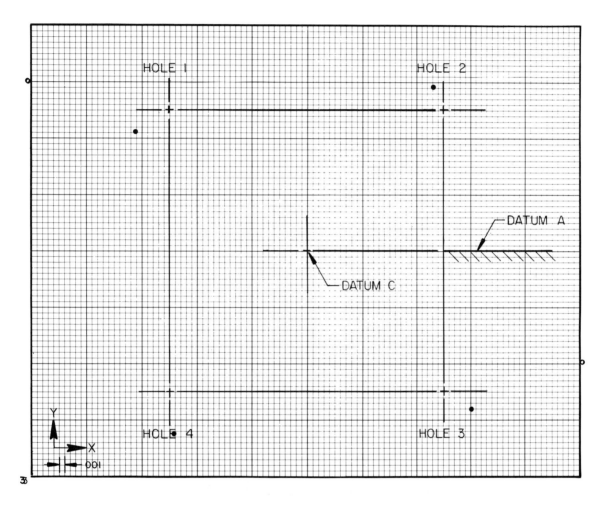

Figure 51. Example 3 DATA GRAPH.

Step 3 requires generation of the TOLERANCE ZONE OVERLAY GAGES for both the hole locations (Figure 52) and any additional tolerance allowed as datum C departs from MMC size (not shown).

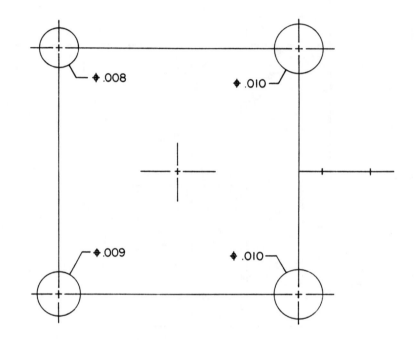

Figure 52. Example 3 TOLERANCE ZONE OVERLAY GAGE.

In Step 4 (Figure 53), the TOLERANCE ZONE OVERLAY GAGES are superimposed over the DATA GRAPH and translated to verify parts acceptance. Notice datum A of the TOLERANCE ZONE OVERLAY GAGE cannot violate the crossed-sectioned area of the DATA GRAPH.

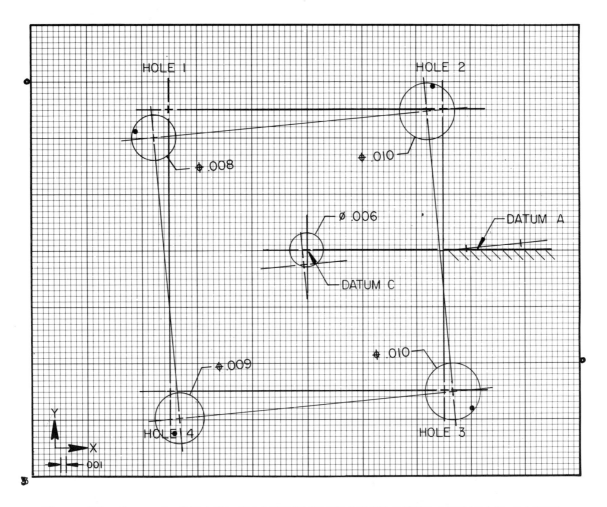

Figure 53. Example 3 DATA GRAPH/TOLERANCE ZONE OVERLAY GAGES.

EXAMPLE 4

Example 4 (shown in Figure 54) requires the graphical verification of non-cylindrical features (.395 wide slots).   Notice the feature control frame associated with the slot width does not contain a diameter symbol preceding the tolerance zone value.   This indicates that the center plane of the slot must be within two planes .005 apart at MMC.

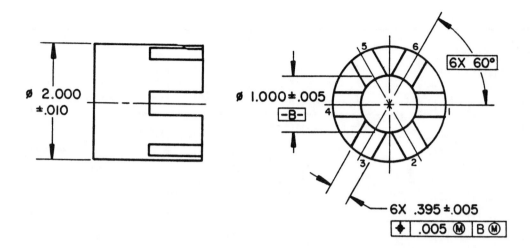

Figure 54. Example 4 Part.

The surface plate inspection process is shown in Figure 55. The feature location deviation is measured for each slot as the part is rotated to the specified basic angle. Deviation above the basic slot locations are specified plus, while deviation below the basic slot locations are minus. The inspection data is tabulated in Table 8.

Figure 55. Example 4 Inspection Process.

Table 8. Example 4 Inspection Data.

INSPECTION DATA

FEATURE NUMBER	FEATURE LOCATION						FEATURE SIZE			POSITION TOLERANCE			
	X AXIS			Y AXIS						MATERIAL CONDITION			
	SPECIFIED	ACTUAL	DEVIATION	SPECIFIED	ACTUAL	DEVIATION	SPECIFIED	ACTUAL	DEVIATION		SPECIFIED	BONUS	TOTAL
1	—	—	—	0°	0°	-.0057	.390	.395	.005	MMC	.005	.005	.010
2	—	—	—	60°	60°	-.002	.390	.397	.007	MMC	.005	.007	.012
3	—	—	—	120°	120°	+.004	.390	.395	.005	MMC	.005	.005	.010
4	—	—	—	180°	180°	+.004	.390	.394	.004	MMC	.005	.004	.009
5	—	—	—	240°	240°	+.002	.390	.395	.005	MMC	.005	.005	.010
6	—	—	—	300°	300°	-.004	.390	.396	.006	MMC	.005	.006	.011
DATUM B	—	—	—	—	—	—	.995	1.001	.006	MMC	—	—	—

The actual feature center lines (shown as dash lines in Figure 56) are plotted on the DATA GRAPH in a manner similar to the inspection procedure. The DATA GRAPH is rotated about datum axis B, and the feature center lines are plotted in relationship to the features basic locations. This procedure will ensure proper location of the inspection data without deviation sign (plus or minus) confusion.

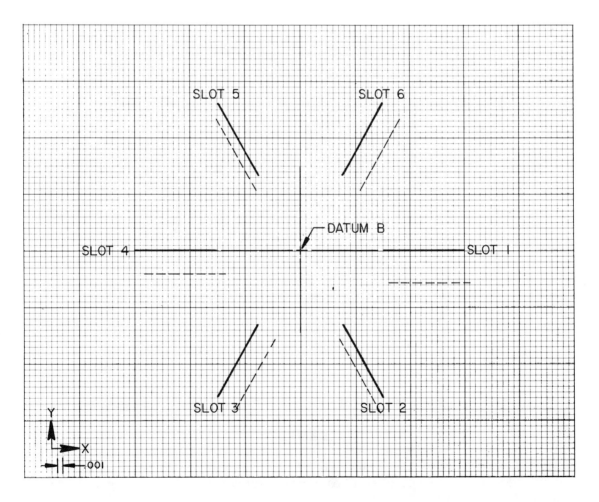

Figure 56. Example 4 DATA GRAPH.

In Figure 57, the TOLERANCE ZONE OVERLAY GAGES for both the feature tolerance zone and the datum departure additional tolerance allowance is superimposed on the DATA GRAPH. The center planes of the features are within the TOLERANCE ZONE OVERLAY GAGE, indicating an acceptable part.

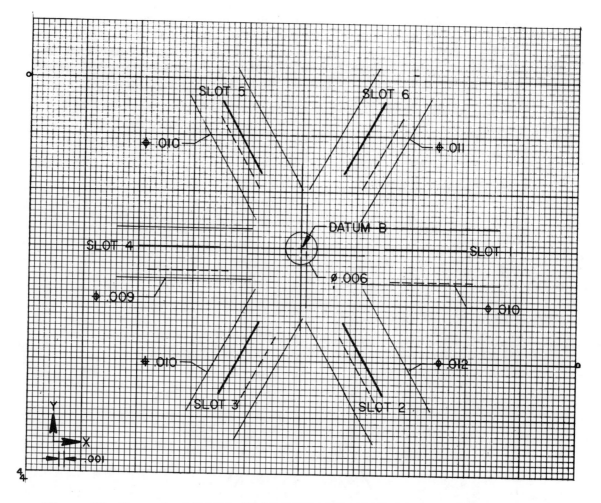

Figure 57. Example 4 DATA GRAPH/TOLERANCE ZONE OVERLAY GAGE.

EXAMPLE 5

In Figure 58, threaded holes are located in relationship to two features of size. Additional product tolerances can be realized as either datum C or datum B departs from MMC.

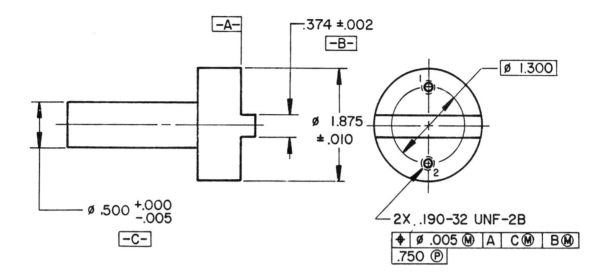

Figure 58. Example 5 Part.

The inspection process is shown in Figure 59.

Figure 59. Example 5 Inspection Process.

The inspection data for Example 5 is shown in Table 9. Notice no "BONUS" tolerance is considered for the threaded hole as it departs from MMC. Any additional tolerances will tend to be negated by the centering effect of the screw thread at assembly.

Table 9. Example 5 Inspection Data.

INSPECTION DATA

FEATURE NUMBER	FEATURE LOCATION						FEATURE SIZE			POSITION TOLERANCE		
	X AXIS			Y AXIS						MATERIAL CONDITION		
	SPECIFIED	ACTUAL	DEVIATION	SPECIFIED	ACTUAL	DEVIATION	SPECIFIED	ACTUAL	DEVIATION		SPECIFIED	BONUS	TOTAL
1	.0000	-.0035	.0035	.6500	.6495	.0005	.190-32	OK	NA	MMC	.005	.000	.005
2	.0000	+.0035	.0035	.6500	.6500	.0000	.190-32	OK	NA	MMC	.005	.000	.005
DATUM B	—	—	—	—	—	.376	.372	.004	MMC	—	—	—
DATUM C	—	—	—	—	—	.500	.500	.000	MMC	—	—	—
O.D	—	—	—	—	—	1.885	1.875	.010	—	—	—	—

Steps 1 and 2 show the DATA GRAPH (Figure 60) with feature locations plotted.

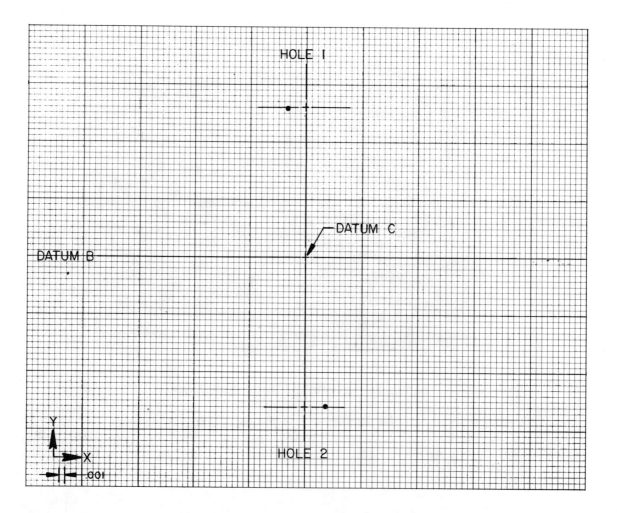

Figure 60. Example 5 DATA GRAPH.

Step 4 shows placement of the TOLERANCE ZONE OVERLAY GAGES on the DATA GRAPH (Figure 61).

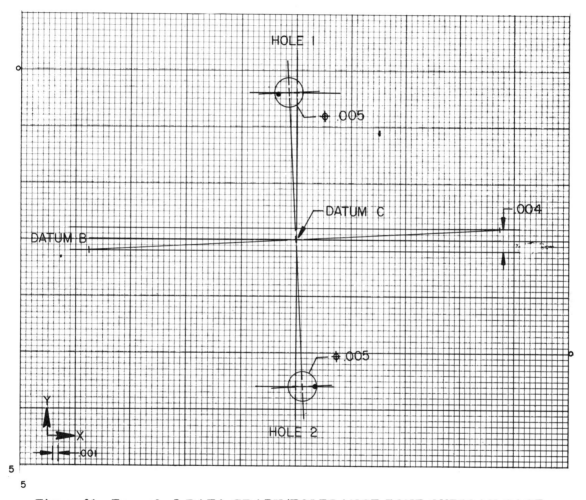

Figure 61. Example 5 DATA GRAPH/TOLERANCE ZONE OVERLAY GAGE.

EXAMPLE 6

Example 6 (Figure 62) shows a part with two patterns of holes located in relationship to the same datum features of size. As discussed in Principle 10, these two patterns of holes must be verified simultaneously since their feature control frames contain the same datums, in the same order of precedence and at the same material condition.

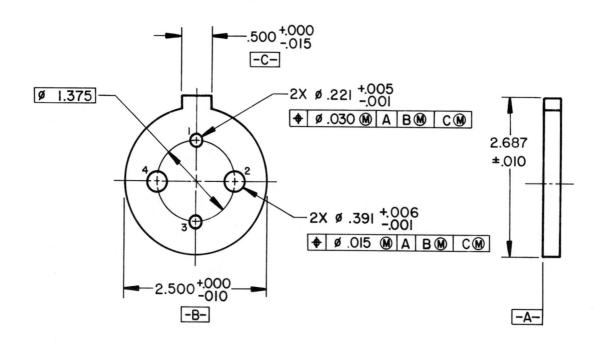

Figure 62. Example 6 Part

Inspection data is shown in Table 10 and plotted on the DATA GRAPH in Figure 63.

TABLE 10. Example 6 Inspection Data

INSPECTION DATA

FEATURE NUMBER	FEATURE LOCATION						FEATURE SIZE			POSITION TOLERANCE			
	X AXIS			Y AXIS			SPECIFIED	ACTUAL	DEVIATION	MATERIAL CONDITION	SPECIFIED	BONUS	TOTAL
	SPECIFIED	ACTUAL	DEVIATION	SPECIFIED	ACTUAL	DEVIATION							
1	.000	-.016	.016	.687	.686	.001	.220	.221	.001	MMC	.030	.001	.031
2	.687	.693	.006	.000	-.006	.006	.390	.391	.001	MMC	.015	.001	.016
3	.000	+.016	.016	.687	.687	.000	.220	.221	.001	MMC	.030	.001	.031
4	.687	.694	.007	.000	+.005	.005	.390	.392	.002	MMC	.015	.002	.017
DATUM B	—	—	—	—	—	2.500	2.500	.000	MMC	—	—	—
DATUM C	—	—	—	—	—	.500	.490	.010	MMC	—	—	—
—	—	—	—	—	—	2.697	2.695	.002	—	—	—	—

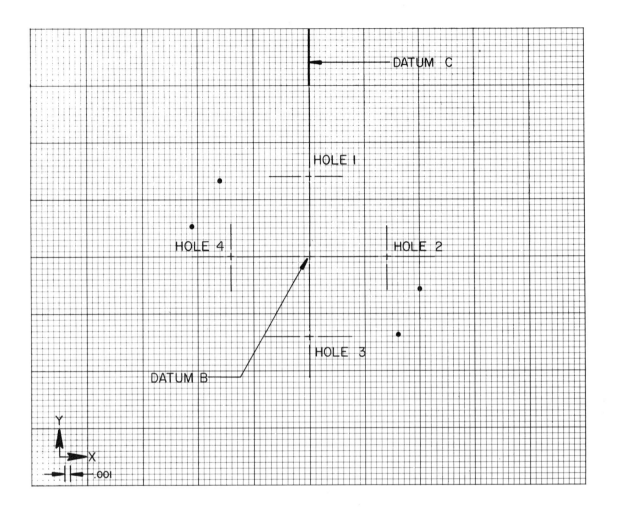

Figure 63. Example 6 DATA GRAPH

The TOLERANCE ZONE OVERLAY GAGES are generated for the feature patterns (Figure 64) and each of the datum features of size (not shown separately). The three TOLERANCE ZONE OVERLAY GAGES are superimposed over the DATA GRAPH and translated and/or rotated within the allowable zones in an attempt to capture the center of the holes within the feature TOLERANCE ZONE OVERLAY GAGE. All four hole centers must appear within the tolerance zone circles simultaneously for part acceptance. As shown in Figure 65, this part does not conform to the drawing requirements. The specified hole-pattern to hole-pattern interrelationship has not been controlled within the necessary limits to allow capture of the four hole centers.

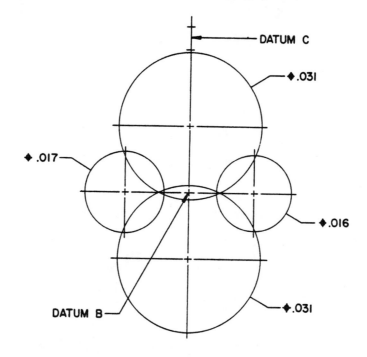

Figure 64. Example 6 TOLERANCE ZONE OVERLAY GAGE.

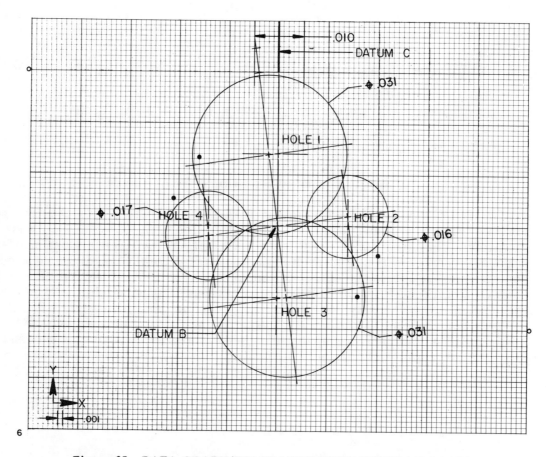

Figure 65. DATA GRAPH/TOLERANCE ZONE OVERLAY GAGES.

Perhaps the hole pattern interrelationship specified by the drawing is not required for function of the part. The drafter could have negated this hole pattern interrelationship by placing a "SEPARATE REQUIREMENTS" notation under each of the feature control frames as shown in Figure 66.

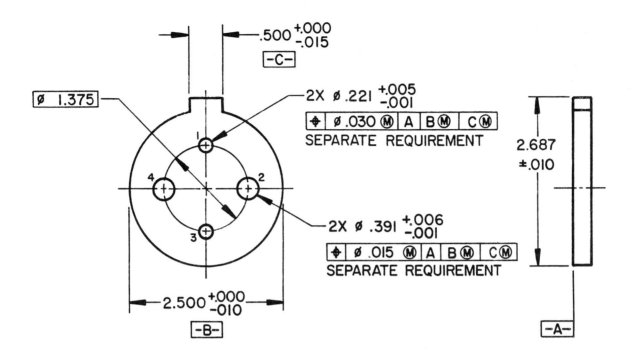

Figure 66. Example 6 Part With Separate Requirements.

With this new drawing requirement, the feature TOLERANCE ZONE OVERLAY GAGE will be regenerated with individual gages for each pattern as shown in Figure 67.

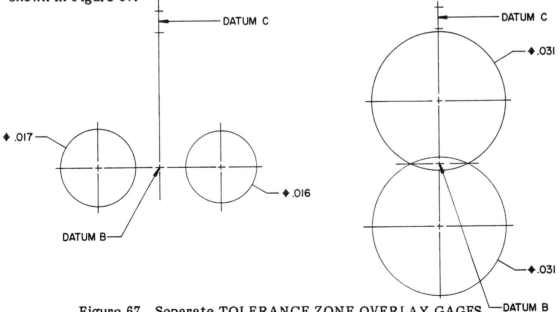

Figure 67. Separate TOLERANCE ZONE OVERLAY GAGES.

Figures 68 and 69 show the separate TOLERANCE ZONE OVERLAY GAGES superimposed over the DATA GRAPH that will allow acceptance of the manufactured part. Notice the individual TOLERANCE ZONE OVERLAY GAGES must be rotated to opposite extremes of the additional tolerance zone allowed by datum C departs from MMC. This was not allowed to occur when we had simultaneous requirements.

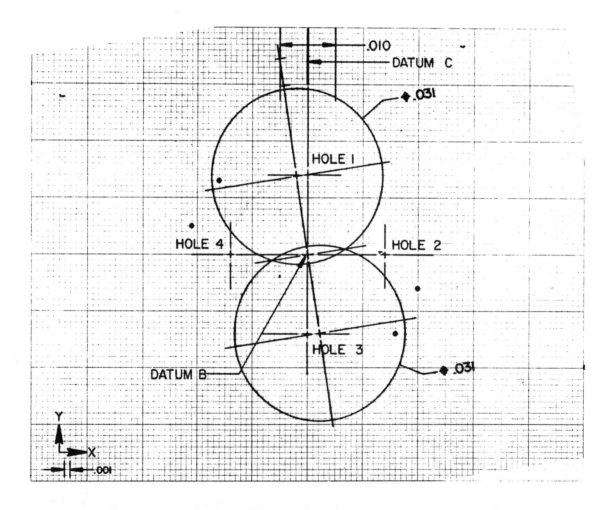

Figure 68. Example 6 DATA GRAPH/TOLERANCE ZONE OVERLAY

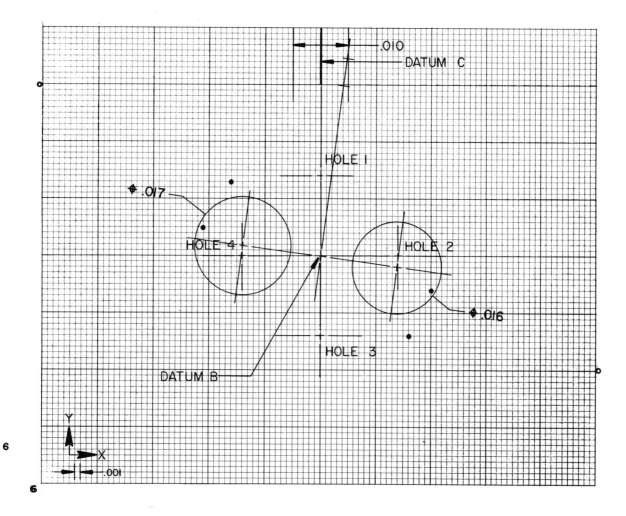

Figure 69. Example 6 DATA GRAPH/TOLERANCE ZONE OVERLAY.

GRAPHICAL INSPECTION ANALYSIS ADVANTAGES

There are many advantages to the Graphical Inspection Analysis concept for hardware verification. A partial list would include the following:

1. **FUNCTIONAL ACCEPTANCE.** Most hardware is designed to provide interchangeability of parts. As machined features depart from their MMC size, location tolerance of the features can be increased while maintaining functional interchangeability. GIA techniques evaluate these added functional tolerances in the acceptance process.

2. **REDUCE COST AND TIME.** The high cost and lead time required for the design and manufacture of a functional gage can be eliminated with the GIA technique. An immediate, inexpensive, functional inspection can be conducted by the parts inspector at his work station.

3. **NO GAGE TOLERANCE OR WEAR ALLOWANCE.** Functional gage design allows 10 percent of the part tolerance to be used for gage manufacturing tolerance. Often, an additional wear allowance (of up to 5 percent) will be designed into the functional gage. This could allow up to 15 percent of the parts tolerance to be assigned to the functional gage. GIA techniques do not require any portion of the product tolerance to be assigned to the verification process. GIA concept does not require a wear allowance since there is no wear.

4. **ALLOWS FUNCTIONAL VERIFICATION OF MMC, RFS, AND LMC.** Functional gages are primarily designed to verify hardware designed with the MMC concept applied. In most instances, it is not practical to design functional gages to verify hardware designed with RFS or LMC tolerance requirements. With GIA techniques all three material modifiers can be verified with equal ease.

5. **ALLOWS VERIFICATION OF ANY SHAPE TOLERANCE ZONE.** Virtually any shape tolerance zone, (round, square, rectangular, etc.) can easily be constructed with graphical verification methods. On the other hand, hardened steel funtional gaging elements of non-conventional configurations are difficult and expensive to produce.

6. **VISUAL RECORD OF TOOLING CAPABILITIES AND PROBLEMS.** Capabilities of machines can be evaluated through the GIA inspection data. For example, the DATA GRAPH for the five hole patterns shown in Figure 26 would indicate a machine problem in the X axis. There appears to be a .004 per inch offset in the one axis. Tooling wear and misalignment can also be detected during the production operation with periodic GIA verified of parts.

7. **VISUAL RECORD FOR MATERIAL REVIEW BOARD.** Material review board (MRB) meetings are postmortems that examine rejected parts. Decisions on the disposition of non-conforming hardware usually are influenced by engineering rank and perseverance rather than engineering information and evaluation techniques. GIA can provide a visual record of verification methods that can be evaluated with tolerance zone overlay to ensure part function.

8. **MINIMUM STORAGE REQUIRED.** Inventory and storage of functional gages can be a problem. Functional gages can also rust and corrode if not properly stored. GIA graphs and overlayes can be stored in drawers similar to drawings.

ACCURACY

The overall accuracy of Graphical Inspection Analysis is affected by such factors as the accuracy of the graphs and layouts, the accuracy of the inspection data, the completeness of the inspection process, and the ability of the drawing to provide common drawing interpretations.

An error equal to the difference in the coefficient of expansion of the materials used to generate the DATA GRAPH and TOLERANCE ZONE OVERLAY GAGES may be encountered if the same materials are not used through out. Papers will also expand with the increase of humidity and should be avoided. Mylar is a relatively stable material and, when used for both the DATA GRAPH and TOLERANCE ZONE OVERLAY GAGE, the expansion/contraction error will be nullified.

The GIA procedure outlined in this paper requires the generation of the TOLERANCE ZONE OVERLAY GAGE using the grid of the DATA GRAPH to determine the proper zone size. Error in the grid printing accuracy will be cancelled with this procedure.

Layout of the DATA GRAPH and TOLERANCE ZONE OVERLAY GAGES will allow approximately a .010 error in the positioning of lines. This error is minimized by the scaling factor selected for the DATA GRAPH. If a 10 x 10 to-the-inch grid is used, with each grid representing .001 inch, a scale factor of 100-to-1 will be provided. With the following formula we can calculate the actual error from line positioning.

$$\frac{\text{Line Position Error}}{\text{Scale Factor}} = \text{Actual Error}$$

The .010 assumed line position error will equal an actual error of .0001 inch. This error can be minimized by consistently working to one side of the line or by increasing the scale factor.

A certain amount of error is inherent in all inspection measurements. The error factor is dependent on the quality of the inspection equipment, facility, and inspection personnel. Most inspection operations will produce an error of less than 5 percent.

Accuracy of Graphical Inspection Analysis operations is dependent on the quality of the inspection report. Inspection reports must contain adequate information to ensure common understanding and provide complete variables data. Drawings that specify properly selected datums and provide datum precedence in the feature control frame will allow common drawing interpretations. When datum features and datum precedence must be selected by the inspector, due to incompleteness of the drawing, the datums and set-up procedure must be included in the inspection report.

Complete inspection data must also be provided to ensure GIA accuracy. For example, the inspection data included in this paper thus far has disregarded the perpendicularity condition of the inspected features. As discussed in Principle 7, position tolerance controls the location of feature through the part's thickness. Figure 70 shows the inspector inspecting both sides of the feature axis. This information would allow the establishment of both ends of the axis on the DATA GRAPH. The two points are plotted on the DATA GRAPH for each hole axis and joined by a line to indicate that they are the axis for a single hole (Figure 71). This procedure creates the effect of a three-dimensional gage at virtual condition (see Principle 9).

Figure 70. Inspecting Perpendicularity.

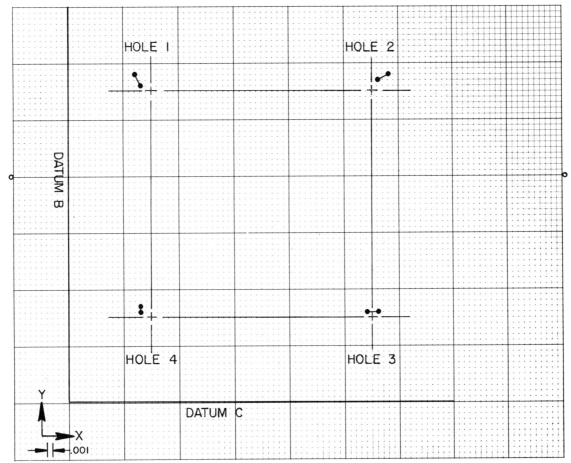

Figure 71. Three-Dimensional Hole Verification.

Figure 71 shows a DATA GRAPH, for the part shown in Figure 15, where the location of the features are plotted at their virtual size (see inspection data in Table 11). The complete axis of each hole must be within the TOLERANCE ZONE OVERLAY GAGE to be an acceptable part as shown in Figure 72. This method of verification is imperative if accurate, meaningful results are to be achieved with Graphical Inspection Analysis. Inclusion of virtual size data has been purposely avoided in the early part of this paper to allow the desired concepts to be presented in a less complicated manner. Table 12 includes virtual size inspection data for Example 7, that will allow the reader to properly evaluate the part at the actual effective feature sizes.

Table 11. Virtual Size Data For Example 1a.

INSPECTION DATA

FEATURE NUMBER	FEATURE LOCATION						FEATURE SIZE			POSITION TOLERANCE			
	X AXIS			Y AXIS						MATERIAL CONDITION	SPECIFIED	BONUS	TOTAL
	SPECIFIED	ACTUAL	DEVIATION	SPECIFIED	ACTUAL	DEVIATION	SPECIFIED	ACTUAL	DEVIATION				
1	.750	.748	.002	2.750	2.751	.001	.515	.516	.001	MMC	.015	.001	.016
		.747	.003		2.753	.003							
2	2.750	2.751	.001	2.750	2.752	.002	.515	.516	.001	MMC	.015	.001	.016
		2.753	.003		2.753	.003							
3	2.750	2.749	.001	.750	.751	.001	.515	.517	.002	MMC	.015	.002	.017
		2.751	.001		.751	.001							
4	.750	.748	.002	.750	.751	.001	.515	.516	.001	MMC	.015	.001	.016
		.748	.002		.752	.002							

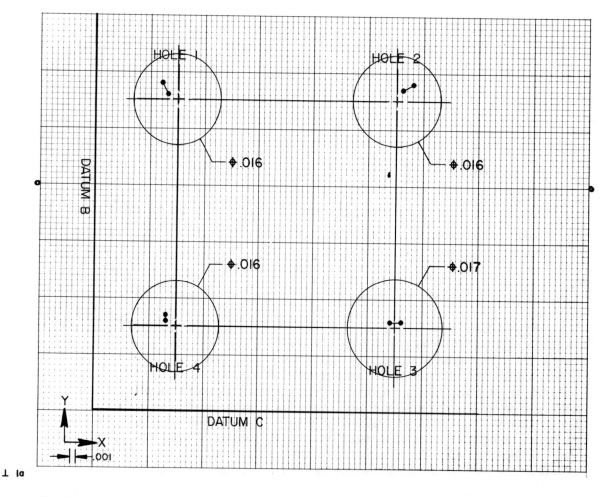

Figure 72. Example 1a With Virtual Size DATA GRAPH/TOLERANCE ZONE OVERLAY GAGE.

300

A datum feature of size which is interrelated by a separate tolerance of attitude or form and is referenced within the same feature control frame, must also be evaluated at its virtual size (often referred to as General Rule 5). The examples previously shown for parts which referenced datums of size did not depict complete part definition since the attitude requirments of the datums feature of size have been ommitted. When we add this additional control required for drawing completeness, we must consider the virtual size of the datum feature when evaluating the allowable additional tolerance that can be gained as the datum feature departs from MMC. Example 7 shows this additional geometric control. The parts inspector will determine the datum feature virtual size (by measurement) and subtract this value from the datum feature virtual condition (calculated from the drawing specifications) to determine the additional tolerance allowance.

EXAMPLE 7

Graphical Inspection Analysis can be performed on the part shown in Figure 73 with virtual size considered for both the holes and datum feature. Figure 74 shows the inspection data from Table 12 plotted on the DATA GRAPH with the TOLERANCE ZONE OVERLAY GAGES in place.

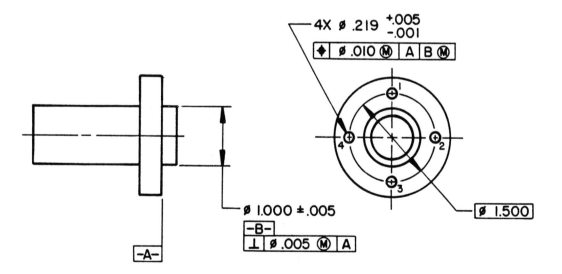

Figure 73. Example 7 Part.

Table 12. Example 7 Inspection Data.

INSPECTION DATA

FEATURE NUMBER	FEATURE LOCATION						FEATURE SIZE			POSITION TOLERANCE			
	X AXIS			Y AXIS									
	SPECIFIED	ACTUAL	DEVIATION	SPECIFIED	ACTUAL	DEVIATION	SPECIFIED	ACTUAL	DEVIATION	MATERIAL CONDITION	SPECIFIED	BONUS	TOTAL
1	.000	.000	.000	.750	.754	.004	.218	.219	.001	MMC	.010	.001	.011
		-.003	.003		.756	.006							
2	.750	.748	.002	.000	.004	.004	.218	.220	.002	MMC	.010	.002	.012
		.746	.004		.006	.006							
3	.000	-.001	.001	.750	.747	.003	.218	.219	.001	MMC	.010	.001	.011
		-.003	.003		.746	.004							
4	.750	.751	.001	.000	.003	.003	.218	.219	.001	MMC	.010	.001	.011
		.753	.003		.005	.005							
DATUM B	—	—	—	—	—	1.005	1.000	.005	MMC			
	PERPENDICULARITY = .007										.005	.005	.010

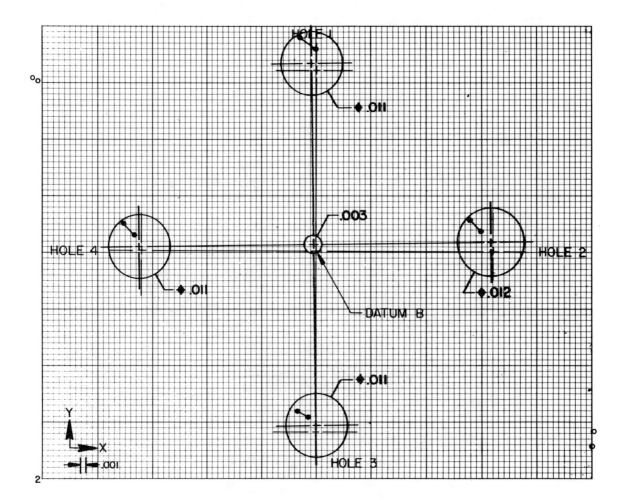

Figure 74. Example 7 DATAGRAPH/TOLERANCE ZONE OVERLAY GAGE.

In this example, datum B must be evaluated at the virtual condition (rather than at MMC as previously done for other examples) due to the attitude control requirement (perpendicularity) in relationship to datum A. As stated in rule 5, if datum B is controlled for attitude to a datum feature that is also specified in the feature control frame for the holes location control, then datum B applies at virtual condition rather than at the specified maximum material condition. As datum B departs from virtual condition, additional tolerance can be gained that will allow greater locational displacement of the hole pattern as a group. This added tolerance is equal to the difference between the virtual condition and the virtual size. In this example the virtual condition is calculated from Figure 73.

Virtual condition = MMC size plus attitude tolerance
= 1.005 + .005 = 1.010

Virtual size is determined from the inspection data in Table 12.

Virtual size = Actual size plus attitude measurement
= 1.000 + .007 = 1.007

The virtual size (1.007) is subtracted from the virtual condition (1.010) to determine the additional tolerance allowed for float of the hole-to-hole TOLERANCE ZONE OVERLAY GAGE.

COMPUTER INSPECTION ANALYSIS

Computer programs can be written to process the X and Y coordinate measurements and tolerance from inspection data, and provide the required rotation/translation of the data to determine functional acceptance of parts. The computer can print out data indicating acceptance or rejection, or produce the data in graphical form, resembling manually prepared Graphical Inspection Analysis. Recommended procedures can be provided for rejected parts to allow rework and ultimate part acceptance.

BIBLIOGRAPHY

Geo-Metrics II, Lowell W. Foster, Addision-Wesly Publishing Company, 1979.

Functional Inspection Techniques, Edward S. Roth, American Society of Tool and Manufacturing Engineers, 1967.

Dimensioning and Tolerancing for Engineering Drawings (USASI Y14.5-1966), The American Society of Mechanical Engineers, 1966.

Dimensioning and Tolerancing for Engineering Drawings (ANSI Y14.5-1973), The American Society of Mechanical Engineers, 1973.

Dimensioning and Tolerancing for Engineering Drawings (ANSI Y14.5M-1982), The American Society of Mechanical Engineers, 1982.

CHAPTER 3

COORDINATE MEASURING MACHINES

STATUS OF STANDARDS FOR
COORDINATE MEASURING MACHINES

Introduction

With the widespread adoption of three-axis coordinate measuring machines (CMMs), a need has been developing within the manufacturing community to establish some standard performance test methods for evaluating these machines. The increasing complexity of part geometries available from four and five axis CNC machining centers has been a powerful driving force for the adoption of the more advanced measurement capability represented by CMMs. At the same time the demand for increased product quality assurance has put pressure on the reliability of the measurement results produced by these measuring machines.

During this time the manufacturers of CMMs both here and abroad were attempting to provide quantitative characterizations of the performance that could be expected from their individual machines. Following widely accepted methods established for machine tools, the primary approach focused on testing of a single axis at a time, with independent tests for several different types of geometric errors. However, CMM users found that this approach did not adequately represent the measurement ability of the machines for a wide variety of part geometries and for the wide variation of operating methods and conditions commonly encountered. The predictable result was a range of specialized methods of testing, often based on an incomplete understanding of the complexity of machine errors in three dimensions, compounded by the relatively large influence of non-machine factors.

Responding to these problems in the mid 70s, the International Institution for Production Engineering Research (CIRP) established a Working Party on Three-Dimensional Uncertainty under the auspices of its Standing Technical Committee on Metrology-Interchangeability (STC-Me). The report[1] of this Working Party in 1978 provided a definition of "accuracy capability" suitable for use in three dimensions, but deferred the question of measurement methods to determine the values of this accuracy in a uniform manner suitable for comparing machine performances. Further, the scope of this effort was limited and did not address the problems that arise from variation in the method of operation or the environment.

In the Fall of 1978, two formal efforts were started with the common goal of establishing a simple unified performance test for evaluating CMMs. In September, under the leadership of Dr. P. Guarnero of Digital Electronic Automation (Torino, Italy), an international organization of CMM manufacturers was formed with separate sections for Europe, Japan, and the United States. The Chairman of the U.S.A. Section of this Coordinate Measuring Machine Manufacturers Association (CMMA) was Richard B. Hook of Brown and Sharpe. In October, the B89 Committee of the American Society of Mechanical Engineers also elected to form Subcommittee B89.1.12 to establish standard methods for performance evaluation of CMMs. Richard B. Hook was selected as the Chairman of this Subcommittee as well, establishing an informal link between the work of CMMA and B89.1.12.

Although technical collaboration took place during 1979 and 1980, differences in perspective between the two groups led to a divergence of approach late in 1980. The CMMA, working primarily through its European section, developed and released its unified accuracy specification in 1982. Its effort addressed three major aspects of CMM performance evaluation:

(a) definitions and measurements of "geometrical accuracies," such as positioning accuracy, straightness and squareness,

(b) master gage measurement methods to define "total measuring accuracy" in terms of "axial length measuring accuracy, volumetric length measuring accuracy, and length measuring repeatability," and

(c) explicit specifications for the form of specification of environmental conditions for the accuracy testing, including thermal parameters, vibrations and relative humidity.

The main thrusts of the CMMA specification were to unify the definition and measurement of accuracy in a manner that was easy to use and related directly to accepted length gages. In introducing the concept of "total measuring accuracy" CMMA explicitly recognizes the need to test coordinate measuring machines as complete systems. Further, they clearly establish the environmental conditions during testing as an important factor in the quantitative results of the tests.

Meanwhile, the ASME Subcommittee B89.1.12 was slowly evolving a somewhat different perspective in addressing the same problems. They also recognized very early the need for testing CMMs as complete systems and the great influence of environmental effects. However, since this group brought together both manufacturers and users, B89.1.12 focused on issues of machine performance involved in the buying and selling of CMMs. The two key issues involved finding suitable functional tests to allow comparisons among different machines and final acceptance testing, and establishing a fair basis for assigning relative reponsibilities for the effects of the environment. All of this needed to be done in the context of the wide spectrum of machines and operating conditions found in the American market for CMMs.

By early 1983 the subcommittee had developed a working document. Due to the complexity of the considerations involved in developing this proposal, and due to the wide impact of this new standard, a trial period of one year as an Interim Standard was recommended. At the end of the year necessary modifications will be made before final approval of the Standard by the appropriate ASME mechanisms. Even release as an Interim Standard requires review and approval through formal ASME procedures, and this is now proceeding. Barring unexpected objections, the trial period of the Interim Standard "Method of Performance Evaluation of Coordinate Measuring Machines" is expected to begin by early Fall 1983. Copies of the document will be available at that time through ASME. During the one year trial period the subcommittee is continuing to develop extensions of the draft standard into major new areas that were not addressed in the basic document.

Fortunately all of the work of B89.1.12 has been done with wide consultation and discussion, making the main thinking of the subcommittee fairly well known within the metrology community. The key features expected to be included are summarized in the following pages.

A third effort to develop a CMM standard is known to be underway in Germany under the auspices of DIN, but little detail has been available on the current state of their proposals. They are reported to be building on the accuracy concepts of CMMA and also to be incorporating a substantive section on environmental influences. At least drafts of this work are expected by Fall 1983.

The System Approach

Virtually everyone involved in efforts to specify or evaluate coordinate measuring machines now has recognized the need to consider the measurement results from these machines as the product of a complete measuring system. Specifically, the same machine used to measure the same part might produce substantially different results if used with a switching probe under direct computer control in a good metrology environment or if used with a hard (passive) probe in a manual mode in a typical shop floor environment. This poses a significant challenge for any attempt to establish a uniform test procedure. Further, the need to test CMMs as complete systems implies a need for the machine manufacturers to specify performance for complete systems, not just for the position sensing portion of the machine.

This problem of how to treat measuring machines as complete systems produced extensive discussions in the B89.1.12 Subcommittee meetings and led to several new developments. Measuring systems were characterized by the combination of "mode of operation" and probe type. The consensus of the Subcommittee was that these two features adequately specify the system for testing purposes. Modes include free floating manual, driven manual, and direct computer controlled. Probe types are passive, switching, proportional, and nulling. (The draft of the Interim Standard contains a complete glossary of terms that define these modes and probe types.) The CMM then is to be tested in the mode and with the probe that will most commonly be used for that machine in the user's application. Further, a standard specification form has been developed by the Subcommittee with a place to indicate the mode and probe to which the performance specifications are applicable. A particular machine, of course, may be used in more than one mode with more than one probe type, but the performance numbers will gennerally differ somewhat among the different combinations.

Another conclusion from the discussions of system evaluation was that the use of parametric testing (straightness, squareness, angular motions) should be deferred for further study since these measurements do not test the system performance and it is difficult to relate the results of these tests to expected performance. Instead, it was agreed that a mechanical artifact should be measured, providing some similarity between the machine testing and the actual measurement of workpieces. It also is important to sample extensively throughout the work volume since the testing seeks to detect systematic errors.

Environmental Effects

One of the most difficult issues to handle in specifying the testing of any system is the sensitivity of the system to external factors. For most dimensional measuring equipment the dominant environmental effects are due to temperature and vibration. The obvious approach is to carefully specify the test conditions to control the effects of the environment. Unfortunately, establishing a quantitative relationship between any particular environmental specification and the effect on any particular machine's performance is virtually impossible. Fortunately the Subcommittee benefitted from the wisdom of an earlier standards effort, "Temperature and Humidity Environment for Dimensional Measurement," (ANSI B89.6.2 - 1973) to point the way out of this problem. The details of this standard applied to thermal effects will be discussed below. The approach of B89.6.2 to the effects of environmental influence on measurement performance served as a model for the treatment of all environmental effects for the B89.1.12 Subcommittee. The essential concept of this earlier standard is that there are two options in treating environmental influences: provide an ideal environment, or perform functional tests of the measuring equipment to assess the actual effect of the environment on the performance of the equipment. The remaining questions are:

> What level of environmental influence is
> acceptable?

> If the environmental influence is not
> acceptable, whose responsibility is it?

Based on the experience of members of the Subcommittee, the environmental influences deemed most significant include temperature, vibrations, electrical supply, and utility air. In each case a simple functional test is described to look for the influence the effect of the environmental factor, and an acceptable level for this effect is described. The user of the equipment is clearly assigned the responsibility for making a good faith effort to provide an acceptable environment. If a machine has an unacceptable level in any of the functional environmental tests, then special tests of the environment itself are recommended to establish whither the user has indeed made a "good faith" effort to meet the CMM manufacturer's nominal needs for an acceptable environment.

In most cases thermal effects dominate the environmental influences affecting a coordinate measuring machine, leading the subcommittee to treat it most thoroughly. The sources of thermally induced errors include deviations of the surrounding air temperature from 68 F (20 C), temperature gradients, radiant energy (e.g. sunlight), utility air temperature, and self-heating in machines with drive motors. Thermal effects take three primary forms: differential expansion between the workpiece and the machine scale system, drift between a workpiece origin and the machine scale system origin, and distortion of the machine structure leading to significant changes in the calibration and adjustment of the machine.

Due to the increasing use of non-steel scales and workpieces, the subcommittee decided that mathematical compensation for the differential expansion effect will be a requirement for all testing. An important consideration in this decision was to keep the computation simple enough for any pocket calculator. The appropriate formulae are presented in detail, with the nominal coefficient of thermal expansion defined as the "effective" coefficient of the scale system as installed. This was based on the observation that different methods of mounting some scale systems cause them to expand with increasing temperature at a rate much different than that of the basic scale material. Following the principles of the B89.6.2 Standard the B89.1.12 Subcommittee also formally recognized the importance of the uncertainty of the nominal differential expansion in the correct evaluation of machine performance. The size of this uncertainty was set at 2 parts per million per degree Celsius (0.000002/C). The practical effect of this is to apply a realistic penalty to the accuracy of measurements performed at temperatures other than 68 F (20 C).

The functional test for thermal effects is a straightforward drift test. This test monitors the repeatability of the machine position over a time comparable to the longest positioning test of the machine. The range of the readings in the drift test is the "temperature variation error." This temperature variation error is added to the uncertainty of nominal differential expansion and taken as a percentage of the working tolerance of each axis to be tested. This combined percentage is called the "thermal error index," and it gives a quantitative measure of the degradation of machine accuracy due to thermal effects.

If the thermal error index exceeds fifty percent, then the thermal environment is deemed unacceptable. For the purposes of acceptance testing, if a user has complied with the supplier's guidelines for the thermal environment, the supplier is responsible for upgrading the machine performance to its specified level. On the other hand, if the thermal environment is unacceptable by its thermal error index and does not meet the supplier's guidelines, the Interim Standard calls for an automatic relaxation of the guaranteed working tolerance for the axis under test. The working tolerance is increased by exactly the amount needed to reduce the thermal error index to just fifty percent.

Due to the very limited knowledge of thermal distortion errors, the subcommittee felt unable to define an adequate test. Instead, as an initial approach to the problem, the burden was placed on the suppliers to evaluate their own machines and specify what change in mean temperature can be expected to degrade their machines' performances to a level where it is prudent to repeat performance testing.

Vibration effects from the environment were treated similarly, but the problems are somewhat simpler. The dominant effect of vibration is to degrade the repeatability of a machine, although there are some systematic effects on measured sizes. The functional testing uses the machine scales or a high resolution displacement indicator to sense relative motion between the machine table and the ram. If the indicated relative motion exceeds fifty percent of the working tolerance for repeatability, the vibration environment is deemed unacceptable. Only in that case is a complete vibration spectrum measured to determine compliance with the supplier's requirements. This supplementary testing is performed at the location of the interface between the user's points of support and the supplier's equipment.

In a similar fashion, the electrical power and utility air are measured in detail only in cases where the coordinate measuring machine's performance is clearly affected. In the absence of definite effects on the performance of the machine, the subcommittee judged the expense of such specialized testing of the power and air supplies to be unwarranted.

Performance Tests

The selection of functional performance tests was a source of lively discussion in the B89.1.12 meetings. The first required test that emerged was a well-defined repeatability test. This test is designed to determine the center of a precision ball repeatedly in the principal mode of operation with the primary probe type at a single location in the work zone. Beyond repeatability a wide variety of mechanical artifacts were considered, along with parametric testing by laser interferometry. Included in the list were ball bars, end standards (gage blocks), step bars, ball and hole plates, and specialized test parts of complex form. At the outset it was agreed that the testing must sample throughout the work zone, not just along a few lines parallel to the machine axes. Most of these artifacts were subjected to evaluation by one or more of the members of the subcommittee to determine their suitability as a general test procedure.

The informal criteria that emerged and influenced the selection of an artifact included:

(a) the usefulness of the test results in predicting how well a machine could be expected to measure actual workpieces,

(b) the fraction of the full three dimensional work zone that could be sampled in a reasonably short test period,

(c) the sensitivity of the test to the known systematic errors in coordinate measuring machines, and,

(d) the number of types of machines which could be tested with a single method.

While all of the test artifacts had certain advantages, most required either a long time to sample the full work zone or required further research to establish the best methods of data collection and analysis.

A compromise was reached by combining two simple tests that separately checked linear displacement accuracy and overall measurement performance. In order to check linear displacement accuracy, either a step bar or a laser interferometer may be used. These measurements are made along three orthogonal lines through the center of the work zone. This is followed by the use of a socketed ball bar based on the method described by Bryan <2>. This type of testing does not directly measure accuracy parallel to each axis independently. Instead it provides a thorough sampling of many combinations of X, Y, and Z errors that occur throughout the work zone of a machine.

In specifying the tests, the subcommittee recognized the great importance of hysteresis as an influence on such performance measurements. However, no suitable universal test for hysteresis could be found that was widely acceptable. Hysteresis generally leads to non-repeatability of the data. Although this often is a separable component of the repeatability that can be easily corrected, it was decided only to advise the potential users of the standard of this problem rather than to specify a separate test for hysteresis.

Using the socketed ball bar provides a means of sweeping out the surface of a (nearly) perfect hemisphere with a physical object (i.e., one of the balls.) The coordinate measuring machine is used to measure the location of the center of this ball at many locations on the hemisphere. The actual measurement data is compared to an ideal hemisphere simply by recording the range of the length of the ball bar computed from the data. Since repeatability has previously been tested, only systematic errors are of interest. The procedure calls for moving the socket defining the center of the hemisphere to several locations in the work zone and repeating the measurements. Three different lengths of the bar also are used. The performance is specified independently for the different lengths.

Open Issues

At the time that the work on the Interim Standard was concluded, the subcommittee identified several issues that were not addressed and warranted further study as possible extensions of the Standard. This list includes the performance of analysis algorithms, probe performance, evaluation of very large machines, evaluation of machines for two dimensional applications, evaluation of machines with rotary axes, parametric testing and the use of ball plates. All of these issues are to be taken up in the future by an "extensions" group in parallel with the continuing review of the comments on the Interim Standard after its official release.

<1> CIRP STC-Me Working Party on 3DU, "A Proposal for Defining and Specifying the Dimensional Uncertainty of Multi-axis Measuring Machines," Annals of the CIRP, 1978, Vol. 27(2), 623-630.

<2> Bryan, J.B., Precision Engineering, Vol. 4:2, July 1982.

The editors wish to thank Thomas Charlton of the National Bureau of Standards, Washington, D.C., for making a substantial contribution to our understanding of the information presented in this status report.

Presented at the CIRP General Assembly and published in the CIRP Annals, Volume XXVI, 1977, p. 403-408

Three Dimensional Metrology

R. Hocken, J. A. Simpson, B. Borchardt, J. Lazar, C. Reeve, and P. Stein, NBS, Washington, D. C.
Submitted by: R. Young, NBS, Washington, D. C./(1)

We present the results of research into the three dimensional measurement process using a classically designed measuring machine. This machine has been retrofitted with laser interferometers to provide a stable metric and is controlled by a minicomputer. Machine motions are programmable in a high level interactive language. Data links are provided to a larger computer for sophisticated data processing.

We have pursued the objective of creating, with the lasers, a machine independent coordinate system based at a point. Measurements made in this reference frame are transformed into the coordinate system of the measured object using the techniques of rigid body kinematics. The error terms inherent to the mechanical design (yaw, pitch, straightness, etc.) are measured over the machine volume, 48" x 24" x 10", on a cubic lattice of spacing two (2) inches. These error terms are stored as matrices and used to correct the data during a measurement. A measurement history on these error terms is being compiled. Real time instrumental drifts due to temperature and other external effects are removed using cross referenced measurement algorithms. Errors that cannot be assessed by calibration, such as axis non-orthogonality, are obtained by measuring the object in different angular positions within the measurement volume. This technique, which we call multiple redundancy, allows the assessment of all metric errors which do not commute with the finite rotation matrix.

Introduction

For several years we at NBS have been concerned with the problem of 3-dimensional measurement as part of our program to provide 3-dimensional calibrations of certifiable accuracy. Our work has been concentrated upon retrofitting a "classical" measuring machine. The machine is a Moore 5-Z(1). We have adapted it for computer control using an interdata 70 minicomputer, added Hewlett-Packard laser interferometers to provide the scales, and added a cooling system to reduce the temperature rise when the machine is in operation. Other features include a temperature controlled environment and a high-speed (1200 baud) data link to a large computer (Univac 1108).

The machine has a measurement volume of 48" x 24" x 10" (120 x 61 x 25 cm). Before retrofitting it had a worst case accuracy of several thousandths of an inch (0.003 cm) along a diagonal of the measurement volume. A schematic diagram of this machine is shown in Figure 1. The workpiece (gage) is mounted on the table which moves 48" (120 cm) to provide the X measurement. Y motion is obtained by movement of the large carriage across the bridge. The Z slide is mounted on the Y carriage. These two movements provide the 24" (61 cm) and 10" (25 cm) displacements respectively. Axes movement is controlled by stepping motors attached to lead screws. The three carriages are mounted upon traditional double-V ways, the X and Y slides with roller bearings and the Z slide with plain ways.

In an attempt to upgrade the performance of this machine, we have developed an extensive formal analysis which to the best of our knowledge is unique to our laboratory(2). The purpose of this paper is to describe this formalism in a broad sense and to illustrate it with specific examples.

Three separable techniques form the basis for our understanding of machine behavior. We begin by examining our machine geometry using the formal structure of rigid body kinematics(3). We then add to this the methods of temporal modeling and production sampling(4,5). Finally, we use the techniques of multiple redundancy(6,7). The first three sections of this paper will explain briefly each of these techniques with illustrative examples from our Moore 5-Z. The rest of the paper will be devoted to relating these techniques to provide a coherent picture of an actual measurement with this system.

Rigid Body Kinematics

The formalism of rigid body kinematics allows one to perform a complete analysis of any machine structure without resorting to complicated and error prone geometric arguments. The technique consists of choosing the minimum number of ideal reference frames (coordinate systems) necessary to characterize a machine and using matrix transformations to relate coordinates in these chosen frames. Ideally these coordinate systems must have their origins at a point but in practice small kinematically mounted "metrology bases" serve to decouple these systems from the distorted machine geometry.

In Figure 2 we show a schematic representation of our machine viewed from above. For simplicity the vertical axis (Z) is not shown even though our analysis is 3-dimensional. Three right handed coordinate systems are used. (Counterclockwise rotations about an axis are defined as positive.) The systems are:

a) The Space System: This is the primary measurement system. Its origin on a small metrology base mounted on the column of the measuring machine. Most measurements are made directly in this system with instruments of small random error. Angles measured in this system will be denoted by Greek letters without superscripts and vectors denoted by $\underline{X}$ without any superscript.

b) The Table System: The origin of this system is on a small metrology base, now rigidly attached to the table. The position of this origin in the space frame, when the probe is at position $\underline{X}_i$, is denoted $\underline{X}_{oi}$. Vectors in this frame are denoted $\underline{X}'$. Of necessity, 3 parameters will be measured in this system.

c) The Object System: This is the coordinate system of the workpiece, usually a gage. It has its origin at some

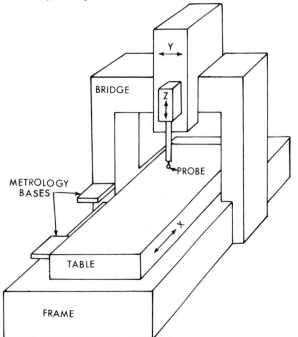

Figure 1 - Schematic diagram of Moore 5-Z

preselected point on the workpiece and all reported co-ordinates are ideally in this frame. Vectors in the object system are denoted by $\underline{X}''$. We assume that goal of any measurement is to obtain a set of N vectors $\underline{X}''_i$ which define positions in this gage and a set of N error vectors $\underline{\sigma}''_i$ which estimate the errors in these vectors.

A typical set of vectors $\underline{X}_i$, $\underline{X}_{oi}$, $\underline{X}'_i$ and $\underline{X}''_i$ are shown in Figure 2. The three coordinate systems are always chosen so that, except for infinitesimal rotations, they are aligned.

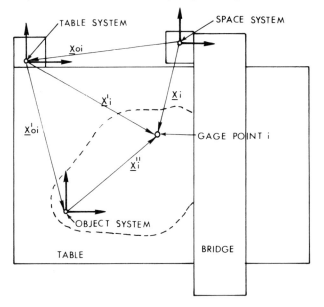

Figure 2 - Schematic diagram of measuring machine viewed from above. Both the table and object system may differ from the space system by infinitesimal rotations.

Any lack of alignment is described by the infinitesimal rotation matrix $\underline{\underline{R}}$ which is

$$\underline{\underline{R}} = \begin{pmatrix} 1 & \Phi & -\Psi \\ -\Phi & 1 & \Theta \\ \Psi & -\Theta & 1 \end{pmatrix}.$$

This antisymmetric matrix retains only first order terms in the infinitesimal angular rotations Φ, Ψ and Θ which are the familiar yaw, pitch, and roll respectively. This matrix will be used to describe the rotations of the three coordinate systems with respect to (WRT) each other. In a well designed measuring machine, the off diagonal terms will be of the order of micro-radians (10^{-6} radians) so that the small angle approximation is quite accurate. The infinitesimal rotation matrices commute, that is

$$\underline{\underline{R}}_i \, \underline{\underline{R}}_j = \underline{\underline{R}}_j \, \underline{\underline{R}}_i, \text{ for small } \Phi, \Psi, \text{ and } \Theta,$$

if we again neglect second order terms in the rotations. Furthermore products of such matrices are also antisymmetric which allows some computational simplification.

Now suppose at time i the machine is moved so that the probe is at point i on the work piece. Let $\underline{\underline{R}}_i$ denote the rotation of the table system WRT the space system at time i and $\underline{\underline{R}}'_i$ denote the rotation of the object system WRT the table system at the same time. The superscripts indicate in which system the measurements are made. Then the vector $\underline{X}''_i$ is related to the vector $\underline{X}'_i$ by the simple equation

$$\underline{X}''_i = \underline{\underline{R}}''_i \, (\underline{X}'_i - \underline{X}'_{oi}),$$

where $\underline{X}'_{oi}$ is the vector to the origin of the object system in the table system. Similarly the vector in the table system, $\underline{X}'_i$ is related to the vector in the space system $\underline{X}_i$ by

$$\underline{X}'_i = \underline{\underline{R}}_i \, (\underline{X}_i - \underline{X}_{oi}).$$

Combining these relationships enables us to simply express vectors in the object, $\underline{X}''_i$, in terms of vectors in the space frame, $\underline{X}_i$. That is:

$$\underline{X}''_i = \underline{\underline{R}}'_i \, (\underline{\underline{R}}_i \, (\underline{X}_i - \underline{X}_{oi}) - \underline{X}'_{oi}).$$

It may appear at first glance that the introduction of the table system is unnecessary. This would be true if the table did not move (as in some machines) or if it were an ideal rigid body, which it is not.

Equation 5 gives us a prescription for obtaining coordinates in the body from measurements made in the other systems as well as rigidly specifying what geometrical characteristics we need to measure. The compactness of the notation and strict adherence to sign conventions can prevent errors that often occur when one attempts to do the same problem using analytic geometry. We will defer further discussion of this equation to a later section.

Measurement Modeling

The second technique we use is related to the visualization of measurement as a production process with a product, numbers, whose quality may be controlled by the methods of statistical sampling. The goal here is to model those aspects of instrument behavior which influence the quality of these numbers and to check this model with quality control techniques.

In a large measuring machine many factors conspire to degrade the quality of measurement. Besides those geometrical factors displayed in Section 1, one of the most troublesome of these factors is a temporal dependence of the measurements. The causes of this time dependence are most likely changing temperature distributions in the machine and drifts in the associated machine electronics. A complete analytic model of such behavior is currently impossible, but an empirical model based on observational data is not. A very limited set of intuitively plausible assumptions makes such an empirical model possible.

Inherent in our discussion in Section 1, though not explicitly stated, was the assumption that the workpiece (object) was an ideal rigid body. Here we make that assumption explicit with the further corollary that, besides being a perfect rigid body, the object is characterized by a fixed set of coordinates of the gage points which, in the object system, are functions only of the workpiece temperature. That is, the numbers computed from equation 5 are invariant in time except for a random error component which is assumed small and independent of the magnitude of $\underline{X}''_i$. Call this error vector $\underline{\sigma}_r$. Then, if on a repeated measurement, the coordinates obtained for a point on the gage differ from the previous value by more than $\underline{\sigma}_r$, the change may have been caused by a drift in some instrument parameter. One can then assume that the drift was linear in time and correct intervening points accordingly. Again this procedure is best illustrated with an example.

Suppose the workpiece is a gage with N gage points which may be located with the same algorithm. The random error in this algorithm, $\underline{\sigma}_r$, may be accessed by repeated location of the same gage point during a time short compared with the time of a complete measurement of the gage (called a "run"). Our technique then is to choose one particular gage point, called the repeated point, and K other points called check points. A run begins with a measurement of the repeated point. This point is remeasured at random intervals throughout a run, and then measured as the last measurement. Between remeasurements of the repeated point the check points are also remeasured. The constraint is then imposed that measurements of the repeated point should agree within the estimated random error $\underline{\sigma}_r$ (i.e. the gage itself does not change).

If they do not, a linearized drift correction is computed and applied to all data. A simple statistical test is applied to confirm that this correction, computed from the repeated point measurements only, reduces the standard deviation of the check points to the predicted σ_r.

The model this algorithm defines is quite simple. Though the machine behavior is governed by many uncontrolled variables, over a time short compared with the machine's thermal time constant, the behavior of the machine is a linear function of time. Extensive experimentation has shown that this is indeed true, an observation that will be proven in a later section. The process of statistical comparison at the check points, analogous to production sampling for part quality, continually checks the adequacy of the model.

Multiple Redundancy

Our third technique is perhaps the most powerful and yet the most difficult to elucidate. It is based on the realization that properly chosen redundant measurements, made with an imperfect instrument can contain sufficient information to also measure and remove the known imperfections in the measurement system itself. In three dimensional measuring machines the type of redundancy we have found most suitable is that of remeasuring the workpiece at different angular orientations WRT the measuring machine. The data then allows us to remove errors due to scale imperfections and axis nonorthogonality. Again we illustrate this procedure with a simple example.

Suppose we had an object with three gage points that we measure in a nonorthogonal coordinate system as depicted in Figure 3. In such a system X coordinates are defined by distances to the Y axis (rather than projections on the X axis) and Y coordinates are defined similarly. Vectors in the non-orthogonal system,

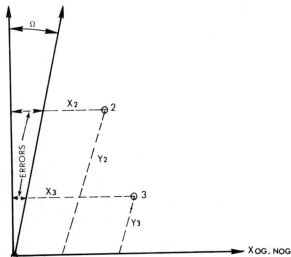

1 Figure 3 - A three point gage as measured in a non-orthogonal coordinant system.

$\underline{X}_{NOG}$, will be related to vectors in the orthogonal system, $\underline{X}_{OG}$, with the same origin by

$$\underline{X}_{NOG} = \underline{A}\ \underline{X}_{OG} \quad \text{where } \underline{A} = \begin{pmatrix} 1 & -\Omega \\ 0 & 1 \end{pmatrix},$$

and Ω is assumed small. The X axes of the two systems are chosen to be aligned for convenience. This choice is arbitrary. Now let us suppose we rotate the object by a finite angle Δ, which corresponds to a rotation of the coordinate systems, and remeasure the vector to the same gage point. The new vector, denoted with the superscript N, is related to the old vector in the orthogonal system by the simple finite rotation matrix $\underline{B}$. That is

$$\underline{X}_{OG}^N = \underline{B}\ \underline{X}_{OG} \quad \text{where } \underline{B} = \begin{pmatrix} \cos\Delta & \sin\Delta \\ -\sin\Delta & \cos\Delta \end{pmatrix}.$$

The coordinates of this new vector in the non-orthogonal frame are simply

$$\underline{X}_{NOG}^N = \underline{A}\ \underline{X}_{OG}^N = \underline{A}\ \underline{B}\ \underline{X}_{OG}.$$

It is easy to see that equation 8 does not correspond to a simple rotation of the original vector in the non-orthogonal frame. Such a vector, call it $\underline{X}_{NOG}^B$, would be given by

$$\underline{X}_{NOG}^B = \underline{B}\ \underline{X}_{NOG} = \underline{B}\ \underline{A}\ \underline{X}_{OG}.$$

The two vectors, $\underline{X}_{NOG}^B$ obtained by a simple rotation of the non-orthogonal coordinates and the measurements of the rotated object $\underline{X}_{NOG}^N$ will be equal if and only if

$$\underline{A}\ \underline{B} = \underline{B}\ \underline{A},$$

ie, if the matrix which describes the imperfection of the measurement system commutes with the finite rotation matrix. For this case it is easy to show that

$$\underline{A}\ \underline{B} \neq \underline{B}\ \underline{A} \quad \text{if } \Delta \neq n\pi \text{ radians, } n = 0,1,2\ldots.$$

Furthermore it can be shown that with just 3 points, (one of which is defined as the origin) measured twice as in this example, it is possible to compute both the coordinates of the points and the angles Δ and Ω. There are 8 measurements and only 6 unknowns so the system is just slightly over constrained. For a real measurement where there are many gage points and measurements at several angles the measurements are termed "multiply redundant" and the angles and coordinates determined by large least squares fits.

The generalization of the above analysis to three dimensions is messy but straightforward. The class of errors that may be computed and removed is exactly the same, that is, those metric errors that do not commute with the finite rotation matrix. Other common errors of this type, besides nonorthogonality, are scale errors. In fact, with sufficient measurements, only one good scale is required for a three dimensional measurement.

It should be noted that though the analysis in Section 1 is made in orthogonal coordinate systems it is equally valid in slightly non-orthogonal systems. This is because the infinitesimal rotation matrix $\underline{R}$ commutes with the non-orthogonality matrix $\underline{A}$ (to first order in the angles).

The technique of multiple redundancy shares with the technique of measurement modeling the assumption of gage stability. In a sense both are self calibrating algorithms in that, through a closed series of measurements, errors in instrument performance may be assessed and removed.

Now with this background and terminology, we are ready to discuss the process we use for three dimensional measurement on our Moore 5-Z.

Machine Calibration

In order to actually make a measurement using the techniques outlined in the preceding sections several more steps are required. The first of these is to actually write out, by components, the geometrical observation equation, equation 5. One then must identify those terms which must be measured, the system they must be measured in, and devise techniques for their measurement. An inspection of the equation reveals that at least 15 independent measurements are required, three (3) for each rotation matrix, and three (3) for each vector, even for this simple model which assumes that the X,Y, and Z positions of the probe can be accurately determined directly in the space frame. Some reduction in this number can be obtained by observing certain machine characteristics. For instance, we assume the table of our Moore 5-Z to be a quasi-rigid body in that it remains rigid WRT the X,Y plane. This assumption merely reflects the fact that distortions of the table are larger in the direction of gravity (Z) and gravity induced changes in X,Y distances only cosine errors. Three measurements, X'_{oi}, Y'_{oi} and ϕ'_i are eliminated by this means.

Table 1 - The geometric parameters needed in 3-dimensions

	Symbol	Description	Measurement Coordinate System	Method
1)	Y_0	Y straightness of X table	space	stored
2)	Z_0	Z straightness of X table	space	stored
3)	ϕ_0	Yaw of X table	space	stored
4)	Ψ	pitch of X table	space	stored
5)	Θ	roll of X table	space	stored
6)	XSP	X straightness of probe	space*	stored
7)	YSZ	Y straightness of Z axis	space*	stored
8)	ZSY	Z straightness of Y axis	space*	stored
9)	Ψ'	Pitch of object	table	on-line
10)	Θ'	roll of object	table	on-line
11)	Z'_0	Z position of object	table	on-line
12)	X_0	Basic table position	space	on-line
13)	Y_0	Basic Y-carriage position	space	on-line
14)	Z	Basic Z carriage position	space*	on-line

*These measurements are transferred to the space system via an intermediate coordinate system at height Z = 0. This system is not described in detail in the text.

Table 1 catalogs those terms remaining which must be measured for an accurate three dimensional measurement. There are 14 in all*, requiring great expense for instrumentation and making the problem of just fitting the instruments on the machine very difficult. Many of the terms are, however, small and if they are sufficiently smooth functions of measuring machine position and are also sufficiently reproducible, they can be measured at regular intervals and used as corrections to the basic X,Y,Z metric. This is the course we chose to pursue.

We concluded that the generalized error term, E, must be a function of the machine position (X,Y,Z) and of such variables as the machine temperature distribution T(X,Y,Z) and the table loading L(X,Y). The load term is the most difficult of these effects to handle. We are currently standardizing the table loading to avoid large changes in its structure. Changes in machine geometry due to self-loading are built into the analysis as functions of the coordinates.

We divide the measurement volume of the machine into a 2 inch (5 cm) cubic array with 1950 lattice sites. Each of the stored error terms is to be measured at all lattice sites. Our algorithm for doing this is, however, basically two dimensional. The instrument is set up for measuring the error term and probe moved through the measurement volume in the X,Y plane at constant Z. This path is depicted in Figure 4.

*The fourteen (14) terms become necessary due to the addition of a 4th coordinate system used to transfer the Z axis measurements into the space system. Since the methodology of this process is the same as described in Section 1 and its addition to the algorithm only an additional complexity, this system (in which only two parameters are measured) has been deleted from the discussion.

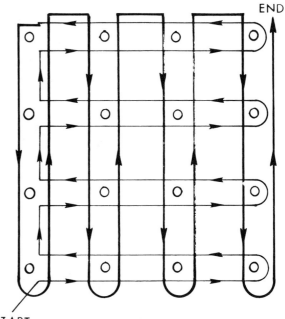

Figure 4 - The measurement path for generation of a machine error surface at constant Z.

After an initial standardized warm up exercise, the machine is moved until the probe is at X = 0, Y = 0 in the X,Y plane (lattice site 1,1,1 for Z = 0), and the interlaced path is swept out as indicated. The error term(s) being measured is, by the end of the run, measured 4 times at each lattice site. Temporal average over a 30 sec interval is performed for each of these measurements. The resulting set of data (1300 points) is divided into what we call cycles (one back and forth run in either the X or Y direction) and a linear temporal drift correction applied within each cycle. A least squares fit is used to "tie in" the whole surface and the standard deviation of the error surface thereby assessed. The stability and reproducibility of the drift correction is discovered by remeasuring the whole surface. Since each run takes some 16 hours, the test that two or more runs should agree is indeed stringent. The extent of this reproducibility is best illustrated by examining the data. In Figure 5, we show typical error surface. This surface is a measure of the roll of the table (at its origin) as a function of X and Y machine positions at Z = 0. The total roll is about 2 sec (as measured in the space frame) and Figure 5 is a result of combining 6 runs (7800 data points, each averaged for 30 sec). The resulting standard deviation of this surface is about ±0.01 sec. Similar surfaces have been prepared for the error terms listed as stored in Table 1. The standard deviations of the angular error surfaces are of the order of hundredths of a second of arc and of the straightness errors of the order of microinches (1 μinch = 25 nm). Our work so far has shown that the five table errors (ϕ, Ψ, Θ, Y_0, Z_0) are independent of the Z probe position, which considerably simplifies the analysis.

The excellent reproducibility of these error surfaces leads us to have considerable faith in both the geometrical integrity and temporal predictability of our machine. That is not to say that the drift is always the same, but rather by careful choice of a measurement algorithm such behavior may be reliably removed. We are now ready to describe a typical measurement of a gage at NBS.

The Complete Measurement Algorithm

The complete measurement algorithm for measurement of a three (or two) dimensional gage is schematized in Figure 6. The algorithm begins with positioning the gage on the table. No attempt

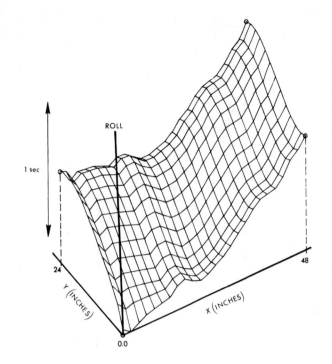

Figure 5 - The roll of the table system WRT the space system on the Moore 5-Z. This surface is the result of 5 runs combined and has a standard deviation of the order of 0.01 second of arc.

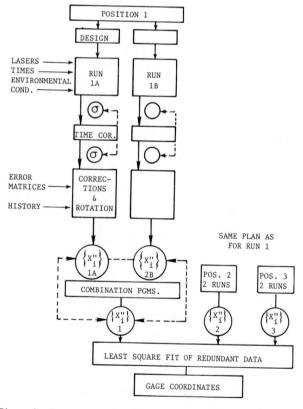

Figure 6 - Measurement flow chart for a multi-dimensional calibration. Each set of 2 runs at a single position has a flow diagram like that shown for position 1. Dashed lines indicate where statistical checks are performed. The last least squares fit of the redundant data is of course a stringent statistical test.

is made to align the preferred (by the user) gage coordinate system with the machine system. Furthermore once the original position (position 1 on the Figure is established) the possible positions for 2 and 3 are constrained since rotations of m π(m = 1,2) yield no new information, in fact π/4 and π/2 rotations are preferred. The operator then designs a typical run algorithm (as described in Section 1) which contains the repeated point and the check points. One such path, for the NBS 2-D ball plate, is shown Figure 7. An attempt is made to randomize the path so that distances measured will not be correlated with time. The machine encoders are set at prescribed values to allow retrieval of the error matrices previously described. The plate is then measured, recording the laser readings, encoder readings and time of each gage point location. Atmospheric conditions and average gage temperature are recorded at the beginning and end of each run.

The resulting data set is then transferred to a large computer (UNIVAC 1108) where the following computations are performed. First the time correction is computed and applied using the standard deviation of the check points to test the applicability of the model. Next the corrections for the machine geometric distortions are made with the error matrices stored in the computer. Finally the laser wavelength is corrected for atmospheric conditions and the gage size corrected for thermal expansion (all results are reported for 20 °C).

The resulting set of coordinates, called a single run, and their standard deviations are stored for future analysis. Such a run may take from 3 to 6 hours.

The process is now repeated and a second such set of coordinates and errors generated. These results are combined with those of the previous run. The standard deviation (rms) of the combined runs is compared to the single run deviation. If there is a

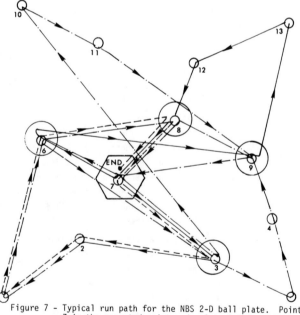

Figure 7 - Typical run path for the NBS 2-D ball plate. Point 7 is the repeated point and points 3,6,8 and 9 the check points.

statistically significant increase in the random error a new run is instituted. This process checks specifically the applicability of the time, atmospheric and thermal expansion correction, as well as giving a good estimate of the gage point location error.

The whole process is now repeated for 2 different orientations of the gage WRT the measurement machine. Rotations of 90° (π/2 radians) and 45° (π/4 radians) are particularly sought after since the former exchanges the axes and the latter mixes the axes.

The final results of these measurements are three complete sets of coordinates and errors in three different coordinate systems. Naturally these coordinates look nothing alike as they differ from each other by finite rotations, by scale errors and by machine axes non-orthogonality. These results are used as the input to a large nonlinear least squares fit which treats these numbers as data and the "coordinates", the angles describing the gage orientations and the angles characterizing the machine axes non-orthogonality as parameters. The constraint that the preferred gage coordinate axes should pass through specific gage points is used in this final fit.

Summary

The algorithm described in the preceding text is a recent synthesis of several different approaches. At the time of this writing it is fully implemented in two-dimensions and nearing completion in three. Only the passage of time will allow complete assessment of its usefulness. Particularly we must establish a history on the multi-dimensional error matrices and perhaps a wear model for the updating of these matrices. At this time the hardware for the pitch, roll and Z-straightness of the gage WRT the table is also in the development stage. (These measurements are only required in three dimensions).

We have however already learned a great deal. The short term reproducibility of the error matrices is quite encouraging and has been shown to be valid over periods of a month or so. We have also created several questions that as yet are unanswered. Specifically though we can obtain single run standard deviations on a large ball plate that are equal to our location error ($\sim$6 μinches, .15 μm). The total error (1) of the complete algorithm after multiply redundant measurements is of the order of twice that. This inclines us to believe that some systematics have been excluded from our analysis and we are pursuing their sources. Scale errors are not currently included in our multiple redundancy programs on the assumption that the properly aligned H.P. interferometers provide an accurate and stable metric. Tests of these lasers against an iodine stabilized helium-neon laser show a frequency stability of parts in 10^8 which would seem to support this assumption. We expect that with continued research and study these problems will be solved.

Bibliography

1) Brand names are mentioned only in the interest of clarity and in no way constitute an endorsement by the National Bureau of Standards.

2) Many other workers have studied three dimensional measuring engines. See, for example, D. L. Lane, UCRL-50679, 1969.

3) Many mechanics texts are applicable. See, for example, Classical Mechanics, Goldstein, Addison-Wesley, 7 ed., 1965, and P. A. McKeown, 23rd CIRP General Assembly.

4) J. M. Cameron and G. Hailes, NBSTN 844, 1974.

5) P. E. Pontius, NBSIR 74-545, 1974.

6) D. C. Brown, AFMTC-TR-58-8, ASTIA Doc. #134278, 1958.

7) C. P. Reeve, NBSIR 74-532, 1974.

Reprinted from *Precision Engineering*, Volume 1 Number 3,
July 1979, pages 125-128. Copyright Butterworth & Company Ltd.

Design of a New Error–Corrected Coordinate–Measuring Machine

J. B. Bryan and D. L. Carter
University of California, Lawrence Livermore Laboratory
Livermore, Calif., U.S.A. 94550

Lawrence Livermore Laboratory is always in the process of upgrading its machining and inspection capabilities. To meet the demands of new programs and to improve the relationship between these capabilities, we must continually investigate new measurement methods and tools. Many times our needs are such that we must design and build our own precision tools, as was the case for the diamond turning machine (DTM No. 3) described previously[1].

The concept presented here for a new error-corrected coordinate-measuring machine is intended to upgrade our inspection–accuracy capability by at least a factor of ten. Because the machine is designed to satisfy the fundamental principles of measurement in the strictest sense, its accuracy can be improved as technology develops, without either total overhaul or the fabrication of a new machine. In a sense, this is then the "perfect" machine! This design has been nicknamed "Ultimat."

Basic Principles — The Key

In the design of a state–of–the–art machine, we must adhere to the fundamental principles of measurement. This is mandatory! Two of the most fundamental principles, explained in detail in an accompanying article in this same issue[2], should be fully understood before applying them to a machine concept. These are the Abbé and the Bryan Principles.

The Abbé Principle

A displacement measuring system should be in line with the functional point whose displacement is to be measured. If this is not possible, either the slideways that transfer the displacement must be free of angular motion or angular motion data must be used to calculate the consequences of the offset.

The Bryan Principal

A straightness measuring system should be in line with the functional point at which straightness is to be measured. If this is not possible, either the slideways that transfer the measurement must be free of angular motion or angular motion data must be used to calculate the consequences of the offset.

The Typical Machine

The Y–Z measuring machine, as it is often called, is the most versatile and most used inspection tool in our Materials Fabrication Division. When using and strictly complying with the Y-14.5 drafting standard, we must rely on coordinate-measuring machines in addition to the more conventional receiver gauges.

The typical Y–Z measuring machine in general use is shown in Figure 1 to illustrate axis orientation and other basic features. The axisymmetric part is centered upon the rotary table or the "C" axis. The rotary table is mounted on the horizontal (Y) slide.

The electronic gauge stylus is typically a ball–tipped, single axis, linear, variable–displacement transducer (LVDT) carried and positioned by the verticle (Z) slide. The axis of the LVDT is typically mounted at a 45–deg angle with respect to the Y and Z axes. A correction is required for the cosine error introduced when the direction of travel of the LVDT is not normal to the part surface. The LVDT travel does not represent the workpiece error and must be corrected by a multiplication factor of $\frac{1}{\cos \theta}$, where θ is the angle the LVDT axis makes with the normal. For example, a 0.010–mm–high spot on either a cylindrical surface or a flat disk will move the transducer 0.014 mm. The θ angle is 45 deg in both cases, and hence the readings must be corrected by the factor 0.707. Data may be taken in the form of circumferential sweeps about the axis of the sample part or longitudinal sweeps through the part pole. Display may be digital or analog (polar or linear charts).

Application of the Principles to Ultimat

Displacement accuracy will be achieved in this new design by laser interferometers operating in helium–shielded pathways. The interferometers are located in strict accordance with the Abbé principle. This design uses the first option of the principle on the Z slide (gauge stylus), i.e., the extension of the laser interferometer axis passes through the center of the stylus ball at its null position (Figure 2). Note that the center of the stylus ball is the "functional point."

On the Y-axis slide, two laser interferometers are separated by 8 in. The difference in readings between these two lasers will be used as a servo input to drive a piezoelectric crystal that supports one end of the Y-axis table. Angular motion or pitch of the table hence is corrected, satisfying the second option of the Abbé principle.

Straightness accuracy is achieved by mounting straightedges parallel to each slide to measure and correct for slideway-straightness errors. For example, the Z slide is supposed to move the stylus along in the Z direction only. Errors in the straightness of travel, however, will cause unwanted movement in the Y direction. A linear variable-differential transformer (LVDT) gauge head that contacts the straightedge detects this movement and corrects it by zero shifting the Y slide. The first option of the Bryan principle is satisfied in this design in that the LVDT is in line with the center of the stylus ball (Figure 2).

The second option of the Bryan principle is satisfied on the Y slide. The straightedge LVDT cannot always be in line with the stylus ball. However, this is permissible, because there is no angular motion of the Y slide (correction achieved by the piezoelectric crystal). When nonstraightness of Y-slide travel is detected, the Z axis is zero shifted in the proper direction to correct the travel.

Application to a "Real" Machine

A real machine deflects and distorts owing to the effects on the structure of changing and moving loads. The base cannot be made infinitely stiff! The metrology or measurement systems then must be independent of the machine base, i.e., the external forces upon the metrology system must be constant. In this new design, the "metrology frame" (Figures 2 and 3) is not influenced by the machine base. The frame is supported on kinematic mounts "inside" the machine base. The plane of the supports is coincident with the bending neutral axis of the machine base, and its influence on the metrology frame is thereby minimized. The metrology frame houses the laser, laser pathways, and remote interferometers and also supports the two straightedges.

The machine base is built of granite (Figure 4). Granite is chosen because of its low coefficient of thermal expansion, 7.2×10^{-6} per °C [4×10^{-6} per °F], and its excellent secular stability. Also, it is relatively inexpensive. Note the simple shapes. The base is supported by three pneumatic isolators.

The metrology frame (Figure 3) can be built of steel. A low thermal coefficient of expansion is desired but is not absolutely necessary because of the temperature-controlled oil shower described later.

The stability of the laser depends on the stability of the media in the pathways. The ultimate for accuracy would be a vacuum. However, even with frictionless sliding seals on the varying-length tubes, the force upon the metrology frame that is due to atmospheric pressure on the cross sections of the tubes varies continuously with barometric change. It is difficult to compensate for this changing force. In this new design, our intent is to maintain helium in the pathways at a pressure slightly above atmospheric pressure. The amount above atmospheric pressure will be held constant by use of a special regulator referenced to atmospheric pressure. The effect of the helium pressure change on the laser wavelength is compensated for by using a Hewlett-Packard* atmospheric compensator referenced to helium instead of air. The consequence of using helium over a vacuum is a degradation of the displacement-system accuracy by 30%. In 0.6 m [24 in] of travel, the worse-case error will be 8nm [0.3 μin].

Both axis slides ride on and are guided by hydrostatic bearings of a hybrid design, a portion of each bearing being evacuated. The evacuated section acts like a vacuum chuck to hold the bearing against the way, in a sense, preloading the bearing. The balance of the bearing surface provides lift. The Y-axis slide-support bearings are externally compensated to enhance the stiffness. Infinite stiffness of the bearing is desired, i.e., no vertical deflection for varying loads. Recent experiments indicate this is possible.

---

*Reference to a company or product name does not imply approval or recommendation of the product by the University of California or the U.S. Department of Energy to the exclusion of others that may be suitable.

The slide drive system is shown in Figure 5. Its placement is shown in the cross sections of Figures 6 and 7. The slide drive system can be thought of as a rack and pinion drive without gear teeth. The capstan is connected directly to the drive motor. The motor tachometer is also part of the same housing. The steel traction bar is squeezed between the capstan and the idler roller. One end of the traction bar is fastened to the slide with a spherical bearing. A coil spring supports the weight of the bar at the opposite end. Both the capstan and the idler are supported on hydrostatic bearings.

This type of drive system has maximum stiffness, minimum sliding friction, maximum linearity of displacement, and no backlash. Other advantages are minimum cost, minimum heat generation, reliability, compactness, and minimum influence on slide straightness. The friction drive system is similar to that used on DTM No. 3.

The thermal environment of the measuring machine and of the part to be inspected is very important. This machine will be showered with approximately 2.5×10^{-3} m^3/s [40 gal/min] of oil that is temperature controlled to 0.0025°C. The shower will be carefully sculptured to maintain machine temperature and to minimize splash. Since all spindle motions are relatively slow compared to a machine tool, very few shields are required. We predict very few parts, requiring high–accuracy measurements, will be encountered that cannot accept the oil environment. If so, however, the oil shower or a portion of it can be switched off. The consequence is a loss of accuracy, which must be considered when making the measurement.

We did consider using two separate measuring systems to inspect simultaneously the inner and outer contours of a closed–end part. One can envision a second gauge head extending up through an annular rotary table into the part to be inspected. This approach requires only one setup and directly provides data on wall thickness of the sample part. We do not know how to do this and still preserve the "perfect machine" concept. The complexity of adhering to the fundamental principles of measurement for both functional points is enormous. An alternative way of measuring wall thickness with two gauge heads is to mount both on a single rigid caliper. This approach does limit the shape of the part (nearly cylindrical or open ended) that can be inspected with any one caliper or bracket. This feature can be added to this machine very easily.

Figure 8 depicts the assembled Ultimat machine. For clarity, the oil shower system and the oil collection pan are not shown. The entire machine is supported by air bearing supports to isolate the machine from ground vibration.

Continuous State-of-the-Art Improvements Possible

This fundamental design we feel can provide the perfect high-precision coordinate-measuring machine, because it is built on basic principles. The design allows for increasingly greater precision capability as new technology and resources develop. Possible future improvements include the following:

- Straightedges can be made straighter.

- Compensation information for straightedge nonstraightness may be stored and used for corrections.

- The laser may be retrofitted with an iodine-stabilized laser for greater stability.

- Further subdivision of fringes would provide increased resolution.

- The helium-filled pathways may be evacuated.

- Compensation information for out-of-roundness of stylus ball may be stored and used for data correction.

- Stability can be improved by better controlling the machine's environment, including the oil shower.

- The upgrading of control systems will always be an ongoing effort.

Acknowledgements

This work was performed under the auspices of the U.S. Department of Energy at Lawrence Livermore Laboratory under contract No. W-7405-Eng-48.

References

1. Bryan, J. B. Design and construction of an ultraprecision 84 inch diamond turning machine, Precision Eng., Vol. 1, No. 1 (1979) 13-17.

2. Bryan, J. B. Abbé principle revisited, Precision Eng., this issue (1979).

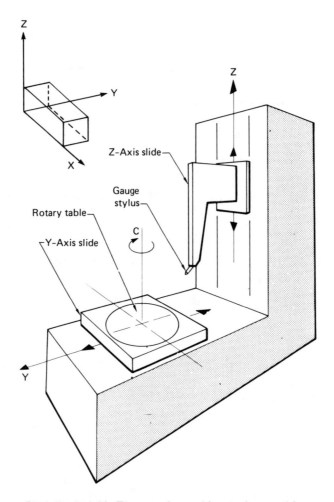

Fig 1 Typical 'Y—Z' measuring and inspection machine (by courtesy of Electronics Industries Association, Washington DC, USA)

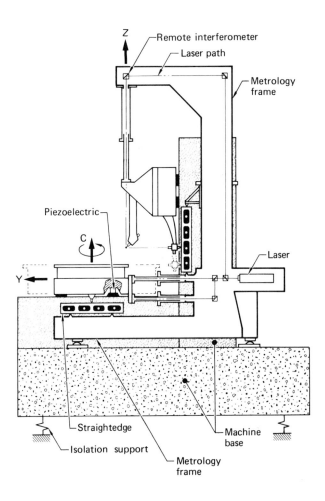

Fig 2 *Conceptual drawing of cross section of Ultimat*

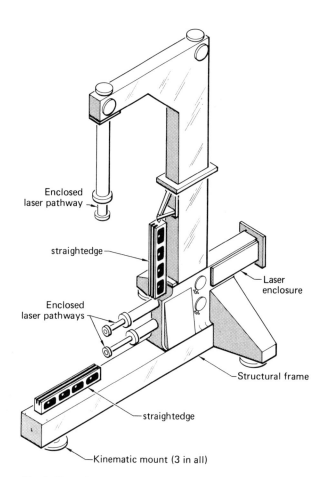

Fig 3 *Metrology system*

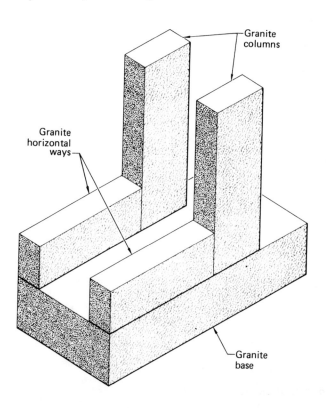

Fig 4 *Granite base and columns, note the simple shapes*

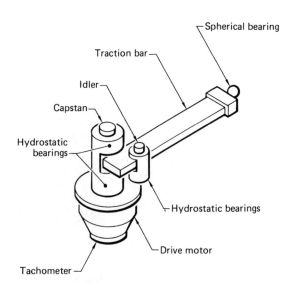

Fig 5 *Slide drive system*

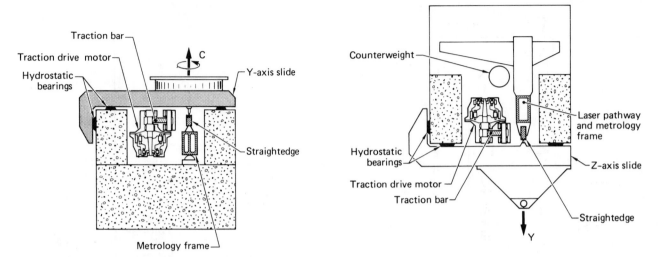

Fig 6 Cross-section of base

Fig 7 Cross-section of column

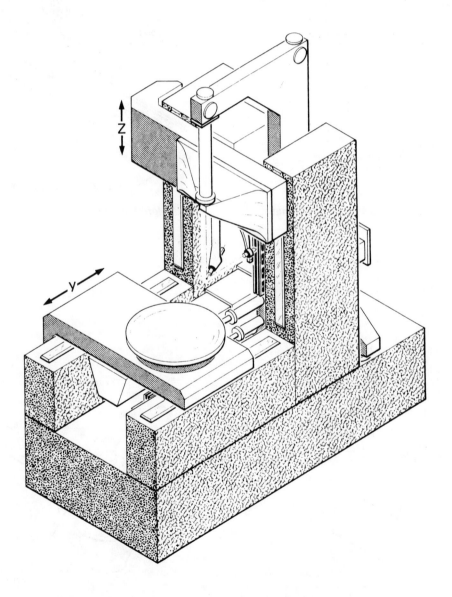

Fig 8 Conceptual drawing of assembled machine

Presented at the SME Precision Machining Seminar, June 1982

An Order of Magnitude Improvement in Thermal Stability With Use of Liquid Shower on a General Purpose Measuring Machine

By J. B. Bryan
D.L. Carter
R. W. Clouser
J. H. Hamilton
Lawrence Livermore Laboratory

Over the past 25 years, LLNL has been developing and contributing to the technology of precision machining and diamond-turning. Our two existing diamond turning lathes in the Materials Fabrication Division routinely produce contoured surfaces within plus or minus 10 microinches, and can produce surfaces within 2 microinches if necessary. The development of these cutting machines has made it necessary to upgrade our inspection tools in order to measure the precision parts coming from these machines.

About 6 years ago the Metrology Group began a project to upgrade our Moore Special Tool Co. #3 measuring machine. The goal was to convert it into a general purpose 3-axis coordinate measuring machine and improve the accuracy capability by a factor of 10 from that of the original machine. That goal has been achieved by the addition of laser interferometry and temperature control by means of an oil shower. The oil showered machine is shown in Figure 1. It has been in use for over a year in the Division's Inspection Laboratory. We believe it is the world's first oil-showered general purpose measuring machine.

Repeatability

The most important characteristic of the upgraded machine is repeatability. A measuring machine cannot be depended upon for accuracy unless the apparent non-repeatability is minimized. The term "apparent" non-repeatability is used because each influence is in itself repeatable if the conditions affecting it are repeated. When the nature of the influence is understood, we are able to bring it under control.

In the field of close tolerance work, experience has shown us that the thermal effect is the largest single source of apparent non-repeatability. Other sources of apparent non-repeatability are sliding oil films, Coulomb friction, and vibration. Figure 2 illustrates the thermally induced drift of the machine in an air conditioned room with the oil shower turned off. Basically this drift test result indicates the repeatability of a measurement taken at different times of the day. With addition of the measurement of room temperature also presented on the chart one can see the influence of temperature upon the repeatability. This illustrates the need for temperature control.

Temperature Control

Once the necessity for improved temperature control is recognized, the question arises as to the best means to achieve it. We feel the best means for achieving high quality temperature control is liquid shower. Our experience in oil showering two diamond turning lathes and three special purpose measuring machines was applied to this general purpose machine.

The primary advantages of liquid shower are its greater heat removal capability and the fact it is easily directed to the critical areas of the machine and workpiece surfaces. The effectiveness of a coolant is measured in terms of a film coefficient ($h = BTU/hr-ft^2 - °F$). The film coefficient in slowly moving room air is about 1.0. This can be increased by increasing the velocity of the air up to a limit at about $h = 10.0$ (above 800 ft./min.) which is approximately the lowest limit for water in natural convection. Liquids in forced convection have film coefficients several hundred times that of slowly moving room air.

Liquids have a higher heat capacity than gases. This characteristic allows removal of heat with corresponding lower temperature differences. The heat transfer characteristics of liquids make accurate temperature control easier and from any initial condition, all machine components, including the workpiece, will reach the desired temperature 10 times faster.

Alternatives to liquid shower for temperature control are conventional room air conditioning, open and enclosed air shower, liquid submersion, and complete isolation from heat sources in an air conditioned room.

We will define conventional room air conditioning as having air velocities between 2 and 20 feet per minute. Even the best room of this type is hopelessly inadequate in controlling the temperature of a machine with varying heat loads around the machine. This kind of room control is expensive to install and is a permanent part of the room.

An air shower of 800 feet per minute is about five times better than conventional room air conditioning. The most common delivery scheme for a high velocity air shower is a ceiling-to-floor circulation. The air enters the workspace through a perforated ceiling from a large plenum above. The exhaust is drawn off near the floor. The workspace may be the entire room, a portion of a room, or a single enclosure.

An open air shower, has the serious disadvantages of operator chill and noise. Air at 68°F with a velocity of 800 ft./min. is roughly equivalent to still air at 55°F as far as human comfort is concerned. This is an unacceptable work environment.

An enclosed air shower system avoids the wind chill problem. A built-in air circulation and temperature control system provides the necessary temperature control and high velocity. Doors allow the operator access to the workspace during part loading, set-up, and maintenance. Air shower noise is also minimized by the enclosed flow system.

Complete submersion of the machine in a tank of temperature controlled liquid has been suggested. This approach is inferior to either liquid or air shower. The reason is that in a liquid at rest, heat transfer occurs by conduction. Natural convection produces bulk movement to keep lower temperature fluid in contact with the surfaces to be cooled. Natural convection is less effective in liquids than it is in still air because of the higher viscosity.

Complete isolation and remote operation of a machine in a special air conditioned room is a technique that is used for diffraction grating ruling machines. It takes weeks to rule a grating on one of these machines. Several days of soak out time after set-up is acceptable. This is not practical for a general purpose measuring machine.

The Measuring Machine

Before proceeding with a discussion of the details of the liquid shower system, it is desirable to have a general description of the measuring machine to which it is applied. Figure 1 shows the upgraded machine.

This Moore Special Tool Co. #3 Measuring Machine was built in 1960 and has been a part of the inspection laboratory since then. It is a plane way machine using the double-vee concept with X & Y slide travels of 18 and 11 inches respectively and a Z slide travel of 16 inches. The Z slide has been converted to hydrostatic bearings.

Displacement of the X and Y slides in the original design was measured by reading the graduated drum attached to the end of the precision leadscrew. The Z axis had no displacement measuring device. Laser interferometry has been added to the machine for displacement measurement of all three axes.

A rotary table was mounted on the X-slide. The machine can therefore be used not only as a 3-axis coordinate measuring machine but also a 'Y-Z' or 'X-Z' measuring machine. In this mode, axisymmetric parts can be rotated on the C-axis to produce circumferential sweeps. Most of our very high precision parts to be measured are axisymmetric.

The electronic gauge stylus is typically a ball or spherical radius tipped, single axis, linear, variable-displacement transducer (LVDT) carried and positioned by the vertical Z-slide. The axis of the LVDT is typically mounted at a 45-degree angle with respect to the Y and Z axes. (It may be used at any other angle however as long as the proper corrections are made). Figure 3 shows the LVDT mounted in the horizontal position. A correction is required for the cosine error introduced when the direction of travel of the LVDT is not normal to the part surface. The LVDT travel does not represent the workpiece error and must be corrected by a multiplication factor of $\cos\theta$, where θ is the angle the LVDT axis makes with the normal to the part surface at the point of contact. Data may be taken in the form of circumferential sweeps about the axis of the part or longitudinal sweeps through the part pole. Display of data is analog in the form of either polar or linear charts as shown in Figure 4.

The Liquid Shower System

Figure 5 shows a schematic diagram of the flow circuit for the liquid shower. Starting at the machine sump, the oil flows through a filter located on the suction side of the pump, through the pump to a heat exchanger, onto the machine and back to the sump. The sump is a part of the collection base. The sides of the collection base extend upward near waist level to catch any splash and still provide easy access to the work area. Figure 1 shows the operator standing on a short stool. The final plan calls for an elevated floor 7" above the supporting floor. This is required since the entire machine is sitting up on the thick support base.

Visible in the lower left are one of the three cylindrical pneumatic isolators, the machine sump, and the suction line leading to the pump which is located in a closet adjacent to the machine. To avoid transmission of vibration from the pump and motors to the isolated machine, the suction pipe is cantilevered from the floor and does not physically contact the sump or machine. The same concept is applied to the plumbing which distributes oil over the machine.

The oil used for the liquid shower is a light weight mineral-base oil (225 Saybolt seconds at 68°). The viscosity is high enough to limit splash and airborne droplets while flowing readily over the surface of the machine. Various pieces of sheet metal and wire screen are used to direct and smooth the flow and minimize splatter. The flow is visible on the column in Figure 6.

Freon solvent is used to remove oil from the part or from the work surface during set-up. A solvent tank is nearby for washing large items. Evaporative cooling has no effect so long as the machine surfaces are separated from the free liquid surface by a layer of flowing oil. Samples of air around the machine show levels of oil in the air to be in the range of .02 to .18 mg/m^3 which is 1/25 of the allowable limit of 5.0 mg/m^3. The particular oil used is free of odor and fumes and has not been a problem.

The oil filter consists of a large (15 gal.) horizontal cylindrical container with a quick-opening hinged front cover. A felt liner material is supported within this cavity, spaced away from the walls by a perforated support, with oil entering the center and flowing outward. The life of the filter is about two years. A new element costs about $20.00. It is worthwhile noting that the oil shower keeps the machine quite clean.

The oil is pumped by a commercial 2 HP integral pump-motor unit. The pump is a constant displacement, rotary screw-type pump, very quiet, and capable of delivering 40 gallons per minute at 20 psi supply pressure. This pressure is adequate to overcome line losses and provide good distribution pressure at the shower nozzles. The suction head of 6 in. Hg through the suction line and the filter is well within the capability of the pump and does not cause cavitation of the oil.

The oil shower pump, filter, and heat exchanger are housed in a 3' x 5' closet only 8 feet from the machine. The closet allows these items to be kept out of sight as well as isolating the room from any pump noise.

Temperature Control System Design

The 40 gallons per minute of circulating oil is heated by the 2 HP pump. Therefore, it is possible to control the oil temperature by cooling only. The shell and tube heat exchanger is used as shown in Figure 5. The oil temperature at the exchanger outlet is sensed by a bare-bead thermister. An on-off controller actuates a solenoid valve to admit chilled water when the oil temperature exceeds 68°F and to bypass it when the temperature is below 68°F. It was found that our chilled water supply temperature varies as much as 5°F. This causes a variation in oil temperature of 0.05°F. A second shell and tube heat exchanger and an on-off control is used to control the temperature of the chilled water. In this case, heat is added intermitently to the chilled water. Plant cooling tower water supplies the heat. The result is that the average temperature of the circulating oil is held constant to $\pm$ 0.01°F.

The temperature oscillations of the oil have a amplitude of 0.08°F and a frequency of about 10 cycles per minute (6 seconds per cycle). At this frequency, the thermal inertia of the machine limits machine distortion. Thermal inertia is defined as the inability of an object to respond instantly to a change in liquid temperature because of its heat capacity.

Measuring Machine Performance

For a drift test to be meaningful, it should include a typical test part in the structural loop, and people in the vicinity of the machine during the test. It should also be setup in a direction most sensitive to temperature changes. For the drift test on the Moore Measuring Machine, we chose an 8 inch long aluminum cylinder. Aluminum, with a thermal coefficient twice that of steel provides a compromise between steel and plastics. Plastics have coefficients five times that of steel. We chose also to have a person working around the machine during the test to provide the additional amount of heat load to the machine that an operator would provide in normal use. The Z-axis of this machine is most sensitive to temperature changes due to geometry. The drift test was therefore set up in the Z-direction.

The drift test was conducted dry (without oil shower) and with oil shower. Figure 7a, taken without oil, shows a total drift of more than 70 μinches with a room air temperature variation of about 1°F. Note the correlation between the machine drift and the room air temperature. In contrast, Figure 7b with oil shower shows an order of magnitude decrease in drift or less than 5 μinches total for the eight hour period.

Soak out time for the oil showered machine is an order of magnitude faster than for a dry machine. Figure 8a shows a 60 microinch setup induced drift for the dry machine. Four hours of soak out time were required to reach equilibrium. Figure 8b shows the same setup with the oil shower on. The soak out time has been reduced to half an hour.

Other Features

The accuracy of measurement was improved by adding laser interferometry to all 3 axes of the machine. A Hewlett-Packard 5501 machine tool laser was mounted in a casting mounted on the rear of the column. The beam is split 3 ways to feed the 3 axes. Figure 9 shows the laser paths and the optical components. The plane mirror measuring system was used to obtain a 0.5 microinch resolution.

The paths of the measuring beam were chosen to minimize Abbe errors. The paths of the X and Y axes beams pass through the center of the working volume, hence minimizing the error due to varying Abbe offset. The vertical Z axis beam passes through the stylus tip, obeying the Abbe principle.

The laser paths are enclosed in either a casting, a pipe or a covered enclosure depending upon the location on the machine. The pathways contain room air and are slowly purged with filtered air to discourage the intrusion of oil. The intensity of the beam is severely lowered when oil is deposited on the optics. Enough oil will cause the system to fail. When the proper techniques of sealing and oil shower management are applied, reliability is very high.

The successful application of laser interferometry to this machine, in terms of accuracy, is due to total thermal control. Adding a laser system would otherwise be counterproductive. Most of the laser mirror and interferometer brackets are external to the machine's massive castings which protect the leadscrews. Exposed brackets quickly distort when exposed to temperature changes. These distortions are likely to cause measurement errors exceeding those of the original leadscrew. The temperature controlled oil shower eliminated thermal distortions.

The LVDT mentioned earlier is an air bearing type. The air bearing eliminates coulomb friction, a requirement when the direction of LVDT travel is not normal to the part surface. This LVDT also uses air pressure to maintain stylus force. The typical stylus force is 0.2 grams. Even when the stylus tip radius is as large as .1 inch at a force of 0.2 grams, the oil does not keep the ball from following the part contour. We have found that the presence of flowing oil on the part is not a problem in contact gaging.

Positioning and holding a part on the rotary table can be difficult in the presence of oil, especially when the parts are small or of relatively low density. The oil tends to get under the part and lift it. In this case a vacuum chuck is used as in Figure 3. Centering is accomplished with an X-Y micrometer stage. We have added micrometer heads that have built-in piezo-electric crystals. The part can first be roughly centered with the micrometer dial and then finely centered 'hands-off' by adjusting the voltage supply to the piezo crystals . Adjustment of tilt is built into the rotary table top.

As stated earlier the Z slide's plane ways were replaced with hydrostatic bearings. This was done to eliminate the Coulomb friction. The weight of the Z slide is held by a roller chain which moves over 2 sprockets to a counter- weight inside the column. One of the sprockets is driven by a handwheel on the side of the machine. The static friction in the original design made it impossible to position the slide with any resolution.

The solution was oil hydrostatic bearings. They have the advantage of eliminating static friction, they are stiff, and are relatively easy to retrofit to the geometry. The leaking oil from the bearings is of no consequence since the machine is oil showered. The bearings are fed with shower oil. A small portion of the filtered shower oil is directed through a high pressure pump to feed these bearings.

The positioning of the Z-slide in the present configuration is done with the bicycle wheel shown in Figures 1 and 6. The bicycle wheel replaces the original handwheel. The Z-slide can be positioned to one microinch in a time period of less than 10 seconds using the bicycle wheel.

Conclusions

Accuracy and repeatability errors in a measuring machine should be significantly less than the errors in the part being measured. An order of magnitude is the goal in many areas of manufacturing. At LLNL, for the past 12 years, the temperature controlled diamond turning lathes have produced parts which had errors less than the measuring machines used for inspection. Now for the first time in 12 years we have a general purpose inspection machine more accurate than our diamond turning machines. It is primarily the result of thermal stability.

There has been, and will be, psychological resistance to oil showered inspection machines. We believe that this initial resistance will be overcome by the need to progress and the lack of practical alternatives.

REFERENCES

1. Bryan, J. B., Donaldson, R. R., McClure, E. R., and Clouser, R. W., "A Practical Solution to the Thermal Stability Problem in Machine Tools", an SME publication #MR72-138, S.M.E. International Engineering Conference, April 1972..

2. Bryan, J. B., McClure, E. R., Brewer, W., Pearson, J. W., "Thermal Effects in Dimensional Metrology", an ASME publication #65-PROD-13, Metals Engineering & Production Engineering Conf., June 9, 1965.

3. American National Standard, "Temperature and Humidity Environment for Dimensional Measurement", American Society of Mechanical Engineers, ANSI B89.6.2-1973.

4. Bryan, J. B., and McClure, E. R., "Thermal Stability-Key to N/C Accuracy", Proc. 7th Ann. Meeting, Numerical Control Society, April 1970, pp. 119-126.

5. Bryan, J. B., Clouser, R. W., and Holland, E. D., "Spindle Accuracy", American Machinist, Dec. 4, 1967, pp. 149-164.

6. Bryan, J. B., "International Status of Thermal Error Research", CIRP Annalen, Vol. XVI, pp. 203-215.

7. McClure, E. R., "Manufacturing Accuracy Through the Control of Thermal Effects", UCRL-50636 (D. Eng. Thesis, Univ. of Calif. at Berkeley, also available through University microfilms).

Acknowledgement

Work performed under the auspices of the U.S. Department of Energy by Lawrence Livermore National Laboratory under Contract No. W-7405-ENG-48.

FIGURE 1: Properly managed oil shower leaves the operator with access to the workspace.

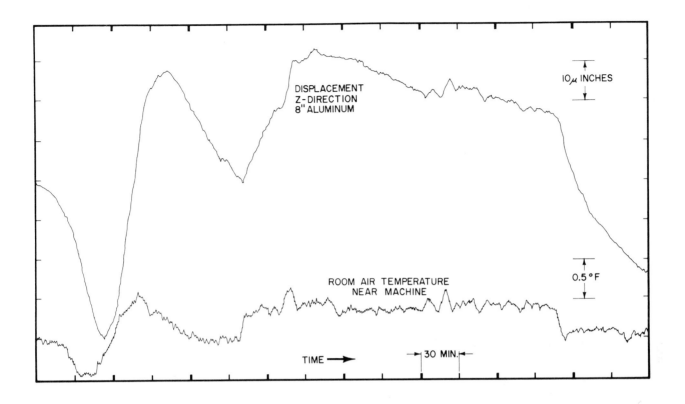

FIGURE 2:   A drift test conducted without oil shower over an 8 hour period shows
how the observed measurement of an 8 inch tall aluminum cylinder changes
with varying room temperature. Note that the room temperature is being
held to ± 1/2°F.

FIGURE 3: The air bearing LVDT bracketed from the Z-axis slide is used to measure size and contour of a ball. The stylus force is adjustable. A force of 0.2 grams is typical.

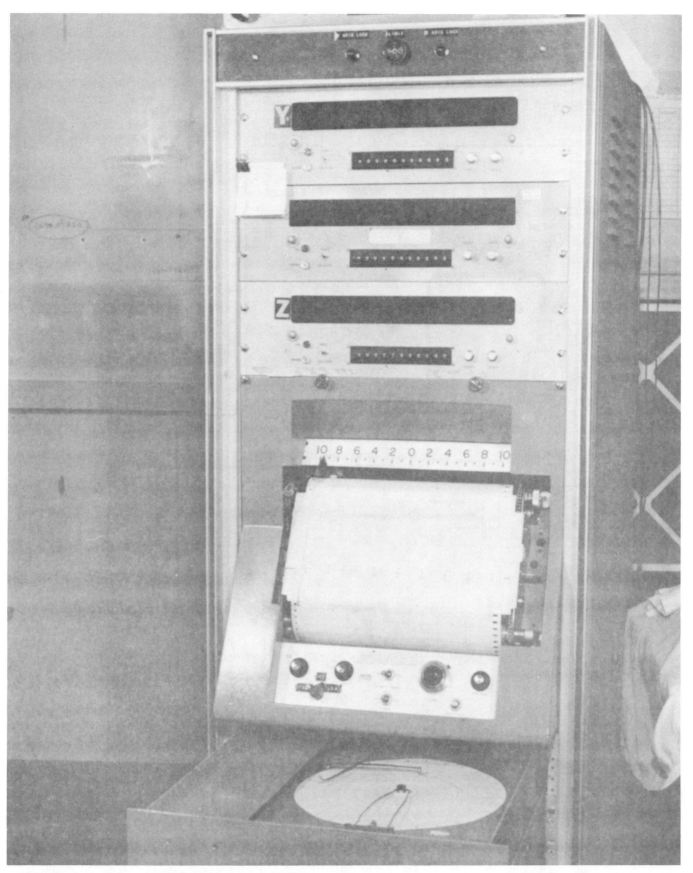

FIGURE 4: The resolution of the digital displacement displays (X, Y and Z) is 0.5 microinch. Recording of the LVDT analog signal is by either the polar or the linear recorder.

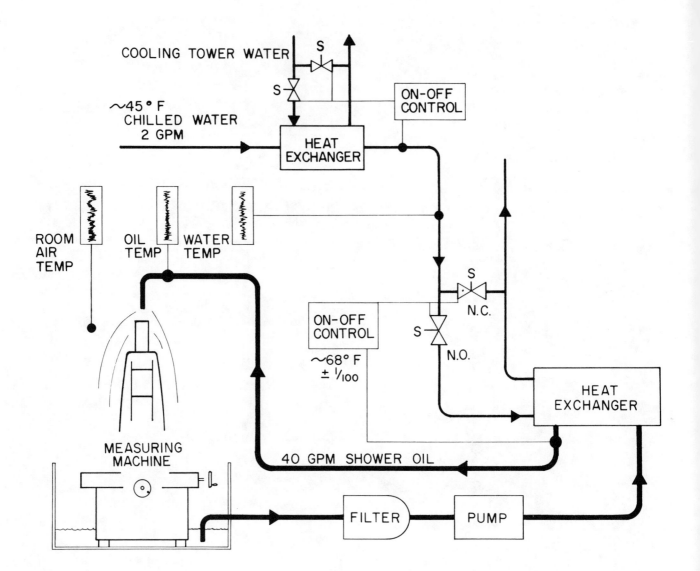

FIGURE 5: The control system maintains the shower oil at a nominal 68°F. The maximum temperature variation, when averaged over 30 seconds, is 0.01°F.

FIGURE 6: This view of the machine shows how the oil flows evenly down and
around the structural column.

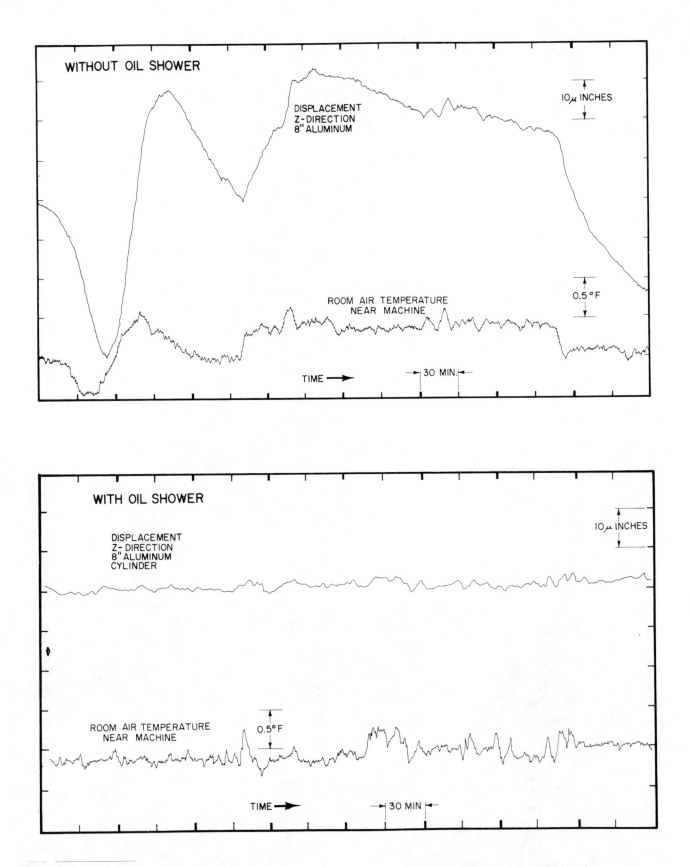

WITHOUT OIL SHOWER

DISPLACEMENT
Z-DIRECTION
8" ALUMINUM

10μ INCHES

ROOM AIR TEMPERATURE
NEAR MACHINE

0.5°F

TIME ——▶   ▶| 30 MIN. |◀

WITH OIL SHOWER

DISPLACEMENT
Z-DIRECTION
8" ALUMINUM
CYLINDER

10μ INCHES

ROOM AIR TEMPERATURE
NEAR MACHINE

0.5°F

TIME ——▶   ▶| 30 MIN |◀

FIGURE 7:   An order of magnitude improvement in thermal stability with the use
of liquid shower is demonstrated with these drift test results.

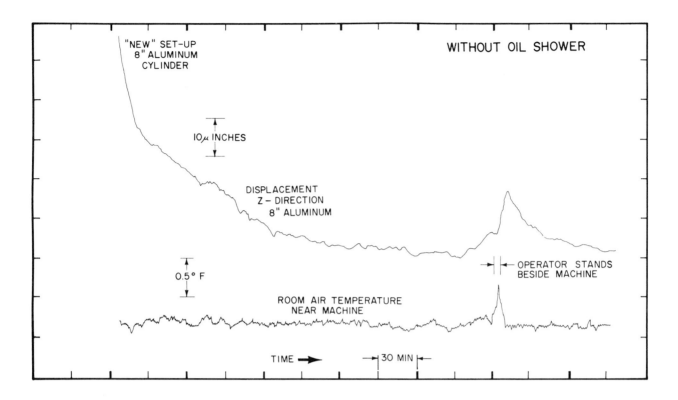

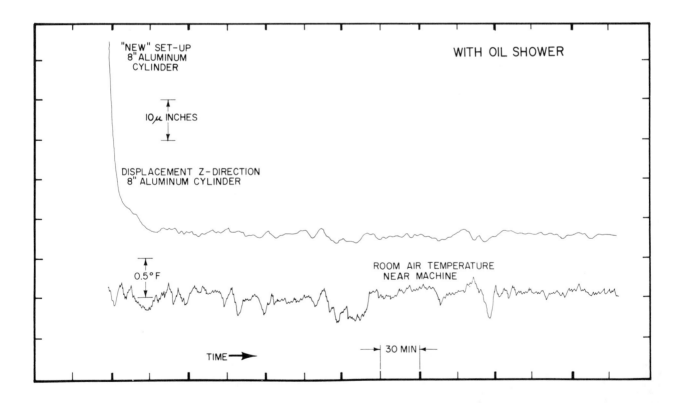

FIGURE 8:   The soak out time with oil shower is an order of magnitude faster than without oil shower.

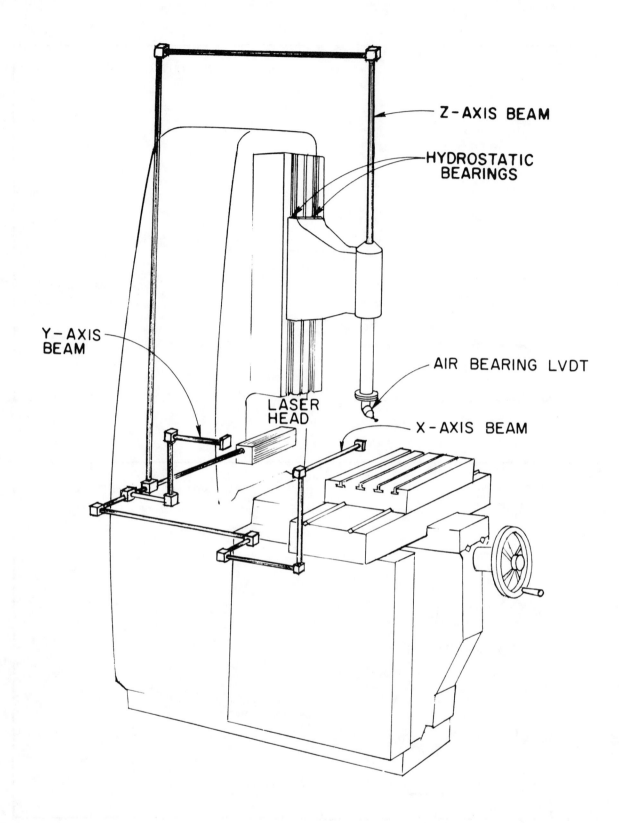

FIGURE 9: Placement of the laser paths satisfies the Abbe Principle in the Z-axis and minimizes the Abbe offset in the X and Y-axes.

FUNCTIONAL GAGING WITH COORDINATE INSPECTION MACHINES

Coordinate inspection machines (see Fig. 2-7) can be converted to functional gaging devices by adding a precision chuck to permit the use of standard gage pins and bushings instead of tapered or flat probes.

CONVERSION

Fig. 5-1 shows the most practical way to convert coordinate inspection machines into functional gages. The calibrated precision chuck forms the basis for this conversion, but the adaptation is slightly different for each machine, since not all have the same axis and spindle geometry.

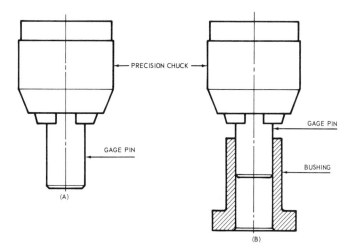

Fig. 5-1. Conversion of coordinate inspection machines to functional gages.

Pins

Once the precision chuck is in place, standard gage pins available in 0.0001 in. increments can be used (see Fig. 5-1A). The pins can be slightly oversize to reflect wear allowances and gage tolerances, if this is desired or required.

Bushings

The use of bushings is also based on the installation of a precision chuck and gage pins to hold the bushings (see Fig. 5-1B). Extra long bushings may be required if a projected tolerance zone is high. If it is impractical to obtain extra long bushings, only a certain portion of a part feature tolerance zone projection can be gaged.

APPLICATIONS

Coordinate inspection machin s modified for functional gaging can be used to check the perpendicularity, size and location of holes, pins, and tapped holes.

Holes

Perpendicularity. Fig. 5-2 shows the method of checking hole perpendicularity by determining the virtual size of the hole. The hole accepts a 0.514 in.

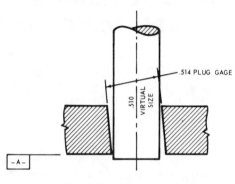

Fig. 5–2. Virtual size check.

diameter plug gage, which means that it has at least a 0.514 in. diameter. However, if the hole is not perpendicular to primary datum surface A, it will not accept a 0.514 in. diameter mating part feature. (This can be readily determined by inserting the 0.514 in. diameter plug gage in the precision chuck mounted on the machine spindle, and attempting to push the plug gage through the hole.) Note in Fig. 5-2 that the largest perpendicular pin that will pass through the hole has a 0.510 in. diameter. This is the *virtual size* of the hole, and means that the hole axis is out of perpendicularity to the primary datum surface by 0.004 in. (0.514 − 0.510 in.).

Location. Fig. 5-3 shows a straight pin being used to check the location of a positionally toleranced hole. Suppose the callout for the hole is 0.510/0.530 DIA, with a 0.010 in. diameter positional tolerance at MMC. The size of the straight pin to be used for gaging location is determined by subtracting the 0.010 in. MMC positional tolerance from the MMC diameter of the hole, which is 0.510 in. Thus, 0.510 − 0.010 = 0.500 in. diameter pin.

The spindle is positioned at the basic hole location using the measuring scales on the machine (datum surface A should be perpendicular to the machine spindle), and the pin is passed through the hole. This procedure is repeated at each hole location. Separate gaging operations to determine hole size with 0.510 in. diameter go and 0.530 in. diameter not-go gages complete the inspection operation.

If the callout for the hole were 0.500/0.539 DIA, with zero positional tolerance at MMC, the same 0.500 in. diameter pin (0.500 − 0.000 in.) would be used to gage hole location, but the zero MMC tolerancing specification would eliminate the need for a separate go gage (0.510 in. diameter) to gage minimum hole size, because the go gage is incorporated in the 0.500 in. diameter straight pin. Thus location and minimum size are gaged at the same time. A separate 0.530 in. diameter not-go gage is still required to gage maximum size.

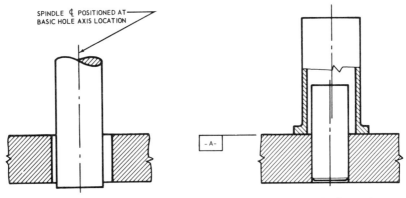

Fig. 5-3. Gaging a hole location.

Fig. 5-4. Gaging a pin.

Pins

Fig. 5-4 shows a bushing used to check the location and size of a dowel pin. Suppose the callout for the dowel pin is 0.5000/0.4996 DIA, with a 0.010 in. diameter positional tolerance at MMC. The inside diameter of the bushing to be used for gaging is determined by adding the 0.010 in. MMC positional tolerance to the MMC diameter of the hole (0.5000 in.). Thus, 0.5000 + 0.010 = 0.510 in. bushing ID.

The spindle is positioned at the basic pin location (datum surface A should be perpendicular to the machine spindle) and the bushing is lowered over the pin. The bushing should touch the top surface of the part. The operation is repeated at each pin location.

Tapped Holes

Fig. 5-5 shows a bushing used to check the location and size of a tapped hole. Suppose the callout for the tapped hole is 1/2–13 UNC 2B, with a positional tolerance of 0.010 in. diameter at MMC and a tolerance zone projection of 0.750 in., which is the maximum thickness of the mating part. The inside diameter of the bushing is determined by adding the 0.010 in. positional tolerance to the MMC diameter of the bolt shank that will fit the tapped hole. Thus, 0.500 + 0.010 = 0.510 in. bushing ID.

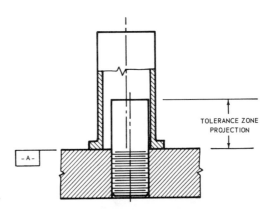

Fig. 5-5. Gaging a tapped hole.

A 0.500 in. diameter go thread gage with a 0.500 in. diameter shank is inserted into the tapped hole. The shank should project 0.750 in. above part surface A. (Datum surface A should of course be perpendicular to the machine spindle.) The spindle is positioned at the basic tapped-hole location and the bushing lowered over the thread-gage shank. The operation is repeated at each tapped-hole location.

For low-volume work, the go thread gage can be moved to each tapped hole and inserted prior to the inspection.

Parts should not be repositioned during the above gaging operations as this would destroy pattern integrity.

Potential N/C Application

The functional gaging capability, because its benefits are not widely known, is not an advertised application of coordinate inspection machines. However, the eventual conversion of these machines to numerical control should bring out their full capability as functional gaging devices. An automated, functionally gaging inspection machine could be programmed to go to the bàsic hole location, insert a straight gage pin into the hole, and electrically record that a particular hole is not functional if the pin contacts the hole at any portion throughout the depth of the hole. The addition of an in-series air-gaging head to such an automated machine would enable the machine to determine hole size as well.

Presented at the CASA/SME Qualinspex II Conference, November 1980

Economic Justification of Coordinate Measuring Machines

By Richard Paolino
Bendix Corporation

ABSTRACT

The problem of justifying the capital expense of a coordinate measuring machine purchase is examined in depth. The first part of the paper describes the advantages of CMMs over hard gages and surface plate methods of inspecting parts, and analyses the reasons for the increased use of CMMs. Degrees of sophistication in CMMs are described and the advantages of each are explained.

Charts and diagrams are used to compute the actual time and dollar savings possible in 100% inspection of a hypothetical part by CMM methods versus surface plate methods. CMM benefits are summarized.

The balance of the paper deals with accounting procedures for determining the Return on Investment. Three methods are described in detail: Discounted Cash Flow Method, Nondiscounted Rate of Return, and Payback in Years. Appendix A and B illustrate in detail the method used to calculate the Discounted Cash Flow Rate of Return of an investment in a coordinate measuring machine.

Detailed examples are given accompanied by appropriate charts and tables which enable the reader to substitute his own calculations for justifying purchase.

Series on CMM Justification

Modern requirements for fast production to high standards of precision makes it difficult for many quality control functions to keep up with the demand for inspection throughput. Machined parts are becoming increasingly more complex with more features and tighter tolerances. Surface plate and height gage techniques are too costly, too slow and often too inaccurate. Finding the necessary space for more surface plates and skilled inspectors is not always easy, is costly and will often not cope with the complexity of the problem and/or the quick turnaround required. Over the past two decades, several inspection techniques have emerged which have greatly helped the Quality Control manager solve his dilemma.

Coordinate Measuring Machines have been in use for almost twenty years now, having started as two axis location devices and having grown gradually into five axis Direct Computer Controlled machines. At first only a few units were purchased each year. Now, annual figures exceed the 1000 mark. Why? The answer is simple, Increased Profits!

There are a number of elements impacting profit and payback when moving from manual, hard gage inspection methods through manual CMM's to Direct Computer Control machines.

Among them are:

1. Compared to hardgage methods, the manual CMM benefits are:
   - Increased inspection throughput.
   - Improved accuracy.
   - Minimization of operator error.
   - Reduced operator skill requirements.
   - Reduced inspection fixturing and maintenance costs.
   - Increased versatility through reduced dedication to specific tasks (as many go/no go type gages).
   - Uniform inspection quality.
   - Reduction of scrap and good part rejection.
2. Additional benefits of CMM with Measurement Pre-Processor or computer assist are:
   - Reduction in calculating and recording time and errors.
   - Reduction in set-up time and fixturing costs through automatic compensation for misalignment.
   - Provision of a permanent record for process control and traceability of compliance to specifications.
   - Reduction in off-line analysis time which can often be longer than the time required for inspection.
   - Simplification of inspection procedures, especially when geometric dimensioning and tolerancing are involved.
   - Possibility of reduction of total inspection time through use of statistical and data analysis techniques (random or systematic sampling).

3. Further benefits of CMM's with Direct Computer Control are:
   - Improved productivity—as a general rule, the machine is 3-4 times faster than above.
   - Improved accuracy and reproducibility—operator influence is eliminated and low gage force improves results.
   - Operable by unskilled personnel.
   - Elimination of monitoring of repeated inspections.

Overall potential profits can be quantified into three areas:
- Inspection Cost Savings
- Reduced Manufacturing Downtime and Scrap.
- Increased Inspection and Manufacturing Capacity.

Inspection Cost Savings:
This area of savings is directly related to inspection time savings. Chart No. 1 delineates a number of inspection operations performed and the *approximate* time required to perform them using—
- Standard Open Plate Techniques and Hand Gages.
- Coordinate Measuring Machines.
- Coordinate Measuring Machines with Measurement Pre-Processors.
- Coordinate Measuring Machines with Computers.
- Direct Computer Controlled Coordinate Measuring Machines.

CHART NO. 1
Labor Time Required (in minutes)

Operation	Surface Plate Method	CMM with D.R.O.	Cordax with M.P.P.	Computer Assisted Cordax	DCC Cordax
Sizing					
Hole	3	1.5	.5	.5	.3
Boss	2	1.5	.5	.5	.3
Length	6	.5	.3	.3	.15
Height	6	.5	.25	.25	.15
Thickness	6	.5	.25	.25	.15
Bolt Hole Diameter	12	.6	2.0	2.0	1.5
Center Distance	6	1.0	.5	.5	.3
Relationships:					
Angle Between Faces	15	8	5	2	.8
Angle Between Bores	24	12	6	2.5	1.5
Arc Length	N.A.	N.A.	2.0	2.0	1.5
Roundness	N.A.	N.A.	2.0	1.5	1.2
Squareness	15	10	6	2.5	1.5
Flatness	4	4	2	2	1
Concentricity	30	6	2	2	1
True Position	15	3	.5	.5	.3
Profile (30 pts.)	N.A.	12	12	4	1.5
N.A. (Not Applicable)					
Set-up	Included	Included	3.0	3.0	1.0
Computation	Included (Hand calculator)	Included (Hand calculator)	Included	Included	Included
Hard Copy Recording	Included	Included	Included	Included	Included

This chart shows that incremental time savings can be achieved by increasing the sophistication of the measuring equipment. Set-up time and computation time is significantly reduced through the two axis and 3 axis automatic alignment routines and geometric and trigonometric computation routines of microprocessor and computer assisted or controlled machines. Profiling is virtually impossible using open plate techniques.

Total savings from the use of coordinate measuring machines can be determined by comparing the actual time it takes to perform the inspection required to the estimated time on the chart if a CMM were used. Note: The times on the chart are for estimating purposes only and may vary from one part to another. Analysis I, II and III show three situations where cost savings result from the use of a CMM. Appendix A illustrates a method of calculating Return On Investment and Payback.

ANALYSIS I
INSPECTION COST REDUCTION

Situation: Existing Business—Increase in Productivity in Inspection and Manufacturing Departments Needed.

Type of Manufacturing: Large Machining Shop—Several NC Machines—Many Manual Mills, Lathes, etc. Independent Q.C. Department.

Current Method of Inspection: Layout Tables and Surface Plates—Height Stands and Transfer Gages—Dial Indicator Hand Gages

Sample Part: Cast Housings with through Bores, Bolt Hole Pattern and Angular Face.

Inspection Required:

1. Diameter of All Bores (3) A B C
2. Concentricity of B to A and C to A
3. Bolt Hole Diameter
4. Location of Bolt Holes
5. Location of Face N Relative to M (Angle)
6. Depth of Shoulder in A and C from M

USED IN LABOR TIME COMPUTATIONS

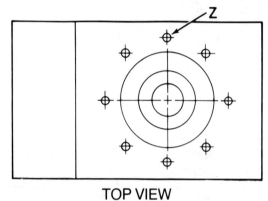

TOP VIEW

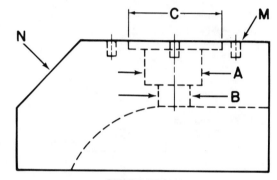

SIDE VIEW

Analysis: 100% Inspection

Determine Time and Cost Required for Manual Inspection and CMM Inspection.

Assume:
1. No New Manual Gages Required
2. One Operator Used for 100% Inspection
3. CMM with Data Assist Package
4. Manual Inspection Time includes Time for Setting Up and/or Mastering Gages
5. Cost/Hour includes Fringes at 33⅓% ($10/hour)
6. Quantity/Lot = 50 units

Manual Inspection:

Inspection	Time	Total Time	Cost	Type of Gage
Bore Size A	.3 min/part	.2.5 hr.	$25.00	Dial Bore Gage
Bore Size B	.3 min/part	.2.5 hr.	$25.00	Dial Bore Gage
Bore Size C	.3 min/part	.2.5 hr.	$25.00	Dial Bore Gage
Concentricity B to A	30 min/part	.25 hr.	$250.00	Surface Plates & Transfer Gage
Concentricity C to A	2 min/part	1.67 hr.	Incl. above	Same Set-up as above
Bolt Hole Diameter	12 min/part	10 hr.	$100.00	Vernier calipers
Bolt Hole Location	18 min/part	15 hr.	$150.00	Vernier calipers
Angle Between N & M	15 min/part	12.5 hr.	$125.00	Surface Plate/Transfer/Sine Plate
Depth of Shoulder A	2 min/part	1.67 hr.	$16.70	Dial Depth Gage ◦
Depth of Shoulder C	2 min/part	1.67 hr.	$16.70	Dial Depth Gage
Sub Total	.90 min/parts	.75 hrs.	$750.00	
Recording Data (Optional)	.1 min/part	8.33 hrs.	$83.30	
Total	1.67 hr./part	83.33 hrs.	$833.30 or	$16.67/part

At this rate of inspection, one inspector could measure 20.74 lots of parts per year at 1728 hours/man year.

CMM Inspection

Inspection	Time/pt	Total Time	Cost	Type Probe
Bore Size A	.5 min.	.415 hrs.	$ 4.15	Ball probe
Bore Size B	.5 min.	.415 hrs.	$ 4.15	Ball probe
Bore Size C	.5 min.	.415 hrs.	$ 4.15	Ball probe
Concentricity B to A	2 min.	1.67 hr.	$16.70	Ball probe
Concentricity C to A	2 min.	1.67 hr.	$16.70	Ball probe
Bolt hole Diameter	2 min.	1.67 hr.	$16.70	Taper probe
Bolt hole Location	Included above		Included above	
Angle Between N & M	2 min.	1.67 hr.	$16.70	Ball probe
Depth of Shoulder A	.25 min.	.21 hr.	$ 2.10	Ball probe
Depth of Shoulder C	.25 min.	.21 hr.	$ 2.10	Ball probe
Set-Up	2 min.	1.67 hr.	$16.70	
Sub Total	12 min.	10 hrs.	$100.00	
Recording Data			Included above	
Total	12 min/part	10 hrs.	$100.00 or $2.00/part	

At this rate of inspection, one inspector could measure 172.8 lots of parts per year, at 1728 hours/man year.

To get the Cost Savings per year, it is necessary to calculate the number of inspection hours required to inspect all lots per year.

Total lots/year = 150

Cost Comparison

	No. Lots	No. Inspectors/year	$/Year
Manual	150	7.2	$149,760
CMM	150	.87	$ 18,096
Savings		.63	$131,644

ANALYSIS II

Reduced Manufacturing Downtime and Scrap

One common application for a CMM is first part inspection on NC and/or CNC machines. Manufacturing Managers have two choices. They can run parts before Q.C. completes verification, taking a chance on a programming error that would cause scrapping the parts or, have machines sit idle at $35 to $50 per hour while the first few parts are verified. The current trend is to let the machines sit idle.

Total Savings and therefore, increased profits, come from inspection time savings *plus* reduced manufacturing downtime; or, if parts are run while awaiting verification, the savings come from Reduced Inspection costs *plus* Reduced Scrap Cost minus Increased Manufacturing Downtime.

The following Analysis shows one way to calculate cost savings resulting from the use of a CMM.

ANALYSIS II
FIRST PART INSPECTION SAMPLE INSPECTION

Assume:
1. Same Part as ANALYSIS NO. 1
2. Opportunity Cost of Not Running NC Machine is $35/Hour
3. CMM with Data Assist
4. Manual Inspection Cost/part — Same as Analysis No. 1
5. Cost/Hour = $10
6. Quantity/Lot = 50
7. 2 Parts Measured to verify NC Program
8. Hard copy record of Data required
9. Sample Size = 13 (2½% AQL Level — General Inspection Level III)

Determine Cost Savings Per Lot

	Time/Part	Time for NC Verification Time	Cost	Op. Cost	Sample Inspect. Cost	Total Cost/Lot
Manual	1.67 hrs.	3.34 hr.	$33.40	$116.90	$217.10	$367.40
CMM	2 hrs.	.4 hr.	$ 4.00	$ 14.00	$ 26.00	$ 44.00
Savings						$323.40/

Cost Comparison

	No. Lots	No. Inspectors/year	Total Cost
Manual	150	2.17	$55,110
CMM	150	.26	$ 6,600
Savings		1.91	$48,510

ANALYSIS III

Increased Inspection and Manufacturing Capacity

This is a combination of the first 2 situations and results in a business opportunity. By significantly reducing the amount of time it takes to measure parts, the total part cost is reduced, and increased profits or a reduced selling price results. With the availability of a CMM it is also possible to perform more inspections, resulting in greater control of materials and processes. The benefits of this increased availability, however, are difficult to forecast.

Increased manufacturing capacity, on the other hand, is a tangible benefit that can be forecasted and dealt with. Increased capacity means cost savings in reduced subcontracting or additional business at incremental profits. Through use of available capacity, in either case, there is a positive cash flow that can be used to offset the cost of a coordinate measuring machine. Appendix B explains the financial benefits that can be derived from such a situation.

Intangible Benefits

CMM operators need not be trained on such an extensive variety of gages as inspectors in the forty's and fifty's. This is a significant benefit since the demand for trained, qualified inspectors far exceeds the supply.

With Measurement Pre-Processors and computers performing the mathematical calculations and recording data, computational and transcription errors are virtually eliminated.

Most coordinate measuring machines can now operate in either English or Metric units eliminating the need for dual gaging systems.

SUMMARY OF CMM BENEFITS

Savings as calculated in Analysis I, II and III present several opportunities.

1. Better use of inspectors time.
   Typically such savings provide the inspector the time to perform more inspections than previous methods. This generally results in improved manufacturing processes and/or improved incoming or in-process inspection. The manufacturing process is improved through early determination of problem areas such as machine drift, machine wear, tool wear, etc. Scrap is also reduced when in-process inspection prevents bad parts from being processed further or when incoming inspection prevents received parts that are out of spec from being processed. All of the above come about without increased effort on the part of the individual inspectors or shop personnel.

2. Increased Capacity
   Since the Inspection Department is more productive and the Manufacturing Department has less downtime (waiting for 1st piece inspection), it is possible for the company to seek additional business or to bring more work in house. This additional business in turn results in increased profits, again without increased efforts on the part of inspectors or Machinists. See Appendix B.

3. Reduced Cost Per Part

By reducing the cost per part, it's possible to get increased profits per part or to reduce the selling price per part while maintaining the same absolute or relative profit per part, the Reduced Selling Price should also result in increased sales and profits.

RETURN ON INVESTMENT

A number of methods are used in quantifying the financial benefits of an investment. Among these are the Discounted Cash Flow, The Nondiscounted Rate of Return and Payback in years.

Discounted Cash Flow Method

This method of calculating Return on Investment (ROI) is currently being used by many companies. It involves the development of the annual increase or decrease in cash expected to flow from a specific investment of funds. Cash flows will differ radically from after tax profit in early years of the project. This situation occurs because a cash outlay is ordinarily required prior to the actual realization of profits.

Discounting the future cash flow results in a rate which can be directly compared to any other investment where a compound return is expected, such as a 10.5% interest rate on a 5-year Bank Savings Certificate.

All companies are dependent upon some type of financing such as sale of bonds, bank loans, sale of stock, reinvestment of profits, etc. and thus desire a rate of return on investment equal to or greater than their cost of capital. Thus the minimum return on investment after tax and before interest for a no-risk, long-term investment varies from company to company, but a range of 10.0% to 15.0% would be considered normal. However, no risk types of investment opportunities are rare and the return on investment requirements in real terms will fluctuate from 15% to 35% depending upon the subjectiveness of the data used in calculating the return on investment. When several investments are being considered, the rate of return can be used to establish priorities.

Nondiscounted Rate of Return

This method takes the average savings or income per year from an investment and divides it by the total investment cost. It does not take into consideration the time value of money and therefore may result in a different rate of return from the Discounted Cash Flow rate of return.

Payback in Years

This method is somewhat similar to the Nondiscounted Rate of Return in that it does not consider the time value of money. It differs, however, in that its prime concern is how long it will take to pay off the investment. In this case, one must calculate the return for each year and compare the returns to the investment cost. If the return is the same each year, the payback is equal to the Total Investment divided by the annual return. Since this method does not concern itself with the returns from years beyond the payback point, it may result in an investment decision different from the Discounted Cash Flow Method and the Rate of Return Method.

Of the three methods described, the discounted cash flow method can be used as the basis for determining the Discounted Return on Investment, the Nondiscounted Return and Payback in Years.

Appendix A and B have been provided to illustrate the method used to calculate the Discounted Cash Flow Rate of Return of an investment in a Coordinate Measuring Machine.

APPENDIX A

SITUATION TWO: *INCREASED SALES VOLUME THROUGH INCREASED CAPACITY*

RETURN ON INVESTMENT AND PAYBACK USING DISCOUNTED CASH FLOW

1. Basic Considerations

   A bid for $500,000 in N/C machining work requiring sample inspection is being prepared by manufacturer A. However, manufacturer A's current inspection methods are too time consuming and expensive for him to be competitive without reducing his normal profit margin. If manufacturer A is awarded the contract, he expects to receive similar orders over the next five years without having to invest in additional metalcutting equipment.

2. Steps to be Followed:
   - Determine the type and price of the CMM required to do the work.
   - Estimate the differential costs (i.e., the difference between the cost of inspecting using current inspection methods and the cost of inspection using the desired CMM).
   - Estimate the total cost of the order under consideration, taking into account the savings above. (This should include all direct and indirect manufacturing costs, engineering (if applicable) and commercial expense).
   - Determine the sales value of the new order under consideration.
   - Calculate the increased investment in inventories and accounts receivable resulting from the added business. (That is, Inventory Investment =

$$\frac{\text{Cost of Sales}}{\text{Annual Inventory Turnover Rate}}$$

Accounts Receivable
Investment = $\dfrac{\text{Sales}}{\text{Annual Accounts Receivable Turnover Rate}}$

EXAMPLE: Using the following assumed data
   1. *Investment* = CMM with computer assist $50,000
   2. *Differential Cost* = Existing Inspection

Costs	$ 50,000
CMM Inspection Costs	20,000
Annual Savings	$ 30,000

   3. *Total Cost of New Contract*

Manufacturing Direct Cost of Sales	$350,000
Less CMM Differential Cost	30,000
Adjusted Manufacturing Cost of Sales	$320,000
Plus Commercial Expense	75,000
Total Cost	$395,000

   4. *New Bid Price of Contract*

Total Cost	$395,000
Profit Before Tax 15%	75,000
Adjusted Bid Price	$470,000*

*The new bid price is equal to a 6% reduction over that which would have been quoted before the change in inspection equipment.

5. *Inventory and Accounts Receivable Investment*

   A. Inventory Investments =
   $$\frac{\$320,000\text{-}\$100,000^{**}}{3\text{ Turns per yr.}} = \$73,333$$

   B. Accounts Receivable Investment =
   $$\frac{\$470,000\text{-}\$100,000^{**}}{10\text{ Turns per yr.}} = \$37,000$$

**This example assumes $100,000 in advance payments are received from the customer to cover materials and expense tools. Disregard if no advances are received.

6. *Inflation Rate*

   7.0% per year for each of the next five years. (For the purpose of this example, inflation at a rate of 7.0% per annum has been used in increasing both cost and sales, as this tends to give a purer return on investment. However, many firms may wish to use constant dollars due to the subjectiveness involved in estimating the future rate of inflation.)

RETURN ON INVESTMENT
DISCOUNTED CASH FLOW METHOD

YEARS (Full 12 Months, Starting from Point Zero)

	DIFFERENTIALS	0	1	2-4	5	TOTAL
1.	Net Sales					
	Manufacturing Cost (increase) decrease					
2.	Direct Labor					
3.	Direct Material					
4.	Manufacturing Burden (includes indirect labor,		30	34	39	171
	fringe benefits, depreciation, tools, supplies, etc.					
5.	Other Factory Costs					
6.	Total Manufacturing Cost (increases) decreases		30	34	39	171
7.	Engineering Expense (increases) decreases		0	0	0	0
8.	Commercial Expense (increases) decreases		0	0	0	0
9.	Income (loss) before federal tax		30	34	39	171
10.	Income (loss) after federal tax (line 9 (a) 52%)		15	18	20	89
11.	Investment Tax Credit		5			5
12.	Depreciation (CMM depreciation included in line 4)		5	5	5	25
13.	Residual Value—Net Book Value				25	25
14.	Capital Outlay (purchase price of CMM plus installation)	(50)				(50)
	(increases) decrease in:					
15.	Receivables (Net of Customer Advance Payments)					0
16.	Inventories (Net of Customer Advance Payments)		10	2	(16)	0
17.	Total Asset Changes (Lines 12 through 16)	(50)	15	7	14	0
	Increase (Decrease) in:					
18.	Current Liabilities		(1)	—	1	0
19.	Total Investment Changes (Lines 17 and 18)	(50)	13	7	15	0
20.	Net CASH FLOW IN (OUT) (Lines 10, 11 & 19)	(50)	34	25	35	94
	RETURN ON INVESTMENT (DCF Method)					
	Line 20 discounted at rates in Table 1					
21.	Discount factor (a) Rates 50% lower than Actual	1.00	.67	.94	.13	
22.	Discounted Cash Flow	(50)	23	24	5	
23.	Cumulated Discounted Cash Flow	(50)	(27)	(3)	2	
24.	Discount Factor (a) Rates 55% higher than Actual	1.00	.65	.86	.11	
25.	Discounted Cash Flow	(50)	22	22	4	
26.	Cumulated Discounted Cash Flow	(50)	(28)	(6)	2	

27. RETURN ON INVESTMENT (DCF METHOD) Rate (1) Line 21 + (Line 23 Cum DCF $\quad$ X 5% = $\dfrac{45\% \ (2 \times 5\%)}{2+2}$ = 52.5%

(Line 23 Cum DCF + Line 26 Cum DCF)

28.	GROSS PAYBACK IN YEARS	2.0

APPENDIX B

SITUATION ONE: *COST REDUCTION*

RETURN ON INVESTMENT AND PAYBACK USING DISCOUNTED CASH FLOW

1. Basic Considerations

   At the instigation of the Director of Quality Control, Company B recently reviewed its product inspection costs. It was determined that the ratio of inspection to machining cost was very high. The Chief Inspector suggested that the efficiency of the detail inspection department could be improved if the company invested in a Coordinate Measuring Machine.

2. Steps to follow:
   - Obtain prices and performance data for the CMM being considered, including installation cost
   - Estimate the differential costs (the difference between the cost of detailed inspection currently being used and the alternate method under consideration)
   - Estimate the effect on inventory investment and current liabilities caused by the cost reduction
   - Prepare the discounted cash flow analysis

I. EXAMPLE: Using the following assumed data for one of the alternatives:

1. CAPITAL INVESTMENT = CMM with computer assist = $50,000.00

   A. Existing Inspection Cost:

Inspection Labor	$37,476.00
Fringe Benefits	12,524.00
Total Manual Inspection Cost	$50,000.00

   LESS:

   B. Inspection Costs using CMM:

Inspection Labor	$11,250.00
Fringe Benefits	3,750.00
Depreciation	5,000.00
Total CMM Inspection Cost	$20,000.00

   C. Differentials:

Inspection Labor	$26,226.00
Fringe Benefits	8,774.00
Depreciation	(5,000.00)
Total Differential Cost	$30,000.00

2. INVENTORY INVESTMENT

$$\frac{\text{Savings}}{\text{Inventory Turnover Rate}} = \frac{\$30,000.00}{3 \text{ Turns Annually}} = \$10,000.00$$

3. CURRENT LIABILITIES

$$\text{Inspection Labor} \ldots \frac{\$26,226.00}{52 \text{ Turns per yr.}} = \$ \ 504.00$$

$$\text{Fringe Benefits} \ldots \frac{\$8,570.00}{12 \text{ Turns per yr.}} = \$ \ 714.00$$

$$\text{Depreciation} \ldots \frac{(\$5,000.00)}{0} = \text{-0-}$$

Net Effect on Current Liabilities .... $ 1,235.00

5. ANNUAL INFLATION FACTOR   7% per annum

RETURN ON INVESTMENT
DISCOUNTED CASH FLOW METHOD

YEARS (Full 12 Months, Starting From Point Zero)

	DIFFERENTIALS	0	1	2-4[1]	5	TOTAL
1.	Net Sales		470	539	616	2703
	Manufacturing Cost (Increases) Decreases					
2.	Direct Labor		(95)	(109)	(125)	(547)
3.	Direct Material		(70)	(80)	(91)	(401)
4.	Manufacturing Burden (Includes Indirect Labor, Fringe Benefits, Depreciation, Tools, Supplies, etc.		(150)	(172)	(197)	(863)
5.	Other Factory Costs (O.T., N.S. Premium, etc.)		(5)	(6)	(6)	(29)
6.	Total Manufacturing Cost (Increases) Decreases		(320)	(367)	(419)	(1840)
7.	Engineering Expense (Increases) Decreases (Part Programming)		(3)	(3)	(4)	(16)
8.	Commercial Expense (Increases) Decreases		(75)	(86)	(98)	(431)
9.	INCOME (LOSS) BEFORE FEDERAL TAX		72	83	95	416
10.	INCOME (LOSS) AFTER FEDERAL TAX (Line 9 (a) 52%)		37	43	49	215
11.	Investment Tax Credit		5			5
12.	Depreciation (CMM Depreciation Included in Line 4) *10 yrs.		5*	5	5	25
13.	Residual Value—Net Book Value—*$50K-$25K Depreciation				25*	25
14.	Capital Outlay (Purchase Price of CMM Plus Installation)	(50)				(50)
	(Increases) Decreases in:					
15.	Receivables (Net of Customer Advance Payments)[2]		(37)	(2)	43[3]	0
16.	Inventories (Net of Customer Advance Payments)[2]		(73)	(8)	97[3]	0
17.	Total Asset Changes (Lines 12 through 16)	(50)	(105)	(5)	170	0
	(Increases) Decreases in:					
18.	Current Liabilities		26	3	(35)	0
19.	Total Investment Changes (Lines 17 and 18)[2]	(50)	(79)	(2)	135	0
20.	NET CASH FLOW IN (OUT) (Lines 10, 11 & 19)	(50)	(37)	41	184	220[4]
	RETURN ON INVESTMENT (DCF METHOD)					
	Line 20 discounted at rates in Table 1					
21.	Discount Factors (a) Rates 40% Lower Than Actual	100.0	.71	1.13	.19	
22.	Discounted Cash Flow	(50)	(26)	46	35	
23.	Cumulated Discounted Cash Flow	(50)	(76)	(30)	5	
24.	Discount Factor (a) Rates 45% Higher Than Actual	100.0	.69	1.04	.16	
25.	Discounted Cash Flow	(50)	(26)	43	29	
26.	Cumulated Discounted Cash Flow	(50)	(76)	(33)	(4)	

RETURN ON INVESTMENT (DCF METHOD): Rate (1) Line 21 + (Line 23 Cum. DCF X 5% = 40.0% (5 x5%) = 42.8%

$$\frac{}{5 \pm 4}$$

(Line 23 Cum DCF + Line 26 Cum DCF)

GROSS PAYBACK IN YEARS (Year in which Line 20 accumulates to zero) 3 Yrs., 1 Mo.

(1) Years 2, 3 & 4 are the same in this example to reduce the amount of copying required, therefore must be multiplied by 3 and added to the other columns in order to arrive at the total.

(2) These are balance sheet items and the largest investment occurs in the first year with increases caused by inflation only.

(3) Year 5 is the final year of the project thus the liquidation of these two assets results in a positive cash flow.

(4) Line 20 Total Column must always be equal to the total Column of Line 10 plus Line 11.

Table A

Table For Approximating Discounted Cash Flow Rate of Return

ESTIMATED PROJECT LIFE

		OVER 20	20	15	10	9	8	7	6	5	4	3	2	1
	1/4	400	400	400	400	400	400	400	400	400	400	400	400	390
	1/2	200	200	200	200	200	200	200	200	200	199	198	196	155
	3/4	133	133	133	133	133	133	133	133	133	133	133	122	60
1		100	99	99	99	98	98	97	97	96	94	92	80	0
1	1/4	81	80	80	80	80	80	80	79	78	76	70	52	0
1	1/2	69	68	68	68	68	67	67	66	64	60	53	30	
1	3/4	59	58	58	58	58	57	56	55	53	49	40	13	
2		51	50	50	50	50	49	48	47	44	40	29	0	
2	1/4	45	44	44	44	44	43	42	41	37	32	20		
2	1/2	41	40	40	39	39	38	37	35	32	26	12		
2	3/4	37	36	36	36	35	34	33	31	27	20	6		
3		34	33	33	32	31	30	29	27	23	15	0		
3	1/4	31	30	30	29	28	27	26	23	19	11			
3	1/2	29	28	28	27	26	25	23	20	15	7			
3	3/4	27	26	25	24	23	22	20	17	12	3			
4		26	25	25	22	21	20	18	15	9	0			
4	1/4	24	23	23	21	19	18	15	12	6				
4	1/2	23	22	22	19	18	16	13	10	4				
4	3/4	21	20	20	18	16	14	11	8	2				
5		20	19	19	16	15	13	10	6	0				
5	1/2	19	18	16	14	12	10	7	3					
6		17	16	15	11	10	8	4	2					
6	1/2	16	15	13	9	8	5	2						
7		14	13	12	8	6	3	0						
7	1/2	13	12	11	6	4	1							
8		12	11	10	5	3	0							
9		11	10	8	2	0								
10		10	8	6	0									
12		8	6	3										
14		7	4	1										
16		6	2	0										
18		5	1											
20		4	0											
30		2												

ESTIMATED YEARS OF PAY-BACK (NON-DISCOUNTED BASIS)

This table is used by reference to estimated years of payback and the Project Life used in the Financial Analysis worksheets.

It can be used to estimate the initial discount rate used for computing the Discounted Cash Flow Rate of Return.

Table B

Table of Present Values of $1 (Discount Factors) and Cumulative Discount Factors

RATES	1	2	3	4	5	6	7	8	9	10	11	12	13	14	15	16	17	18	19	20	MAXIMUM CUMULATIVE DISCOUNT FACTORS
5%	.95	.91	.86	.82	.78	.75	.71	.68	.64	.61	.58	.56	.53	.51	.48	.46	.44	.42	.40	.38	
	.95	1.86	2.72	3.54	4.32	5.07	5.78	6.46	7.10	7.71	8.29	8.85	9.38	9.89	10.37	10.83	11.27	11.69	12.09	12.47	20.00
10%	.91	.83	.75	.68	.62	.56	.51	.47	.42	.39	.35	.32	.29	.26	.24	.22	.20	.18	.16	.15	
	.91	1.74	2.49	3.17	3.79	4.35	4.86	5.33	5.75	6.14	6.49	6.81	7.10	7.36	7.60	7.82	8.02	8.20	8.36	8.51	10.00
15%	.87	.76	.66	.57	.50	.43	.38	.33	.28	.25	.21	.19	.16	.14	.17	.11	.09	.08	.07	.06	
	.87	1.63	2.29	2.86	3.36	3.79	4.17	4.50	4.78	5.03	5.24	5.43	5.56	5.73	5.85	5.96	6.05	6.13	6.20	6.26	6.67
20%	.83	.69	.58	.48	.40	.33	.28	.23	.19	.16	.13	.11	.09	.08	.06	.05	.05	.04	.03	.03	
	.83	1.52	2.10	2.58	2.98	3.31	3.59	3.82	4.01	4.17	4.30	4.41	4.50	4.58	4.64	4.69	4.74	4.78	4.81	4.84	5.00
25%	.80	.64	.51	.41	.33	.26	.21	.17	.13	.11	.09	.07	.06	.04	.04	.03	.02	.02	.01		
	.80	1.44	1.95	2.36	2.69	2.95	3.16	3.33	3.46	3.57	3.66	3.73	3.79	3.83	3.87	3.90	3.92	3.94	3.95		4.00
30%	.77	.59	.46	.35	.27	.21	.16	.12	.09	.07	.06	.04	.03	.03	.02	.02	.01				
	.77	1.36	1.82	2.17	2.44	2.65	2.81	2.93	3.02	3.09	3.15	3.19	3.22	3.25	3.27	3.29	3.30				3.33
35%	.74	.55	.41	.30	.22	.17	.12	.09	.07	.05	.04	.03	.02	.02	.01						
	.74	1.29	1.70	2.00	2.22	2.39	2.51	2.60	2.67	2.72	2.76	2.79	2.81	2.83	2.84						2.86
40%	.71	.51	.36	.26	.19	.13	.09	.07	.05	.03	.02	.02	.01								
	.71	1.22	1.58	1.84	2.03	2.16	2.25	2.32	2.37	2.40	2.42	2.44	2.45								2.50
45%	.69	.48	.33	.23	.16	.11	.07	.05	.04	.02	.02	.01									
	.69	1.17	1.50	1.73	1.89	2.00	2.07	2.12	2.16	2.18	2.20	2.21									2.22
50%	.67	.44	.30	.20	.13	.09	.06	.04	.03	.02	.01										
	.67	1.11	1.41	1.61	1.74	1.83	1.89	1.93	1.96	1.98	1.99										2.00
55%	.65	.42	.27	.17	.11	.07	.05	.03	.02	.01											
	.65	1.07	1.34	1.51	1.62	1.69	1.74	1.77	1.79	1.80											1.82
60%	.63	.39	.24	.15	.10	.06	.04	.02	.01												
	.63	1.02	1.26	1.41	1.51	1.57	1.61	1.63	1.64												1.67

Note: It is often necessary for a number of discount rates be tried before arriving at the two which yield a positive cumulative number for the lower rate and a negative number for the higher rate.

Reprinted from *Manufacturing Engineering*, October 1981

CMM Support Packages — Making the Most of the Machines

Enhanced data processing capabilities that are easy to use solve the skilled worker shortage problem for many CMM applications

Zeroing in an electronic probe on a reference sphere. Up to 10 tips can be precalibrated.

GARY S. VASILASH
Associate Editor

NOT ONLY ARE skilled machinists and toolmakers in short supply, but so too are trained individuals in the inspection and QC department. The solution to the shortage (in addition to basic training programs) is advanced electronics and ancillary equipment. Servodrives are replacing handwheels on lathes; wire EDM is sinking dies with unparalleled accuracy; computer-driven coordinate measuring machines are making highly precise readings. The trick now is to develop controls that don't require advanced programming training.

Bendix Automation & Measurement Div., Dayton, OH, is a pioneer in the field of equipping coordinate measurement machines with computers. In 1968 the company produced its first CNC machine and in 1978 its first microprocessor unit. 1981 marks the introduction of its first distributed microprocessor systems, a family of data processing support packages for its Cordax® CMMs, including the newly introduced Series 2000 horizontal arm machines. Key to the system is the proprietary Measurement Processor (MP), a dedicated microprocessor-based, self-diagnosing computer, keypad, and LED display housed in a single unit.

Says Fred Witzke, Cordax product manager, in describing the systems, "We have separate dedicated microprocessors for computation, operator communication, and also microprocessor servocontrol in machines equipped for power operation. This means we use each computer to fullest advantage, and all can work simultaneously."

Witzke explains that while most major CMM builders are utilizing the same Hewlett-Packard HP-85 or HP-9835 desktop computers that Bendix A&M does, the use of the computers differs. That is, other builders typically use the computer to activate servodrives, perform algorithms, initiate printouts, and so on. Bendix A&M uses it primarily as a communication channel to the operator and output devices. It acts as what Bendix calls an "operator friendly" interface for the development and execution of inspection programs and routines.

Explaining the "operator friendly" concept, Witzke says that the operator is presented with a series of operational decisions and instructions on a CRT. "It asks him what he wants to do, tells him how to do it, and with his okay, does it. It also records the operations so that all similar parts can be routinely inspected in the future," says Witzke.

Three Levels

The new data processing support packages are available in three levels of sophistication. The most basic, Level 1, utilizes the MP-2 Measurement Processor, that automatically performs complex tasks such as part leveling and axis alignment, which significantly reduces setup time. It provides automatic calculation of the data required to find centers and to determine diameters and polar coordinates, and performs other measurement and data handling requirements.

The Level 2-equipped CMMs utilize the HP-85 computer and either an MP-10 or MP-15 measurement processor. The HP-85 can be programmed in BASIC to provide inspection routines beyond those provided in the Bendix menu of measurement and instructional routines. The routines are displayed on the HP-85's CRT. The operator simply makes pushbutton selection from the menu, so data input is typically performed by doing no more than following the instructions displayed for the selected routines. Routines are built into the function library in the microprocessor; there aren't separate tapes for the various routines.

The MP-10 is designed for manually operated machines and can be retrofitted to any Cordax CMM. The MP-15 is a step up: it utilizes servos, a microprocessor-based servocontrol, and a microprocessor-based remote joystick control with which complete programming can be performed. Both the MP-10 and MP-15 are available with an optional 80 or 180 cps printer and dual floppy disc storage.

The most sophisticated are the Level 3 MP-20 and MP-25 measurement processors. Level 3-equipped CMMs can perform fully automatic, direct computer controlled (DCC) measurement.

Not only does this level utilize a more powerful HP-9835 computer with 128K-byte memory as the operator interface, an 80K-byte memory measurement processor for performing calculations and handling the function library, and the microprocessor-based joystick control (this only if it is a servodriven MP-25 package), but it also utilizes an additional microprocessor for controlling and monitoring the CMM during DCC

Level 3 system for coordinate measuring machines makes use of a measurement preprocessor, an HP-9835 A computer, and other equipment. The system facilitates programming and speeds measurement operations.

Microprocessor-based remote control unit permits CMM programming directly at the work site. Joystick control drives machine servos as required during the programming phase.

operation. Level 3 equipment can store and retrieve up to 10 datum reference frames; 29 features including points, lines, planes, circles, cylinders, and spheres; and nine user-defined inspection reports.

What Level 3 Does

Describing the use of the Level 3 measurement processor, Bendix A&M Engineer John Everhart says, "When the user writes a program with the Level 3 'operator friendly,' it actually writes the program in BASIC for him. To do it, the user puts his first part on the machine and performs the first measurement operation. By doing that he has written a debugged, fully automatic program and he's done his first inspection."

The processor establishes part coordinates, not machine coordinates. As a result, this teaching operation can be performed on one area of the table, then the following parts (of the same type) can be moved for subsequent operations. The system will seek out the part, establish its location, and perform the required measurements.

The measurement processors, as mentioned, generate all of the data and perform the algorithms. This means that the DCC controller in the MP-25 system is free to work in real time.

There is no hesitation during movement, as can be the case when a single computer is performing work on the data, communicating with the operator, and activating the drives. The DCC controller constantly monitors the electronic probe (up to 10 tips can be precalibrated) and shaft, and drives the combination at two programmable speeds. The standard speeds are 5 ips (127 mm/sec) and 0.1 ips (3 mm/sec), a seek or touch motion. If the probe encounters an unexpected obstruction, the machine will stop, awaiting further instruction; if the shaft hits something, the drives are disengaged and the machine shuts down.

Says Everhart, "The DCC system will take care of the electronic probe better than any manual system since it is constantly monitoring the probe. I would say that any time a user loses a probe with a DCC system it's because one of his operators has manually driven it into the part."

Justification

According to Richard Paolino, director of marketing for Bendix A&M, the company has been devoting a great deal of time to developing the support systems. He says that the best way to get the most out of a mechanical measuring machine is through electronics.

Asked about justification of a Level 3 measurement processor, Paolino responds by noting the shortage of skilled people and the difficulties that occur when a worker is trained, then decides that he can make more money elsewhere. It is essential that a company make the most of the trained people it does have, and the measurement processor is one way of doing it, Paolino asserts.

Says Paolino, "When a skilled operator runs through the first measurement operation on a part, he'll probably do it in the most efficient way. Once the machine is taught through the first run, a lower-skilled individual can put the part on and push the correct buttons to start the operation. You don't have your most expensive person sitting there watching the part being run." And if that skilled worker leaves, the program remains in memory.

Paolino stresses that the systems are designed to be run by measurement personnel, not programmers, as it's often difficult and inconvenient to try to teach a computer programmer how to measure a part. The systems are simple enough to use so the operator doesn't have to learn how to program a computer.

While these capabilities might seem as though the systems would be much more expensive than other builders' equipment, Paolino says that the packages are cost competitive. ∎

CHAPTER 4

PRODUCTION GAGING SYSTEMS

Reprinted from *VDI International Magazine* March 1982

Users' Report: In-Process Gauging

1 Introduction

Gauging is a necessity when machining close-tolerance workpieces. Normally gauging is carried out as a dimensional check within the framework of quality control. But when accuracy demands as well as the cost of the machined workpiece are high, the gauging operation is advanced in the process sequence so that it takes place as close as possible to the machining cycle (Fig. 1).

True in-process gauging has established itself as a standard technique in precision grinding applications. In turning, boring and milling applications gauging can only be carried out after the actual machining operation. With reference to the workpiece measured this is post-process gauging, and with reference to the next workpiece to be

Gauging the workpiece or the tool during the machining cycle is gaining in importance on lathes and machining centers. In most applications gauging is effected immediately after machining the previous workpiece and is therefore strictly speaking a post-process operation. In-process action, however, is obtained by feeding back measured deviations as tool corrections. This report gives details of selected applications which point the way to a wider use of this method in general machining applications.

Le contrôle des pièces ou des outils en cours d'usinage gagne de l'importance, notamment en ce qui concerne les tours et les centres d'usinage. Dans la plupart des applications, le contrôle s'effectue immédiatement après l'opération d'usinage. Ainsi, à proprement parler, il s'agit plutôt d'un procédé post-opératoire. Cet article présente quelques applications choisies en fonction des possibilités qu'offre ce procédé en vue d'une utilisation plus poussée dans d'autres domaines de l'usinage.

El control de piezas o herramientas en fase de fabricación va adquiriendo más importancia, particularmente en lo que concierne a tornos y centros de mecanizado. En la mayor parte de las aplicaciones, la verificación se efectúa inmediatamente después de la operación de mecanizado. De esta forma, y propiamente hablando, se trata más bien de un procedimiento postoperatorio. Este artículo presenta algunas aplicaciones seleccionadas en función de las posibilidades que ofrece este procedimiento, con objeto de una utilización más avanzada en otros campos de la fabricación.

machined it can be called pre-process gauging.

Recently in-process gauging has gained in importance because of the growing interest in extended automatic operation of numerically controlled lathes and machining centers in linked lines, flexible manufacturing cells or flexible manufacturing systems. In most machining set-ups of this type an extended unmanned shift – and in exceptional cases a completely unmanned third shift – is only possible by employing in-process gauging to initiate corrective measures before tool wear, workpiece variations (including clamping inaccuracies, Fig. 2) and other error sources lead to rejects and shutting down of the line.

2 Case Study: Wheel Hub

Because of the preponderance of linked lines of automatic

Fig. 2: Compensating errors in workpiece alignment is another major function of probes on machining centers. A system of this type is employed by ASEA for unmanned operation of a linked line. (Burkhardt & Weber)

◄
Fig. 1: Probe mounted in the spindle of a machining center for checking the dimension of a critical bore. (Deckel)

lathes in the automotive industry, in-process gauging has found numerous applications for components such as wheel hubs (Fig. 3). In this example taken from Volkswagenwerk, a 3-spindle NC lathe (Heynumat 10 F/3 developed by Heyligenstaedt) machines the critical cylindrical surface (57 mm diameter, h 9 tolerances) on three workpieces at the same time. The in-process gauge (a Samsomatic unit) automatically adjusts the cutting edge of the turning tools so as to compensate for tool wear during the production process.

"Permissible dimensional tolerance of the workpiece, the required correction step for the cutting tool, the limiting value for the total correction range for the tool, and the control tolerance are preset on the in-process gauge", explains Dr.-Ing. H. J. Ehmer, Manager of Technology Planning in Volkswagenwerk's Central Planning Division in Wolfsburg. "In our experience, the optimum value for the control tolerance is determined by the following limits: On the one hand it should be larger than 50% of the repeatability (determined by the dispersion of errors) of the machine, and on the other hand it should be smaller than one-third of the required workpiece tolerance."

During the gauging operation – which is effected in a gauging station when the loading device picks up the machined workpieces – all three components are checked against the h9 tolerance of the cylindrical surface of 57 mm diameter. If the gauged value is outside the control tolerance, a tool correction is effected after every second gauging operation at the earliest. (If tool correction is effected immediately after detection of a deviation, hunting would occur, i. e. repeated corrections would become necessary.)

If the gauge detects that the limiting tolerance of the workpiece is being approached or has been exceeded – or the tool is broken – no further tool correction is carried out. After completing the current machining operation, the machine is switched off for tool change. Switching off for tool change is also initiated if the total correction range of the tool has been reached before the limiting workpiece tolerance has been approached or exceeded.

Dr. Ehmer points out that downtime due to tool changing is reduced to a minimum by fitting two identical tools to each of the swivel toolholders which are mounted on a precisely adjustable carrier. After a predetermined number of workpieces has been machined (checked by a subroutine of the Siemens Sinumerik 6 T control system), or the permissible workpiece tolerance has been exceeded (checked by the in-process gauge), the stand-by tools are swivelled into operating position. In this case all adjustment steps previously effected by the control unit of the gauge are reset to zero. The same number of workpieces can now be machined before the swivel tool head has to be replaced.

3 Case Study: Differential Gear Unit

A similar tool correction system is in operation in the new General Motors Engine and Transmission Plant at Aspern, near Vienna. The workpiece is a differential gear unit with two close-tolerance bearing journals (Fig. 4). This component is machined in a two-section transfer machine (Heller). In one of the stations, both bearing journals are simultaneously precision turned by two slides fitted with in-process gauges (Feinprüf).

"In-process gauging was applied in this case to minimize subsequent grinding time", explains W. Siegfried, Production Planning Department, Adam Opel AG, Rüsselsheim. "The bearing journal diameters are required to be within ± 0.02 mm (i. e. within the range 35.261–35.300 mm). When the control tolerance is exceeded, the turning tools are automatically adjusted. And when the permissible workpiece tolerance is exceeded, the transfer machine is switched off."

Optimum machining results are ensured by adjusting the

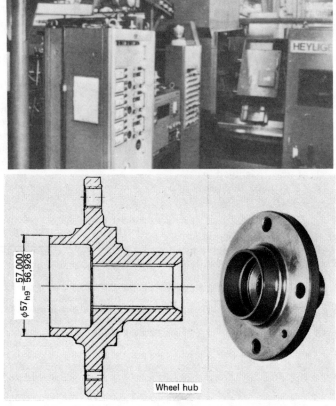

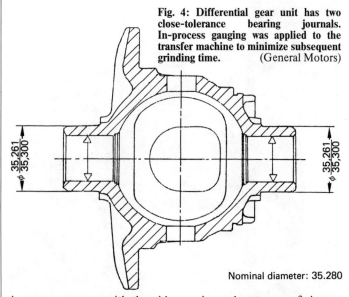

Wheel hub

Fig. 3: A 3-spindle NC lathe (top) is one of the machines in a linked line for machining wheel hubs (bottom right). An in-process gauge automatically adjusts the cutting edge of the turning tools to compensate for tool wear, thus ensuring an h 9 tolerance on the bore of 57 mm diameter (bottom left).
(Volkswagenwerk)

Fig. 4: Differential gear unit has two close-tolerance bearing journals. In-process gauging was applied to the transfer machine to minimize subsequent grinding time. (General Motors)

Nominal diameter: 35.280

in-process gauges with the aid of a master. This master is ground and polished to within a few microns. Gauge adjustment is carried out before every shift.

Mr. Siegfried reports that major advantages of inprocess gauging in this application are: eliminating downtime due to manual gauging and cutting tool correction; increasing utilization rate of the machine; eliminating rejects.

Fig. 5: Machining aero engine disk rotors on a vertical turning lathe fitted with a touch trigger probe (top). Since the probe can check the complete cutting edge, deposited or crumbled cutting edges can also be detected (bottom). (Fiat Aviazione)

Fig. 6: Workpiece gauging ensures IT 6 tolerances (± 0.02 mm) on a family of bevel-gear pinions for trucks machined on a linked line comprising two NC lathes at Renault Véhicules Industriels. (Georg Fischer)

4 Case Study: Machining Aero Engine Components

High demands are placed on dimensional accuracy when machining aerospace engine components. The exotic materials employed not only make the cost of the unmachined workpieces high but also call for difficult machining operations. An interesting example is machining disk rotors for aero engines at Fiat Aviazione in Torino. The material employed is Waspalloy, a high-strength alloy containing nickel, chromium, molybdenum and titanium.

Machining is carried out on a vertical lathe (Comau) fitted with a continuous-path numerical control unit (Elsag) and a touch trigger probe (Renishaw) with three-dimensional deflection. Before the machining cycle – which comprises two semi-finishing and two finishing operations – is started, the touch probe emerges from its protective housing (Fig. 5, top) and checks the cutting edge of the turning tool. The measured values are entered in the numerical control unit as tool corrections. Since the probe can check the complete cutting edge, deposited or crumbled cutting edges can also be detected (Fig. 5, bottom). An alarm signals these tool defects and possible tool breakage.

"Setting-up time for this workpiece has been reduced by a factor of 5 through the use of the touch trigger probe", reports a spokesman for Fiat Aviazione S.p.A., Torino. By fitting the probe to the ram of the vertical lathe, dimensional check of the workpiece can be effected on the machine itself. Here the probe replaces a conventional toolholder and is stored in the magazine. However, because of the heat generated during the long turning operation (10 h), a prerequisite for accurate measurement is the use of a heat exchanger for maintaining the coolant at constant temperature.

5 Case Study: Truck Bevel-Gear Pinions

Machining a family of bevel-gear pinions for trucks at Renault Véhicules Industriels to IT 6 tolerances (± 0.02 mm) on a linked line comprising two NC lathes (Georg Fischer) is another example of the potential offered by in-process or feedback gauging (Fig. 6).

Because of the high cutting speeds (220 m/min at feed rates of 0.25 mm/rev.) during finishing, the cycle time lies under 3 min. Workpiece loading and gauging take up only 15 s. To ensure reliable broken tool detection, gauging of the heavy workpiece (25 kg) is effected before starting the machining cycle for the next workpiece.

Workpiece gauging on the machine is one of the two measures taken to ensure unmanned operation of the machines. The other is to fit multiple tooling, thus enabling tools to be automatically changed when tool life has expired. Since only four tools are needed for the machining operation, three sets of tools can be fitted to the 12-tool revolver.

After changing a set of tools, the first workpiece is machined so as to be oversized. The deviations are then measured and become effective as corrections for the next workpiece. As a result, this workpiece is within the required tolerance of ± 0.02 mm at the first try. The oversized workpiece is automatically ejected and later on loaded into the linked line for finishing.

6 Outlook

In-process – or more precisely, indirect in-process or feedback – gauging brings distinct advantages to close-tolerance machining operations such as precision turning, boring and milling. Already used in numerous transfer machines and linked lines carrying out precision boring or turning operations in the automotive industry, this method is a prerequisite for extending the operation of NC lathes and machining centers.

Recently, the development of touch trigger probes which can replace normal tools in a tool magazine has opened new possibilities for this gauging method. In conjunction with numerical control units, the probe can automatically compensate for workpiece misalignment, tool length variations and errors due to tool wear (Fig. 7). ■

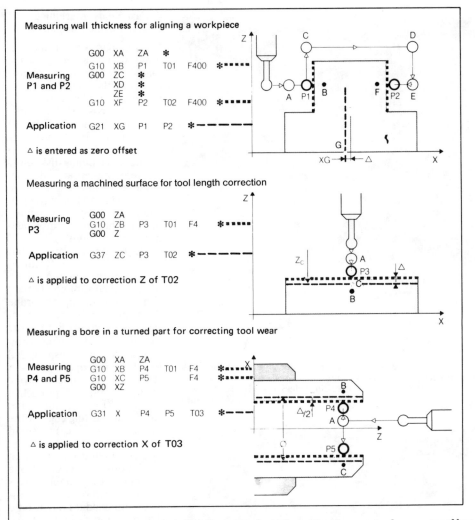

Fig. 7: Special software on numerical controls enables the control system to process values measured by a probe and thus compensate for workpiece misalignment (top), tool length variations (center), and errors due to tool wear (bottom).

(NUM-Güttinger)

Reprinted from *VDI International Magazine* November 1982

Progress Report: Measuring Machines

Coordinate measuring machines have achieved a high technological standard both in hardware and software. Dimensions, form and position of geometric elements can be measured quickly and precisely by standard software. In addition, special software is available for specific applications such as statistical evaluation of measured data, curve and gear measurement as well as probing curved surfaces. A recent trend is that towards the use of coordinate measuring machines on the production floor.

Fig. 1: Numerically controlled coordinate measuring machines are increasingly being placed directly next to machining centers. Automated measuring sequences enable routine checks to be effected on multiple workpieces precisely by merely pressing a button. (Zeiss)

Table 1. Coordinate Measuring Machine Applications and System Design

Application Area

- Metrology department
- Workshop
- Production shop
  Island application
  Integrated application: Transfer line, flexible manufacturing system

System Design

- Hardware: Basic machine, probe system, probe changers, rotary table, workpiece handling devices
- Software: Basic system software, application-oriented software, interfaces to DNC, CAD/CAM, CAI.

1 Introduction

Coordinate measuring machines are finding growing applications in metalworking plants. Originally designed for high-precision measurements on relatively few workpieces in the metrology department, they are now increasingly used both in the workshop and on the production floor (see Table 1). The reason behind this shift is the demand for accurate checking of dimensions, form and position of geometric elements of workpieces after machining. In many cases precise checks are needed for evaluating long-term trends and thus compensating for the influence of thermal effects or tool wear on machining accuracy.

Basically, the coordinate measuring machine represents a multifunction measuring system designed to maintain a high level of accuracy. A single measuring setup, and hence a single workpiece clamping, suffices to provide data on deviations in dimensions, form and position. This contrasts with conventional methods which involve three different setups and need three workpiece clampings.

In recent years measuring machine hardware has been extended to cover a wide range of workpieces. The majority are designed as portal or bridge-type machines with a fixed granite measuring table. The cantilever or horizontal arm type is normally employed for small workpieces. Measurement is usually carried out in all three co-ordinates. Three-dimensional probes are common, and many are suitable for dynamic measurement.

Standard measuring software available with modern measuring machines enables automatic calibration of the probe, determination of the workpiece coordinate system, calculation of geometric elements (either using a minimum number of points or by means of statistical evaluation of multiple points), as well as automatic evaluation and recording of the complete measurement. Automatic compensation of workpiece misalignment is a feature offered on many machines.

Special software is available for numerous specific applications including statistical evaluation of measured data, curve measurement, gear measurement and probing curved surfaces (Table 2).

2 Wider Applications

"Recent applications show that numerically controlled measuring machines can be located next to machining centers", states H. J. Neumann of Carl Zeiss, Oberkochen (D). "Fast feedback of measuring data significantly reduces idle times of the production equipment. Another advantage of NC measuring machines is their application in unmanned production lines. During the manned shifts unskilled operators can carry out checks on all workpieces, including combinations of different workpieces (Fig. 1). After changing the probes and loading the workpieces, the automatic measuring sequence can be initiated by simply pressing a button."

According to Mr. Neumann, Zeiss has greatly extended their range of measuring machines. These include models designed for fully automatic monitoring with the aid of robots as well as measuring installations suitable for large measuring ranges such as those needed for checking automotive bodies

Des progrès considérables ont été réalisés récemment dans le domaine des machines à mesurer par coordonnées. Le logiciel d'application raffiné assure le contrôle dimensionnel entièrement automatique et l'impression des résultats pour différents types de pièces d'usinage. Le présent rapport traite des derniers développements en hardware et en software.

Recientemente ha habido progresos muy notables en el campo de las máquinas de medición de coordenadas. La dotación lógica (software) de hábil aplicación, asegura las verificaciones totalmente automáticas de dimensiones y la representación gráfica de los resultados para diferentes tipos de piezas de trabajo. El presente informe trata de los últimos adelantos en la dotación física (hardware) y en la dotación lógica (software).

accuracy. Measuring robots, for example, are designed for high measuring speeds and therefore have to take points on the fly, i.e. not at constant speed, thus reducing their accuracy. Our machine range has five vertical models up to 900 x 1650 x 600 (Fig. 3) and two horizontal models of 400 mm cube and 400 x 650 x 400 mm. Accuracy is ± 3 μm for the smaller models and ±8μm for the larger models."

Fig. 2: Multi-column measuring setup for simultaneously checking tolerances at three points of an automotive body. (Zeiss)

and railway waggons (Fig. 2). "Coordinate measuring is now successfully used in many forming areas, whether they involve complex castings, forgings or sheet metal structures", notes Mr. Neumann.

Ing. Sergio Naurelli, Marketing Manager, Measuring Systems and Robotics Division, Olivetti Controllo Numerico S.p.A., S. Bernardo d'Ivrea (I), states that today's market demands machines of different types: "Users should bear in mind, however, that there is always a trade-off between measuring speed and

3 Sophisticated Software

As Mr. Naurelli points out, a decisive factor in applying a measuring machine is the available software. Olivetti employs SCAI (Software for Automatic Control on Inspector) for speeding up measuring operations and automatically printing out inspection records listing all out-of-tolerance values. "In addition to giving automatic workpiece alignment in space, this software enables measured results to be

Table 2. Basic Types of Measuring Machines, Probe Types, Operating Modes, Programming, Software, Accessories

Machine Types	1. Cantilever or horizontal arm 2. Portal or bridge type	**4. Scanning** Scanning enables contours to be digitized at closely spaced intervals for form measurement. The scanning device digitizes probe deflection in each measuring axis.
Probe Types	**1. Hard probes** A hard probe (or trigger probe) is a rigid sensor designed to engage the workpiece being measured. Disadvantages: Hard probes are slow in operation and can only be used on manual machines. Although machines using hard probes are cheaper, they have limited applications.	
		Programming Methods — **1. Teach-in (self-learning)** Basically corresponds to the manual measuring sequence. Once the first part has been measured, the program is complete.
	2. Soft probes A soft probe (or touch-sensitive probe) has the capability of sensing motion or deflections when it contacts the workpiece being measured. Advantages: Soft probing eliminates some of the error factors due to operator inconsistency in manual machines, thus leading to higher repeatability.	**2. Part programming** The measuring path is programmed by methods similar to those employed for NC machining.
		Application Software — **1. Standard programs** Geometric elements can be measured for dimensions, form and position. Evaluation can be made for straightness, squareness, roundness, cylindricity, rectangularity, inclination, parallelism, position, symmetry, coaxiality, concentricity, runout, etc.
Operating Mode	**1. Manual measurement** Operator moves probe from point to point and takes readings from digital display.	**2. Statistical evaluation** Statistical analysis of measurement data and their organization in statistical form, i.e. graphical representation in diagrams, histrograms, etc.
	2. Computer aided measurement The probe is moved manually by operator but the measured data are processed by pre-programmed software routines so as to obtain dimensions and tolerances.	**3. Curve measurement (2-dimensional)** For measuring cams and other complex contours. **4. Gear measurement** For measuring gear parameters such as flank profile, base diameter, pressure angle, tooth trace, helix angle, cumulative pitch error, difference between adjacent pitches, radial runout. For spur, helical and bevel gears.
	3. Numerically controlled (NC or CNC) measurement All axes of the measuring machine are equipped with servo controls (i.e. drives with feedback loops) so that the machine is controlled by the part program. The numerical control moves the probe in all three axes to each nominal location. Application-oriented software in the computer carries out required mathematical computations and prints out the results (dimensional, form and position values and tolerances).	**5. Curved surfaces (3-dimensional)** For measuring models, turbine blades, etc.
		Machine Accessories — **1. NC indexing tables** For simplifying and speeding measurement and programming of rotationally symmetrical parts.

Fig. 3: Software capabilities are a decisive factor when studying the application of coordinate measuring machines. Olvetti's SCAI provides component alignment in space, dimensional tolerance print-out, statistical processing and other features for 4 simultaneously controlled axes.

statistically processed", adds Mr. Naurelli.

An extensive software library is offered by DEA, Digital Electronic Automation S.p.A, Moncalieri (I), one of the pioneers of coordinate measuring machines. "Our latest development is SAGE, an interactive computer graphics system for statistical analysis of inspection data on the shop floor with visual presentation of trends", reports M. Ercole, Measuring Robot Sales Support Manager of DEA. "This system enables the user to qualify machines and route work, optimize toolchange frequency, isolate causes of machine downtime, and reduce scrap by monitoring dimensional trends and preventing production of defective parts. Results are presented in full-colour tables,

bar charts, and other graphic displays."

The system provides real-time statistical analysis of inspection data and full monitoring of the inspection process on an interactive basis that is easy to learn and use. It is capable of presenting graphic data in 16 different ways.

4 Measuring Robots

DEA's statistical analysis software is particularly useful when applied to the company's measuring robot (Fig. 5). According to Mr. Ercole, the robot called "Bravo" first came on the market roughly a year ago and has already been applied for fully automatic in-line dimensional inspection in a flexible manufacturing system designed for producing seven different motorcycle

gearboxes (Fig. 6). The robot has a maximum traversing speed of 33 m/min and an acceleration capacity of 3.3 m/s^2. Axial length measuring accuracy is $10 + 6L/1000$ μm with software error compensation, and $10 + 16L/1000$ μm without compensation. (L is the measuring stroke length in mm.)

When applied to inspection of as-cast aluminium cylinder heads, the DEA robot can check 165 points in less than 3 min, thus giving a time saving of 93% compared with conventional measuring methods. The computer numerical control fitted to the system

enables the robot to carry out multi-point measurement of plane and spatial geometric elements. Part misalignment is automatically compensated. Checks can also be effected for a full range of form and attitude errors. Since the CNC system can control two robot arms at the same time, it provides an economic solution to the problem of reducing the overall cycle comprising inspection and part handling phases. (The two robots are programmed to operate on opposite sides of the same part.)

"The measuring robot has two major fields of application", explains Alain Louveau,

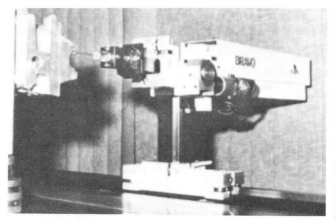

Fig. 5: DEA's measuring robot is designed for fully automatic in-line dimensional inspection. It is suitable for monitoring 100% of medium volume production runs.

Fig. 4: Measuring robots are a new approach to reducing the time delay between the machining and measuring operations in transfer lines or in flexible manufacturing systems.   (Renault - SEIV Mesure)

Commercial Manager of SEIV Mesure, Evry (F), a subsidiary of Renault Automatismes. "One is the FMS or flexible manufacturing system and the other is the transfer line. In the first case small and medium batches are handled and in the second high-volume production runs."

Mr. Louveau emphasizes that SEIV's robot – called AMT (automate de mesure tridimensionnel) – has been designed to provide full integration into the manufacturing process, whether based on FMS or transfer line techniques. All phases of operation, i.e. loading the workpiece, initiating the measuring cycle, transferring the measuring results to the production system, and unloading the workpiece, are fully automatic. Fig. 4 shows one solution based on a roller conveyor for

automated handling of the workpieces fitted to pallets.

In the case of a measuring robot integrated in an FMS environment, two points are particularly important: First, there must be a dialog between the robot and the supervisory computer system. Second, the robot has to be equipped to provide real-time data on workpiece quality so that the central computer can analyze the results and initiate corrective measures if necessary. The AMT robot employs powerful software derived from PROMESUR and specifically adapted to effecting on-line diagnostics and real-time machine compensations.

Another approach to high-speed measurements on the production floor is the PAG (Programmable Automatic Gauge) machine developed by Imperial Prima S.p.A., Torino

Table 3. Accuracy Specifications for Coordinate Measuring Machines
(Source: CMMA, Coordinate Measuring Machine Manufacturers Association)

Geometrical Accuracies	Determined by independent measurement because they make major contribution to overall accuracy of machine

1. Straightness of axes

Definition: Deviation from a straight line in two orthogonal planes for each axis of movement. Six measurement parameters are therefore needed: Straightness of X axis measured in Y and Z directions; of Y axis in X and Z directions; of Z axis in X and Y directions.

Measurement is effected against a suitable straightness reference, e.g. laser beam (min. 10 points measured in each direction over full travel of each axis). Typical deviation record:

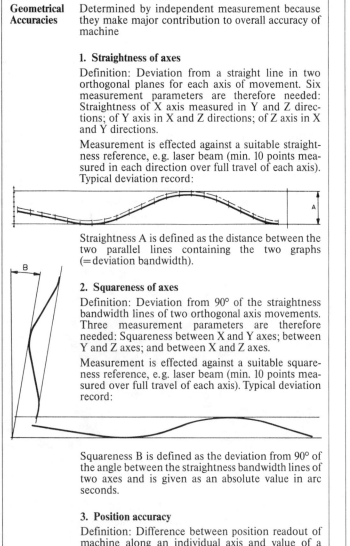

Straightness A is defined as the distance between the two parallel lines containing the two graphs (= deviation bandwidth).

2. Squareness of axes

Definition: Deviation from 90° of the straightness bandwidth lines of two orthogonal axis movements. Three measurement parameters are therefore needed: Squareness between X and Y axes; between Y and Z axes; and between X and Z axes.

Measurement is effected against a suitable squareness reference, e.g. laser beam (min. 10 points measured over full travel of each axis). Typical deviation record:

Squareness B is defined as the deviation from 90° of the angle between the straightness bandwidth lines of two axes and is given as an absolute value in arc seconds.

3. Position accuracy

Definition: Difference between position readout of machine along an individual axis and value of a reference length measuring system. Three measurement parameters are therefore needed: Position accuracy of X axis, of Y axis and of Z axis.

Measurement is effected along one measuring line for each machine axis located approx. at center of measuring travel of remaining two axes. For this purpose a suitable reference length measuring system, e.g. laser interferometer, is aligned to each machine axis within a permissible deviation of 1 arc minute (min. 20 points measured over full travel of each axis). Typical deviation record:

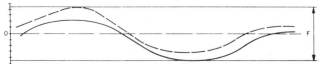

Position accuracy F is defined as the distance between the two parallel lines containing the two graphs for the two directions.

Total Measuring Accuracy	Determined by utilizing the entire measuring machine system as applied to master gauges.

1. Axial length measuring accuracy

Definition: Difference between the reference length of gauges, aligned with a machine axis, and the corresponding measured results from the machine.

Three reference gauges are measured in each of the three axes X, Y and Z. Gauge lengths approx. 1/3, 1/2 and 3/4 of full travel of respective axis (up to a max. of 1000 mm).

Length measuring accuracy G is defined as the absolute value of the difference between the calibrated length of the gauge block and the actual measured value.

2. Volumetric length measuring accuracy

Definition: Difference between the reference length of gauges, freely oriented in space, and the corresponding measured results from the machine.

Three reference gauges are measured, their lengths corresponding to approx. 1/3, 1/2 and 3/4 of the full travel of the longest axis (up to a max. of 1000 mm).

Volumetric length measuring accuracy M is defined as the absolute value of the difference between the calibrated length of the gauge block and the actual measured values.

(I). The novel features of this coordinate measuring machine (Fig. 7) are its automatic probe changer and its completely enclosed design. "Even in the case of complex workpieces, the lathe or machining center operator can handle the PAG measuring machine", says Ing. Luigi Lanotte of Imperial Prima. "In fact, the PAG not only looks like but also operates like a machining center. Pallets and probes are coded so that the risk of errors in conjunction with different part programs is eliminated."

It is claimed that the machine's enclosure eliminates the problem of the adverse effect of the production shop's environment, in particular dust, smoke and steam. Measuring accuracy is stated to be $6 + L/120\,\mu m$. The rotary indexing table has a repeatability of $\pm 2\,\mu m$. Accuracy of each axis (measured with Renishaw probe against high accuracy gauges aligned along each axis) is $4 + L/150\,\mu m$. (L is the measuring length in mm.) Maximum traversing speed is 20 m/min, and maximum useful table load is 100 kg.

5 Form Measurement

"Form measurement repre-sents an important field of application for coordinate measuring machines", states Dipl.-Ing. R. Rottstock, European Sales Manager of Ernst Leitz Wetzlar GmbH, Wetzlar (D). "For this purpose we have developed a scanning technique enabling up to 50 measuring points to be processed per second", explains Mr. Rottstock. "This contrasts with one measuring point every two seconds with conventional single-point probing. The main application is in measuring contours which cannot be described by simple geometric formulae, e.g. cams and cutting dies (Fig. 8)."

The diagram shows the scanning plot of a cutting die compared with a master. Deviations from the reference are indicated by small, fine lines or bristles at right angles to the reference contour and enveloped by the tolerance limits. The diagram is based on 600 measuring points which were taken in 60 s. Evaluation time is 80 s. With conventional single-point probing, a minimum of 27 measuring points would be needed for the five circular and six straight-line sections of the complete contour. This would involve a measuring and evaluation time of roughly 150 s.

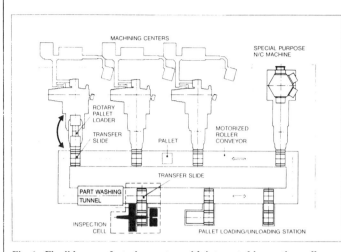

Fig. 6: Flexible manufacturing system with integrated inspection cell comprising DEA's measuring robot. The FMS is designed for a family of seven different motorcycle gearboxes. Average inspection time is 56 s including component recognition, misalignment compensation and checking 20 geometric elements.

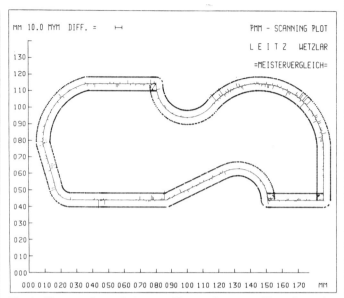

Fig. 8: The scanning technique enables continuous probing of complex contours, such as those needed for cams, based on closely-spaced equidistant measuring points. This diagram shows the scanning plot of a cutting die compared with a master. Deviations from the reference are indicated by the small, fine lines at right angles to the reference contour and enveloped by the tolerance limits. Here 600 measuring points are taken in 60 s. (Leitz)

According to Mr. Rottstock, the scanning method enables a large number of points to be measured. If evaluation of these points is effected on the machine, the total cycle time is approximately the same as for conventional single-point probing, but three times as many points are acquired. "And if evaluation is done remote from the machine, the measuring cycle is reduced by 60% (60 compared with 150 s)", he adds. "The scanning method has the further advantage of providing a graphical representation of the results, thus enabling a qualitative evaluation of the form."

Form measurement is the underlying objective of the coordinate measuring machines developed by FAG Kugelfischer Georg Schäfer & Co., Schweinfurt (D). The latest addition to the company's range is model MGF 10.2 with traverse paths of 600 x 400 x 900 mm and providing a maximum measuring range of 1300 mm. An optical incremental linear measuring system gives a resolution of 1 μm or 0.01°. An inductive probe is used for form measurements, whereas coordinate measurements are effected with an adjustable 3D probe. Straightness is measured at a speed of 0.25 or 1.5 mm/s and roundness at 0.1 rev./s. A total of 1024 points are continuously measured in both cases. Three-dimensional sensing is effected point by point, the maximum speed of travel being 25 mm/s.

6 Microprocessor-Based Controls

Several new microprocessor-based controls have been developed by coordinate measuring machine manufacturers to provide the user with operator-friendly techniques. An interesting approach is that taken by Ferranti Ltd., Dalkeith (GB). As shown in Fig.10, the Micro-900 display is provided with eight soft keys, i.e. keys with variable functions, which handle all the commands needed for man-machine dialog. The soft keys change their functions throughout the measuring sequence. At each stage their function is clearly defined on the display, immediately adjacent to the keys.

Among the commands available are: full 3D alignment,

Fig. 7: Specifically designed for three-dimensional inspection on the shop floor, this completely enclosed machine features a probe changer automatically controlled by the measuring program. Developed by Imperial Prima, it was shown at the IMTS '82 by Bendix.

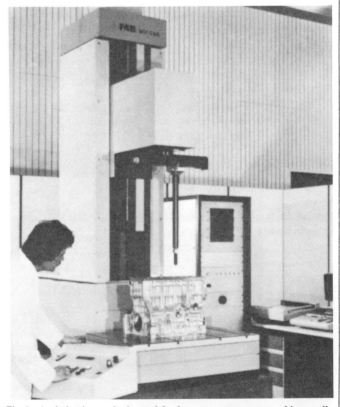

Fig. 9: An inductive probe is used for form measurements on this coordinate measuring machine which has a maximum measuring range of 1300 mm. Straightness and roundness measurements involve 1024 points. Three-dimensional sensing is effected point by point. (FAG Kugelfischer)

375

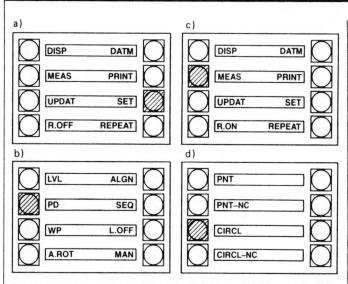

Fig. 10: **All commands needed for man-machine dialog are handled by eight soft keys on the Micro-900 inspection machine display. The microprocess-based unit was shown for the first time at Microtecnic '82 in Zürich. The illustrations show how the soft keys change their functions for (a) setting, (b) probe diameter input, (c) measuring routine selection, and (d) measuring circle mode.** (Ferranti)

During the plain-language dialog the operator answers with only a single key stroke. Default answers are highlighted on the screen. Once a part program file has been created it becomes a permanent record of the measurements to be carried out including nominal dimensions and tolerances involved. The program then leads the inspector through the part, displaying features to be measured on the screen. A special software package is available for incorporating CAD/CAM quality assurance functions for on-line applications.

Linking of coordinate measurement with CAD/CAM is also a feature of the computer numerical control system fitted to the SIP-560M machine built by SIP, Société Genevoise d'Instruments de Physique, Geneva (CH). "Our LOGISIP software gives the system a number of distinct advantages", states Fréderic Aubert, a SIP engineer. "In addition to providing man-machine dialog, it automatically controls the machine, its indexable spindle and the indexing table. Workpiece mis-

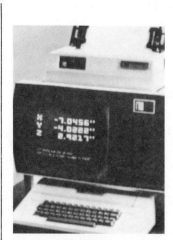

Fig. 11: **User-friendly prompting, plain-language dialog with single-stroke operator answers and highlighted default answers are features of the Numerex coordinate measurement software.**

alignment can be corrected, and the probe can be calibrated. Measurement results can be analyzed and basic geometric calculations carried out, e.g. calculation of distances, intersections, angles, symmetries, etc. Measured and contour points can be stored, and results can be graphically displayed." ∎

choice of three working planes, multi-point measurement of circle centers and diameters, probe diameter or radius compensation, datum translation or transfer, sub-datum facility to match drawing dimensioning, axis rotation. A repeat function simplifies inspection of similar features. And a "break-in" facility enables a second component to be inspected in manual mode without loss of original component references.

User-friendly prompting is also a feature of the system introduced by Numerex Corporation, Minneapolis (USA).

Presented at the SME QualTest-I Conference, October 1982

Automatic Inspection Machines

By Alberto Imarisio
Marposs Gauges Corporation

INTRODUCTION

In high production manufacturing facilities, the use of automatic gauging machines to perform final inspection operations are an economical means to qualify parts for assembly or shipment. The variety of gauging machine designs are as varied as the vast array of parts produced. However, there are enough similarities between all gauging machines that a simple definition can be attempted.

A gauging machine automatically transports parts to one or more measuring stations so that they can be inspected for quality. This inspection is performed during the automatic cycle of the machine and does not involve direct human intervention to judge quality. Movement of the parts through the machine is accomplished by electro-mechanical, hydraulic or pneumatic mechanisms. The machine can be either manually or automatically loaded. Several measurements are usually taken, by this type of final inspection equipment, due to the great efficiency of their operation. In cases where full automation is incorporated into the cycle of the machine, some way of identifying good and bad parts must be accomplished. This may mean a marking station and/or the use of a reject chute to collect scrap parts.

In the past, the main purpose of an automatic gauging machine was simply to final inspect parts at the end of a manufacturing process. The present industry trend is to use them also for automatic inter-operational inspection for the purpose of realizing a direct process control. Inter-operation machines are normally simpler than final inspection machines. Fewer measurements are required because they are related to only one or two metal removal operations. In addition, most measurements are taken statically which simplifies the inspection process. By monitoring the quality output of a machine immediately after the operation is completed, direct process control can be achieved. If part size is approaching an out-of-tolerance limit, proper compensation feedback can be given. If size is out of tolerance, the machine can be stopped before any more scrap is produced. Electronic gauging machines are the most suitable means to achieve direct process control and overall quality control. This is because they can provide measurement output information which can be easily interfaced with statistical analyzers, printers and computers.

The individual components of either a final inspection or inter-operation gauging machine deserve a closer examination to appreciate their basic function and how they work together to make a complete system.

1. Part Transportation: Each different part requires a detailed study and engineering to assure efficient and safe movement through the machine.

2. Measuring Station: Depending upon the complexity of the gauging task and the physical characteristics of the part, one or more measuring stations are required to inspect the part.

3. Gauging Amplifier: Information received from the measuring station(s) is processed by the gauging amplifier and compared to a desired and predetermined standard.

4. <u>Logic System</u>: This system controls all of the machine
    movements while also keeping track of good and bad part
    conditions that were determined by the gauging amplifier.

5. <u>Marking Station</u>: Some way of identifying good and/or
    bad parts must be accomplished.

6. <u>Reject Station</u>: Bad parts can be moved out of the machine
    into a scrap chute to avoid any possibility of using them
    in a final assembly.

7. <u>Segregation of Good Parts</u>: In some instances there may
    be different classifications of good parts. Segregating
    by size can be an effective way in which to improve and
    facilitate assembly.

The development of microprocessor-based circuitry into the design of today's
automatic gauging machine amplifiers has opened a whole new world of advantages
to the user:

-   Increased measuring capacity while reducing the physical
    size of the equipment.

-   Precise and clearer measurement information is available
    to the operator because it is displayed as a written
    message on a screen.

-   Easier resetting operations allow for faster and more
    economical checking procedures.

-   Automatic or semi-automatic mastering of the system is
    easily performed.

-   Self-diagnostic systems allow for easy and fast identi-
    fication of faulty circuits reducing downtime and
    improving maintenance efficiency.

-   Retooling is less expensive and faster because in most
    cases it will not require hardware changes, but only
    software reprogramming.

Considering all of the above factors, the realm of automatic gauging machines and their applications is indeed a broad area. The focus of this paper will examine one specific automatic gauging machine used as a final inspection system in an automotive plant. It represents the latest state of the art in automatic gauging machines.

Figure 1

This machine (Figure 1) performs the final inspection on crankshafts used in an automobile engine. A total of 85 measurements are taken simultaneously during a 16 second gauging cycle. These measurements are checked by two gauging stations. The first station gauges the five main bearings, post end, flange, pilot hole and thrust faces for a total of 57 measurements. The second station inspects each of the pin bearings for the remaining 28 measurements. All measurements are taken dynamically while the crankshaft is turned approximately two revolutions. Figure 2 shows a schematic diagram of the position of each gauging point and a list of measurements taken at each gauging station.

GAUGE STATION #1

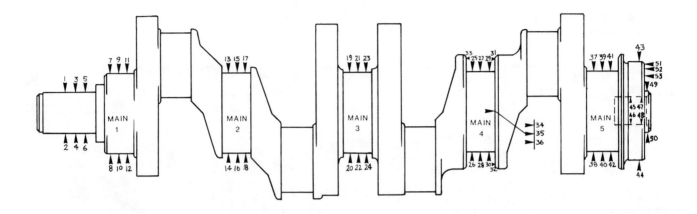

Measurements Performed:

Diameter, Maximum Diameter, Taper, Ovality and Straightness of
Main Bearings 1 – 5; Post End, Gear Fit, Oil Seal, Hub Diameters;
Bushing Bore Diameter, Taper and Concentricity to Main Bearings
1 & 5; Thrust Wall Width and Concentricity to Main Bearings 1 &
5; Total Indicated Runout of Mains 2, 3 & 4 with respect to Main
Bearings 1 & 5, and overall T.I.R.

GAUGE STATION #2

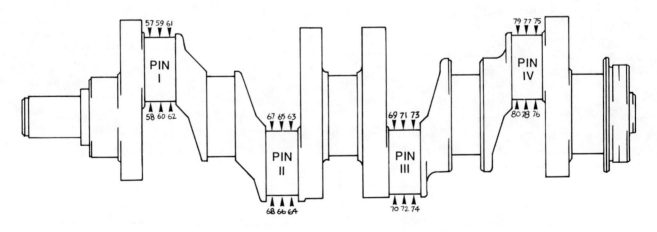

Measurements Performed:

Diameter, Maximum Diameter, Taper, Ovality and Straightness of
Pin Bearings I – IV

Figure 2

Figure 3

Each part checked is loaded by the customer's automation system onto the gauging machine's lift and carry mechanism (Figure 3). After two idle stations, the part reaches the first gauging station. Here two centers lift the crankshaft from the nests of the lift and carry bars. A driver mechanism engages the end of the post end diameter while the frame supporting the gauging fixture is lowered onto the part. When each of the 53 contact gauging points are in proper position, the part is rotated and the gauging cycle begins (Figure 4). All measurement data from each of the contact points are sent to the microprocessor-based amplifier where they are stored in memory.

Figure 4

At the end of the second revolution the gauging fixture is raised, the driver disengaged and the centers opened so that the part rests again on the lift and carry nests.

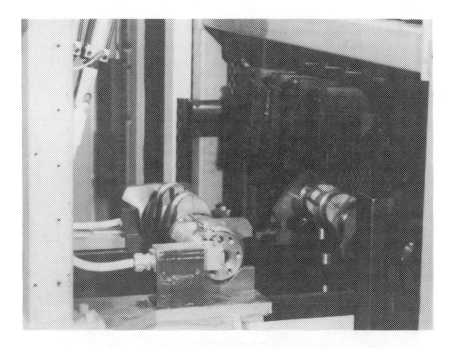

Figure 5

After the crankshaft has been carried through the next five idle stations by the lift and carry mechanisms, it reaches the second gauging station (Figure 5). Although the gauging cycle is very similar to the one just described for the first station, the gauging structure is completely different because each gauging assembly related to an individual pin bearing is independent from the other three. This is due to the cranking movement necessary during the rotation of the part. Again the part is rotated for approximately two revolutions, and measurement data is sent to the gauging amplifier.

| Figure 6 | Figure 7 |

After an idle station, the following two are marking stations. Here selective number of colored dots are stamped onto the part to identify the size class of each main bearing (Figure 6), or each pin bearing (Figure 7). Only good parts are marked.

Following the marking stations, parts judged to be out of tolerance that must be scrapped or reworked, are taken out of the main stream of the machine by a reject station (Figure 8).

Figure 8

In the reject station crankshafts are lifted and turned 90° by a lift-turntable mechanism which unloads them onto a reject chute magazine. Good parts are allowed to continue to the last station of the gauging machine where the customer's automation unloads them.

The brain of the measuring system is the gauging amplifier which is located in a separate cabinet on top of the pushbutton console (Figure 1). All signals coming from the 77 contact gauging points are fed into a multiple analog amplifier where all of the dynamic measurement computations like T.I.R. or minimum and maximum diameters are performed. The resulting information is then fed into a microprocessor-based, amplifier-controller. Here the results of all 85 measurements are compared with each related set of tolerance limits. This measurement data is also memorized in order to control the machine's marking and reject stations so that they can be activated at the proper time and sequence. Due to the transfer sequence, each part is gauged and marked/rejected at different times within the machine's cycle. In fact, the amplifier-controller must keep track of several parts which move simultaneously through the machine.

The microprocessor amplifier is equipped with interface circuitry so that it can be connected with a Marposs Statistical Analyzer and an industrial printer. In this way, complete documentation of part quality and statistical evaluation of production can be accomplished. These devices are not permanently connected to the automatic gauging machine. They are conveniently mobile so that they can also be utilized and plugged into other inspection machines within the plant.

DESIGN CONSIDERATIONS

This condensed description of an automatic crankshaft gauging machine shows that complex multiple measurements can be confined into a relatively small space with many advantages. The nature of those measurements, the close tolerance required and the conditions that this machine must work in (a normal manufacturing environment) require very special engineering and design techniques. A close-up examination of some of these special features will provide a clearer understanding of how it was designed to work in a difficult production environment while also limiting the inevitable possibility of human error.

The structure of the machine was engineered to completely separate the part transportation system from the gauging structures. In fact, the base of the hydraulically-operated, lift-and-carry system has floor anchors which are completely independent from those supporting the gauging structures. In addition, special antivibration mounts are used on each of the supporting legs of the gauging structures. Designing the machine in this way helps to insulate the gauging fixture from any possible vibrations from the rest of the machine and the surrounding environment. This will help to ensure that any vibration which may influence gauging accuracy is eliminated.

To avoid any possibility of the machine mechanically going out of sequence, special design considerations were incorporated. The sequence of part lifting, clamping between centers, the gauge fixtures moving down and up, as well as the rotation operation are all linked by a single control mechanism. All movements of the gauge structures in each gauging station are cam operated and powered by a single electric motor reducer. Using this engineering technique avoids any out-of-sequence mishaps.

The centers used to hold the part during gauging rotation are used only for mechanical positioning. They are not used by the gauging system as reference axis. The part axis in fact is electronically computed using the gauging points on main bearing #1 and #5. In this way, the accuracy of the measurement results and gauge repeatability are not affected by small particles present on the part holding centers or from any mechanical wear. The "floating" capability of the measuring points will more than compensate for these error conditions.

A full complement of protective guards are used on all exposed areas of the machine. Not only do they protect the safety of the operator and bystanders, but they also ensure the integrity of the gauging operation. The protective guards help to ensure that a part cannot be loaded into the middle of the machine during the automatic cycle. In fact, if a guard is opened during the normal operation of the machine, it stops running and the "manual" mode is automatically triggered. The manual mode can also be called for by the operator when the mastering procedure of the gauge stations are performed. Safety of hands and arms during this procedure are assured and meet all government regulations.

Normally, due to the stability and linearity of transducers and circuitry used by Marposs, only a single mean size master would be required. In this case, the customer desired to use both a minimum and maximum master, so that he could check the high and low tolerances. The minimum master is used for the automatic mastering operation. In this situation a failsafe method has to be employed so that the machine will not master itself to the wrong master or to a part. After the minimum master is in gauging position and the machine put into the mastering mode, all dimensions are automatically zeroed. If by mistake the operator calls for automatic mastering after he placed the maximum master into the machine, the amplifier screen will display a warning that the wrong master is in place and the mastering command will be refused. Similarly, if a crankshaft is in the gauging station and the mastering command is called for, it too will be refused. The machine has special sensing devices that communicate to it exactly what kind of master or part it is inspecting. Using this failsafe method during mastering, the customer is assured that the gauging machine is properly set up for his final inspection operation.

Another typical human error the machine is equipped to recognize is if the part is mounted backwards or improperly seated on the transfer nests. This is true both at the loading station and the gauging stations. One can imagine the physical damage that would occur if the measuring fixture were to try to gauge a part in the wrong orientation. Not only would it be expensive to repair, but the time lost on the machine would not be acceptable.

A frequent occurrence in a production environment is an unground or oversized crankshaft coming down the line with finished parts. Again, to avoid damage to the gauging fixture a number of safety devices are used to monitor these kinds of conditions. All of the mechanical or gauging devices "entering" the part are springloaded into their proper position instead of being positively pushed. The main gauge frame, which supports most of the gauging points, has a template-like guard which is used for oversized part protection. If excessive force is needed to reach the correct gauging position, it will automatically recoil away from the part before any damage can occur. A safety switch is immediately triggered which shuts the machine down so that the obstruction can be manually removed.

If for any reason too many rejected parts are fed into the reject chute magazine, a full magazine switch will stop the machine so that the operator can check the condition causing the rejects.

OPERATIONAL CONSIDERATIONS

Although this machine is truly an automatic gauging device intended to inspect approximately 225 parts per hour, the operator may want to visually monitor the measurements related to a particular part. In order to do this he must first switch the machine into the manual cycle mode. He can now scan the various measurements of the part, on the amplifiers display screen. This can be accomplished even if the gauge fixture has just been retracted. The memory of the amplifier retains the information automatically.

Should a more comprehensive study be needed of several dimensions for many consecutive parts, a different and much more efficient method is available. A Marposs Statistical Analyzer can be plugged into the machine. This device will automatically gather the necessary data while the machine proceeds in its normal production mode. The statistical analyzer not only gathers the measurement data faster than other methods, but it is much more reliable and accurate. This is achieved by the efficient and direct communication between the two microprocessor-based systems.

The versatility of the microprocessor-based gauging amplifier-controller becomes more evident when a system malfunction needs investigation. A built-in diagnostic self-test mode can be used to pinpoint any electronic failures. Simply, should any portion of the circuit be malfunctioning, the amplifier screen will display which card needs to be replaced. Repairs can be done promptly and downtime for the machine will be minimal.

SUMMARY

Recalling the content of this paper we can say that automatic gauging machines can be divided into two main categories; final inspection and inter-operational. All final inspection gauging machines automatically transport the parts to be inspected to one or more measuring stations. A number of individual components are working together to make a complete system. Today's state of the art takes full advantage of modern microprocessor-based circuitry.

In our Case Study we examined a specific automatic gauging machine used for the final inspection of an automobile crankshaft.

- A total of 85 measurements were taken at a production rate of 225 parts per hour.

- Two gauging stations were used and the part was dynamically gauged during rotation.

- Marking was provided as a classification method.

- A reject station segregates reworkable or bad parts from the good crankshafts.

- It is possible to connect a statistical analyzer and printer to the machine.

- Special design characteristics are necessary to allow a gauging machine to perform at high standards in a manufacturing environment.

- Safety of operators and system integrity must be considered.

- Easy maintenance procedures are accomplished because of the use of modern microprocessor-based gauging amplifiers.

With careful attention to the design and function of an automatic gauging machine, it is possible to bring the accuracy of the quality control lab right into the manufacturing environment. However to do this, a close cooperation among manufacturing and quality control departments within the plant and the designer of modern gauging machines must be maintained. Only this combination will be able to succeed and bring quality control where it is needed most: on the plant floor where parts are manufactured.

THE TECHNOLOGY OF AUTOMATIC GAGING

HARRY M. ABRAHAM, Jr.
Vice President, Engineering
AA Gage Inc.
Detroit, Mich.

How gaging probes are actuated

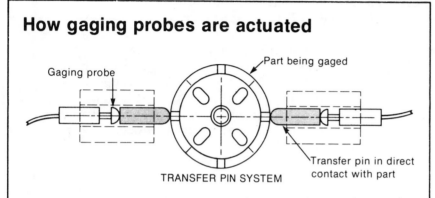

TRANSFER PIN SYSTEM

Whenever feasible, gaging probes are designed to be actuated by an auxiliary probe, or transfer pin, that functions in tandem with the gaging probe. The tip of the transfer pin is machined to the exact shape of the gaging probe.

This technique greatly increases the service life of the relatively expensive probes by eliminating direct contact with the parts being checked. The protection is especially important in automatic-gaging systems where advancing parts would be sliding against or ramming into the probes. The auxiliary tips also help ensure precise linear movement of the probe plunger with respect to the part being gaged.

When the part configuration or the means of conveyance limits the space for probe placement, the probes can be made retractable. This method holds the probe heads back until the part is in gaging position, preventing possible wear or damage to the probes.

Another type of probe actuation uses rocker-arm components to transfer movement from the datum through one or more pivot points to the gaging probe. The ratio of movement from the surface being gaged to the probe plunger can be varied through the mechanical linkage, amplifying the measured tolerance for greater accuracy.

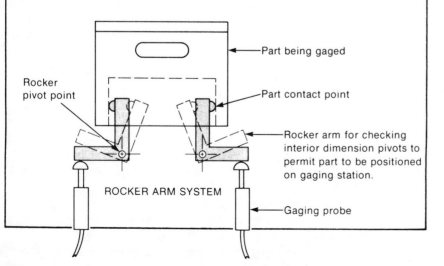

ROCKER ARM SYSTEM

Spurred by product liability and escalating labor costs, more companies are turning to automation for dimensional inspection of manufactured parts. Initital investment in equipment can be substantial, but prudent design planning can keep costs reasonable. The trick is to specify inspection of only essential features.

DIMENSIONAL inspection can be one of the most critical operations in a manufacturing plant. Usually, dimensions are checked manually with micrometers, scales, gages, comparators, or coordinate-measuring equipment.

Inspectors are not infallible, however, and fatigue or carelessness can cause errors. Out-of-tolerance parts may be sent to the assembly line, or the product may fail in service. Manual inspection is also slow and expensive, often contributing to a weak competitive position, especially in selling against foreign-made products.

Thus, for reasons of product safety, manufacturing economy, eliminating the human decision, and legal protection, automation is increasingly being viewed as a practical approach to checking and documenting the dimensional

How a typical automatic gaging system works

Dimensional buildup in a tractor track consisting of 150 or more links could, if the parts are out of tolerance, cause the assembled track to be too short or too long. Thus, the individual links for tractors are 100% inspected by automatic gaging equipment.

①

This part, one of four sizes of forged-steel track links in production, is gaged directly downstream from the production machining system. The automatic gaging machine, using electronic gaging, inspects eight dimensions on the link, compares them against established tolerance limits, and makes a decision as to whether all measurements are in tolerance, out of tolerance, or trending toward out of tolerance.

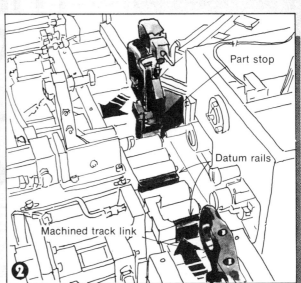

②

The part is released to the gaging station by a reciprocating gate (not shown), which restrains any additional links waiting to be gaged. As the link approaches the gaging area on the inclined conveyor, a bar moves out, stopping the part at the gaging station. The conveyor rolls then drop down, allowing the part to rest on the precision datum rails.

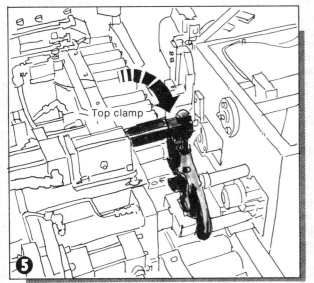

⑤

A clamp swings down and presses onto the top of the part, ensuring that it is properly seated against the base datum rails. The part is now "staged" and is ready to be checked. Before the gage heads are activated, however, a compressed-air blast from the gage spindles cleans the part of chips and coolant.

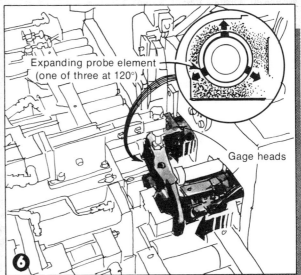

⑥

The gage heads, which can "float" in both the vertical and horizontal directions, then advance into the bores of the part. The heads are designed so there is no rubbing contact as they move into the part. After they are in position, the probe elements (similar to a machine-tool chuck) expand to contact the diameters to be checked. With this one gaging operation, a total of eight checks are made and compared against set limits, and a decision is rendered. Features checked are:

- 1 and 2: diameters of both through holes
- 3 and 4: diameter and depth of the counterbore
- 5: radius of the counterbore
- 6 and 7: height of both holes above the base rail
- 8: location of the holes with respect to each other

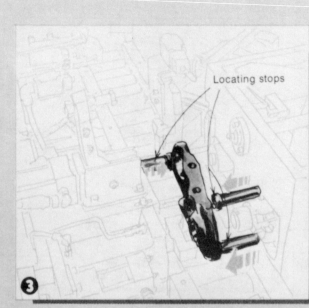

Locating stops

3

Three locating stops move out to a fixed position to establish a plane. The stop bar is retracted.

V-block clamp

4

A V-block clamp advances out from the opposite side, securing the link in the plane of the three locating stops. The clamp also adjusts the part into the proper fore-and-aft position.

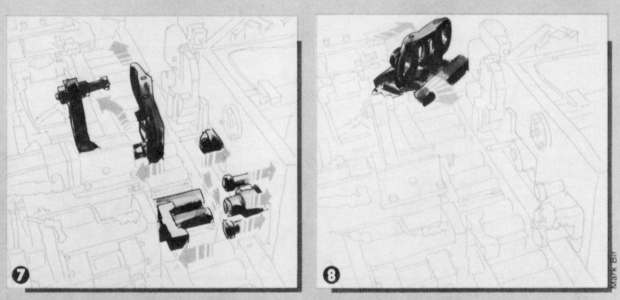

7

After the gaging operation, the gage heads, locating stops, and clamps are retracted, the conveyor rolls are raised, and the link moves along. If all features are within tolerance, the conveyor feeds straight ahead to either an assembly or storage area.

8

If the link has one or more of the checks out of tolerance, a section of the conveyor shunts the link off to another area where it is checked manually for possible salvage.

characteristics of mass-produced parts.

In most cases, the design and configuration of fabricated parts are determined entirely by functional, assembly, or manufacturing requirements. In others, however, some options exist as to the location,

Controlling costs through design

Most of the cost for automatic gaging equipment is not in the gaging operation itself. Rather, it is in the machinery that brings the part to the gaging station, secures it accurately, and moves it out. Consequently, parts should be designed, if possible, so that all dimensional checks can be made from the same clamped position.

Some features are considerably more difficult to check than others. Automatic gaging of these should be avoided if possible:

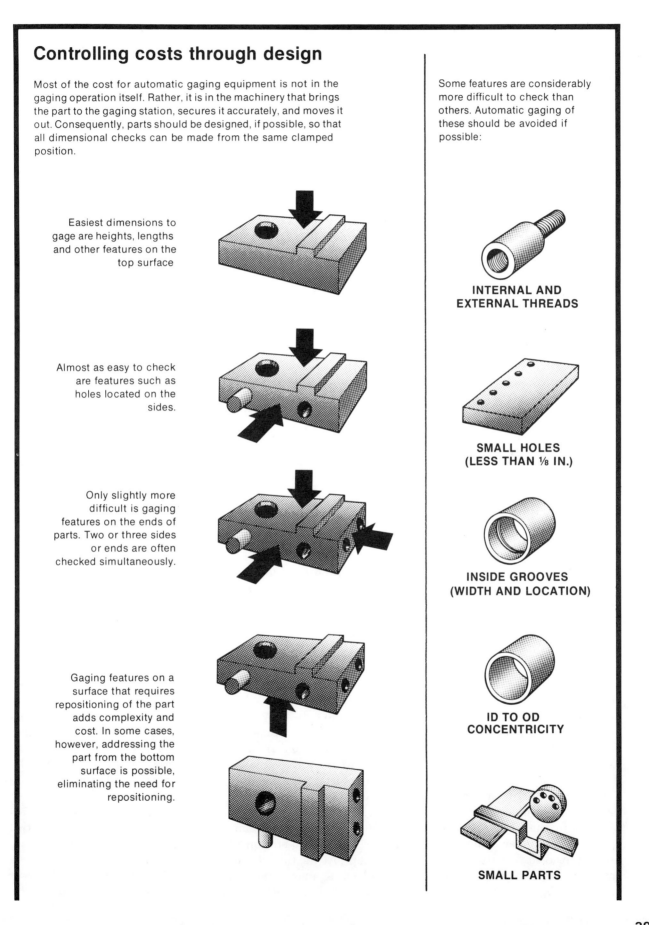

Easiest dimensions to gage are heights, lengths and other features on the top surface

Almost as easy to check are features such as holes located on the sides.

Only slightly more difficult is gaging features on the ends of parts. Two or three sides or ends are often checked simultaneously.

Gaging features on a surface that requires repositioning of the part adds complexity and cost. In some cases, however, addressing the part from the bottom surface is possible, eliminating the need for repositioning.

INTERNAL AND EXTERNAL THREADS

SMALL HOLES (LESS THAN ⅛ IN.)

INSIDE GROOVES (WIDTH AND LOCATION)

ID TO OD CONCENTRICITY

SMALL PARTS

size, or shape of part features. If such parts are destined for automatic inspection, chances are that economies in inspection-equipment needs can be made in the design stage.

By far the largest use of automatic inspection equipment has been in checking automotive production such as rear-axle assemblies, wheel spindles, brake parts, and collapsible steering-column components. Other high-production parts that warrant automatic inspection are, for example, track links for tractors and relay-core spools for communication equipment.

Gaging in action

Automatic inspection—or automatic gaging as it is called—almost always requires custom-built equipment. Although certain standard components and instrumentation are common to a number of installations, each system involves individual engineering, assembly, and break-in, or "debugging," attention. Thus, even a relatively small gaging system—one that checks only one or two features on a part—may cost in the range of $40,000 to $50,000. From there, the cost of automatic gaging can escalate into the area of $500,000 for a highly sophisticated system. Software cost alone for one of the more complex machines can exceed $100,000.

Interestingly, a large portion of the machine cost is usually not in the gaging system at all but in the means necessary to bring the parts to the gage and to move them out after inspection.

The principles of automatic gaging are simple. A part is brought to the gaging station by any of a number of methods and clamped so that its significant features can be measured (usually from a machined

datum surface). Sensing probes then move into or around the part to measure designated features, and electronic responses compare the gaged values to standards. The part is then moved out of the gaging station, often through one of several trap doors or chutes, depending on how the machine is programmed. The exit paths may be simply "acceptable," "salvagable," and "scrap" or, in cases involving selective assembly, for example, parts may be segregated according to the size of a given critical feature.

Staging the part: Before a part can be gaged, it must be delivered to the gaging station in proper orientation. This may be done manually or by means of a vibrator bowl or conveyor. Some of the newer systems use robots for this function. Robots are not particularly fast, however, nor are they especially accurate; moderately priced robots cannot repeatedly locate parts closer than about 0.060 to 0.070 in. Most systems, robot-loaded or otherwise, require a locating device such as a tapered clamp or a hole in the part to position the part precisely.

Taking the measurements: After the part is positioned and clamped, various sensing probes, pins, and expanding-chuck devices move into and over the part to measure the significant features and relay the data to the control instrumentation. The measuring probes need not touch the part directly, however. The delicate probe tips are usually protected from impact and wear from the moving part by transfer pins and other intermediate devices.

Another method used to protect delicate probes from damage involves a two-station arrangement. At the first station, a simple plug moves into, for example, a hole in the part to

verify the presence of a hole of approximately the right size. The part then moves to the second station where a measuring probe checks the exact dimension. If the plug, or "functional probe," detects no hole or one considerably undersize, the part is rejected before it reaches the gage station. Alternatively, the conveyor line may be shut down automatically until the problem upstream—for example, a broken or worn tool—is corrected. This type of probe protection is used most commonly where the gaging equipment is directly downstream from the machining operation.

Other than functional probes, there are two other types of gaging probes: electronic and pneumatic. The primary considerations in choosing between electronic and pneumatic logic are the size and configuration of the feature being checked.

Electronic probes are readily adaptable to most part configurations except inside diameters under ⅝ in. Pneumatic probes can be used to measure any inside diameter larger than the probe itself—about ⅛ in. in diameter is the smallest.

A typical pneumatic gaging spindle has a central air passage and diametrically opposed air jets. It gages a hole diameter based on the amount of clearance between the spindle and the wall of the hole. The change in air-flow velocity is interpreted as a dimension on the indicator. Air gaging, the slower of the two types, limits the gaging speed of a system to about 1,800 parts per hour. Electronic gaging can handle parts at speeds three or more times faster.

Another important consideration is the type of control logic to be used. In potentially hazardous areas, air logic is usually favored to minimize or eliminate the use of electricity.

Other factors are the dimensional tolerance required, indicator resolution, and amplification. For example, gaging tolerances of 0.0001 in. requires electronic means. And for very high amplification, air systems become unstable. Only through electronic techniques can high resolutions be utilized with any degree of stability. Based on these criteria, the system could well use a combination of pneumatic and electronic probes.

Speed of gaging is also affected by the closeness of tolerances being measured and by whether the part is gaged while at rest or "on the fly." Parts on a continuously moving conveyor can be gaged at a faster rate than those that are checked while stationary but not to the same degree of accuracy. For measurements involving a tolerance of one ten-thousandth of an inch, even the at-rest parts require a second or two of "stabilization" time for the moving components in the gaging station to recover from any flexing or vibration. Only then can acceptable readings be taken.

Sorting the parts: The exit path of the part is determined by automatic comparison of measured features with the standards programmed into the controlling instrumentation. Some systems are designed to recognize only a simple pass/fail gaging result; others are more sophisticated. For example, an additional notification of "trending toward out of tolerance," can be incorporated into a program to warn an operator of tool wear that will soon require attention.

Other types of systems sort parts according to size to simplify selective assembly. One such system, which gages telephone-relay cores at 3,600 parts per hour, segregates the parts into five sizes according to length. The five groups—

differing by only 0.0002 in.—are then guided into separate bins.

The mechanics of moving the parts after gaging can be handled by gravity roller conveyor—often with a bypass section for out-of-tolerance parts—or by various arrangements of chutes and trap doors controlled by the machine program.

Another variation of an automatic gaging system has been applied to checking a stacked-tolerance assembly of components for an automotive differential. Such assemblies are often specified to be manufactured so that the sum only of the maximum-material values equals the desired total length. The difference, then, for most assemblies, is made up by adding a shim member (usually a washer-shaped part) to reach the desired length. An automatic gaging machine is used to measure each assembly and determine the exact shim thickness needed. The value appears as a digital readout on the panel, and the operator then adds a shim of the indicated size.

Design recommendations

The principal design recommendation applying to designers of parts to be gaged automatically is to try not to do too much. The tendency to specify "check all dimensions to tolerances indicated" must be tempered with reality. Such a specification would almost always result in an unnecessarily complex, expensive piece of equipment. Responsible builders of gaging machines will, in fact, discourage this type of order and ask that only the essential dimensions be checked automatically—and preferably those that can be reached from a single positioning of the part. With some parts, this may involve only one or two dimensions; others may require eight or ten checks.

A recommended goal is to gage the features that could cause perhaps 90% of the problems (in assembly or safety, for example) rather than to increase the complexity—and cost—of the machine two or three-fold to reach near 100%. The increased complexity may lead to the machine being down 50% of the time. The simpler the system, the higher its reliability.

In general, the features that are easiest to gage are lengths, heights, holes larger than ⅝ in. in diameter (both diameter and depth), hole location, and run-out. Features that are difficult—and hence expensive—to gage include threads, inside groove locations, concentricity of an inside to an outside diameter, and holes having diameters smaller than ⅛ in. In addition, small parts are often difficult to orient or to hold for gaging. Finally, parts must be clean, free from burrs in critical areas and, particularly for checking tolerances of one ten-thousandth of an inch or finer, must be stabilized at a known temperature at which the tolerances are to apply.

Automatic inspection is not limited to the checking of dimensions only. For example, valve seats in automotive engine heads can be checked for hardness and for the presence of cracks or other flaws. This inspection involves electronic instrumentation using an eddy-current response to verify hardness or to detect discontinuities.

Another characteristic that can be inspected automatically is porosity of vessel-type parts. Here, the part is sealed and filled with a fluid under pressure. Pressure readings taken before and after a suitable dwell period are then compared to detect any leakage.

Evaluation of Performance of Machine Vision Systems

By C.A. Rosen and G. J. Gleason
Machine Intelligence Corporation

Abstract

A simple method is proposed for comparing the relative performance of machine vision systems designed for programmable automation. Some major performance criteria are the degree of discrimination between patterns and the execution time required to determine the identity, the state, and the position of workpieces.

The method is based on the use of a small set of two-dimensional test figures with readily-reproducible silhouettes. Workpieces with multiple internal regions of interest are simulated by a circular disk with multiple holes, increasing complexity represented by an increase in the number of holes, with a consequent increase in the required processing time for analysis.

The method of measurement is described, the geometric test shapes are illustrated, and typical performance measurements are reported for a commercial machine vision system.

Introduction

Machine vision, based on the processing and interpretation of electro-optical (television) images is being introduced into practice for a number of diverse programmable automation applications. (1-5) These include inspection, material handling, and robot assembly tasks. The first generation of available vision machines are designed to recognize and identify workpieces, their stable states, and to determine their positions and orientations under the following major constraints:

- Each workpiece is supported in one of a small number of stable states. (A workpiece hanging on a hook, able to turn, sway or swing has an infinite number of states and does not satisfy this constraint.)

- A binary image, representing the silhouette of the workpiece can be extracted by enhancing the optical contrast between workpiece and background by suitable lighting techniques.

- Each workpiece is separated (not touching) from every other workpiece in the field of view. No overlapping of workpieces is permitted.

The above constraints define the conditions for present cost-effective vision systems dealing with binary images. The complexity of a visual scene under the above conditions depends on the number of separate entities (or "blobs") that must be analyzed in a given scene, and the number of states that must be discriminated for each workpiece. One work-piece may have several internal regions of interest, such as holes, for which some measured features are important for

both recognition and inspection. Further, to provide visual
sensory information for an industrial robot, each stable
state of the workpiece must be recognized, and position and
orientation determined.

A semi-quantitative classification of scene complexity
for this type of simple vision has been described by Foith,
et al (6) in terms of simple geometric relations between
workpieces in the field of view. These range from the
simplest case of one randomly-positioned workpiece in the
field of view, to the most complex -- in which several
workpieces are either touching or overlapping each other.
They distinguish one intermediate case of wide practical
importance, namely, when several non-contacting non-overlapping workpieces are in the field of view and cannot be
separated by non-overlapping rectangular bounding "windows".
Two cases of such scenes are shown in Fig. 1. In Fig. 1(a)
two workpieces are shown (non-touching) but intertwined such
that the dotted rectangular "windows" overlap. In Fig. 1(b)
a disk with two holes represents three discrete regions of
interest, (three "blobs")-- the disk and the two holes.
These images can be analyzed using connectivity analysis
(3, 7) which automatically groups together picture points to
form the regions of interest, that we have labeled "blobs".
It is plausible to assume that one measure of complexity is
the number of "blobs" that must be processed to analyze the
scene. A second measure is the number and type of geometric

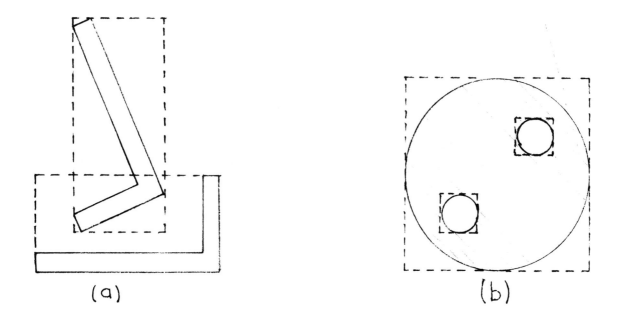

(a)   (b)

Figure 1 - Complex objects requiring Connectivity
or "Blob" analysis.

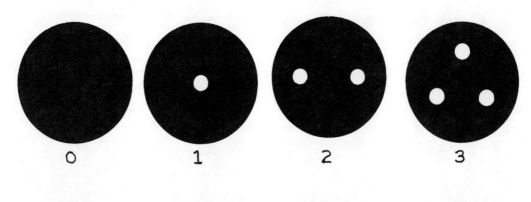

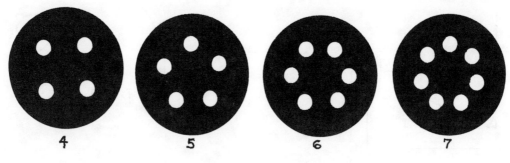

All holes are ¼ inch diameter.
The centers of all holes lie
on a 1 inch diameter circle.

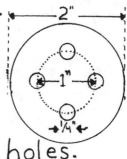

Figure 2 – Set of disks with
variable number of holes.

features that must be extracted to characterize or discriminate each state. Such features may include area, perimeter, second moments, maximum and minimum radii from centroid, number of internal holes (or "blobs"), and others. A third measure of complexity is the amount of total data that must be processed per scene, that is, the total number of bits, whether binary, gray scale, color or range, that describe the object of interest as distinguished from its background.

It is difficult to define an analytical expression to quantify the complexity of the rich variety of scenes that are possible. We propose, instead, a standard reproducible set of two-dimensional test figures which can be used to obtain comparative performance data of different vision systems, such data related to the three complexity measures described above. Undoubtedly, other measures will be proposed and added as we learn more about appropriate classification of imagery and can better define what we mean by complexity.

Proposed Method of Test and Pattern Sets

It is proposed to use two sets of two-dimensional black and white geometric figures as test patterns for evaluating a few major performance factors of vision systems. The first set, shown in Fig. 2, is designed to measure execution time for recognition and measurement of important parameters as a function of complexity due to increase in the number of regions of interest, ("blobs") while total binary bits processed and number of discrimination features remain constant. The first set is composed of eight 2" diameter disks, each with a different number of ¼" diameter holes, ranging from zero to seven holes. The disks can be readily reproduced with high accuracy, photographically, providing high contrast binary patterns. They are of constant total area, therefore, requiring the processing of a constant number of binary bits. There is little change in the general shape of the different patterns in the set, such that the discrimination between each pattern will not be overly sensitive to selection of discrimination features.

The second set, shown in Fig. 3, are designed to present four subsets of patterns with significantly different shapes to test the efficacy of the set of discriminating features selected for differentiation. There are four different shapes, in subsets of two: There are two circular disks, two squares, two rectangles, and two equilateral triangles. All patterns have a constant total area of π square inches. Each subset of two of a kind have one pattern with no holes, and the other with one ¼" diameter hole. Thus, this test should yield some insight into the relative worth of a selected set of features while the area (number of bits processed) remains constant, and the number of "blobs" are two or less.

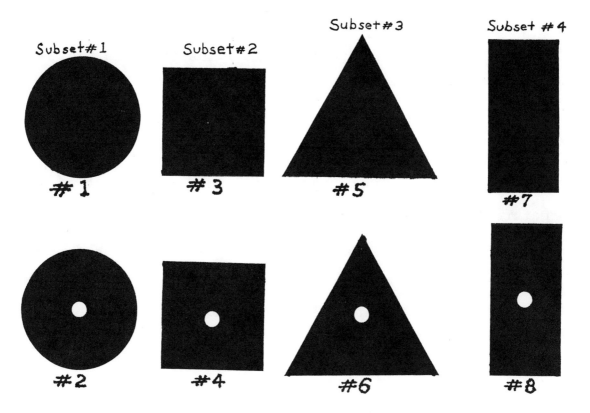

Notes: 1) The area of each pattern is π square inches.
2) Holes are ¼ inch diameter.

Figure 3 - Subsets of Patterns with equal area, varying shapes.

Experimental Results

The performance of a commercial vision system (8) using these two sets of test patterns is illustrated in Fig. 4 and Table 1.

In Fig. 4 the system was trained to adequately discriminate randomly-positioned disks with varying number of holes, using a nearest-neighbor classification. (7) A set of eleven features (Fig. 4) for one series of measurements was used. A single distinguishing feature, the total area, was used in another series of measurements to provide a basis for comparison. The 2" diameter disks were randomly-positioned in a 4" x 4" field of view, imaged by a 128 x 128 element solid state camera (GE TN2200).

Curve "A" indicates recognition times increasing from approximately 700 milliseconds for the disk with no holes to 850 milliseconds for the 7-hole disk, when the eleven features are used. Because the identification is based on eleven measurements, the disks are identified with high reliability. Further, complete data as to position and orientation of each important region (the whole disk and each hole) is available at the output terminals through the use of an appropriate communication protocol.

Curve "B", the reference curve, shows recognition time of an approximately constant 250 milliseconds for all samples. This test used only the net area feature, i.e., it counted the number of bits representing the disk area, without the hole area, thus permitting rather poor discrimi-

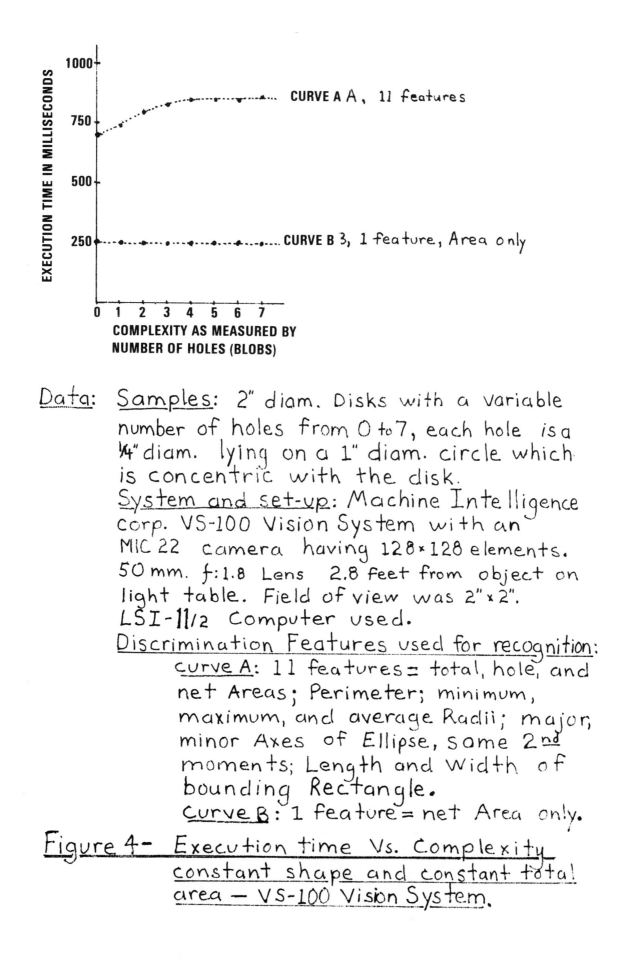

Data: Samples: 2" diam. Disks with a variable
         number of holes from 0 to 7, each hole is a
         ¼" diam. lying on a 1" diam. circle which
         is concentric with the disk.
         System and set-up: Machine Intelligence
         corp. VS-100 Vision System with an
         MIC 22 camera having 128×128 elements.
         50 mm. f:1.8 Lens 2.8 feet from object on
         light table. Field of view was 2"×2".
         LSI-11/2 Computer used.
         Discrimination Features used for recognition:
              Curve A: 11 features = total, hole, and
              net Areas; Perimeter; minimum,
              maximum, and average Radii; major,
              minor Axes of Ellipse, same 2nd
              moments; Length and Width of
              bounding Rectangle.
              Curve B: 1 feature = net Area only.

Figure 4- Execution time Vs. Complexity
           constant shape and constant total
           area — VS-100 Vision System.

Pattern No	Shape	Max Radius Mean	S.D.	Hole Area Mean	S.D.	Net Area Mean	S.D.	Perimeter Mean	S.D.
1.	○	1.06	.027	0	0	3.12	.034	6.66	.057
2.	◎	1.06	.027	.051	.001	3.055	.034	6.63	.042
3.	□	1.60	.037	0	0	3.125	.028	7.40	.127
4.	▣	1.61	.037	.055	.0007	3.085	.031	7.465	.094
5.	△	2.47	.063	0	0	3.16	.038	8.55	.117
6.	◬	2.43	.053	.054	.001	3.06	.023	8.46	.114
7.	▭	2.01	.036	0	0	3.09	.030	7.84	.186
8.	▭	2.00	.031	.055	.001	3.049	.025	7.86	.175

The mean and standard deviation measurements are given in square inches, at least 10 samples for each of the 8 patterns were used.

Area of each pattern equals π square inches.

The field of view was 4"x4", using a "Silicon Vidicon Camera" having 240 x 240 elements, on the VS-100 Machine Vision System.

Table 1- Discrimination Measurements

nation based on small differences of net area.

Table 1 shows the key measurements made on the second set of patterns. The eight patterns (Fig. 3) were used as training samples, each shown to the vision system at least ten times in random positions and orientations in the field of view. A silicon vidicon camera, with a resolution of approximately 240 x 240 elements was used with a commercial vision system (8). The field of view was 4" x 4" and the area of each pattern was π square inches. The system computed the mean values and standard deviations for each of the discriminating features used to differentiate and recognize each pattern. Shown in the table are the mean values and standard deviations for four selected features, namely, the net area, the hole area, the maximum radius from the centroid and the perimeter. Analysis of these data reveals the following conclusions:

(1) No single feature can differentiate all eight patterns. For example, the patterns in any subset consisting of two identical shapes except for one pattern having a hole, cannot be discriminated by the maximum radius, the net area, nor the perimeter features. The only feature able to discriminate for the presence or absence of a hole is the hole area feature. At the same time the hole area feature, by itself, cannot discriminate between the four subsets.

(2) The net area feature is not useful as a discriminating feature for all the patterns, since we have designed the experiment to maintain constant area for each shape, and the

only differences are due to presence or absence of a single hole. The resolution of the system is too low to compensate for the noise and optical distortions which yield standard deviations comparable to the size of the ¼" D. hole. The major value, therefore, of this column of data is to inform the application engineer that more resolution elements are needed if this particular feature is to be of value for discrimination.

(3) The <u>maximum</u> <u>radius</u> feature is revealed as a useful discriminator between each of the four subsets. Together with the <u>hole</u> <u>area</u> feature, all eight patterns can be discriminated. For example, for the triangles and rectangles, the difference of the means, adjusted for three times the standard deviation spread is:

$$\underset{\text{Max. Radius-Triangle}}{\Big[2.47 - (3)(.063) \Big]} - \underset{\text{Max. Radius-Rectangle}}{\Big[2.01 + (3)(.036) \Big]}$$

$$+ 2.281 - 2.118 = .163$$

This <u>positive</u> difference is $\frac{.163}{2.118}$ = 7.6% of the measured means and is reasonably significant. Shown in Table II are the calculations for all the subsets using <u>maximum</u> <u>radius</u> and <u>perimeter</u> as the measured features.

(4) The <u>perimeter</u> <u>feature</u> is also a useful feature for discriminating three of the four subsets, but is marginal for discriminating between the square and rectangles. A calculation of the adjusted means indicates considerable overlap:

Discrimination between	Maximum Radius	Perimeter
○ □	32.8%	2.8%
○ △	107.5%	20.5%
○ ▭	71.8%	6.8%
△ □	35.6%	5.6%
△ ▭	8.1%	-2.5%
□ ▭	11.9%	-6.7%

Table II - Discrimination Factor.

$$\overset{\text{Perimeter-Rectangle}}{\left[7.84 - (3)(.186)\right]} - \overset{\text{Perimeter-Square}}{\left[7.40 + 3 (.127)\right]}$$

$$= 7.282 - 7.781 = -.499$$

This negative difference indicates considerable overlap and, therefore, this feature measurement is not useful as a discriminator between the rectangle and square. It is, however, useful for the other discriminations.

The combined use of <u>maximum</u> <u>radius</u> and <u>perimeter</u> measurements do, in fact, aid in assuring the discrimination of all the subsets, and the <u>hole</u> <u>area</u> measurements then can discriminate the two members in each subset.

Conclusions

A method for comparing some important elements of performance of machine vision systems is proposed and described. Based on the use of several sets of standardized, reproducible two-dimensional geometric patterns, execution time as a function of complexity, and discrimination capability can be compared. Initial experimental results are shown for one system.

BIBLIOGRAPHY

1. Dodd, G. G. and L. Rossol, "Computer Vision and Sensor-Based Robots". Plenum Press, New York; (1979)

2. Rosen, C. A. and D. Nitzan, "Use of Sensors in Program-mable Automation", COMPUTER, Vol. 10, No. 12 (1977)

3. Gleason, G. and G. J. Agin, "A Modular Vision System for Sensor-Controlled Manipulation and Inspection". Proc. of 9th Int. Symp. on Industrial Robots. March, Washington, D.C. (1979)

4. Ward, M. R. and L. Rossol, S. W. Holland, and R. Dewar, "Consight" a Practical Vision-Based Robot Guidance System". Proc. of 9th Int. Symp. on Industrial Robots, March, Washington, D.C. (1979)

5. Foith, J. P., H. Geisselmann, U. Lubbert, H. Ringshauser, "A Modular System for Digital Imaging Sensors for Industrial Vision: Proc. of 3rd CISM IFToMM. Symp. on Theory and Practice of Robots and Manipulators", Sept. Udine, Italy. (1978)

6. Foith, J. P., "A TV-Sensor for Top-Lighting and Multiple Part Analysis, 2nd/FAC/IFIP Symp. on Information Control Problems in Manufacturing Technology, Stuttgart, FRG, October (1979)

7. Agin, G. J., "Computer Vision Systems for Industrial Inspection and Assembly", COMPUTER, Vol. 13, No. 5, May (1980)

8. Model VS-100 Machine Vision System, Machine Intelligence Corporation, Palo Alto, California

Vision Methodology as it Applies to Industrial Inspection Applications

By Gary G. Wagner
Inspection Technology, Inc.

Inspection Technology, Inc. is a high technology, non-destructive testing and inspection system design and manufacturing company. ITI presently uses microprocessor based electronics, low light level camera based with X-ray intensifier fluoroscopic devices, and visible light and laser non-contact metrology systems to detect and analyze surface and internal defects and characteristics on a real time and automatic basis.

One of ITI's product lines is based on video image analysis for on-line inspection of parts. VIDOMET, or video metrology, is ITI's name for high speed, non-contact video image analysis inspection systems. This product is analogous to replacing the human inspector who uses his eyes for vision, brain for storage and is taught an inspection technique with a camera, video memory, and microcomputer. The camera image is temporarily stored in a solid state memory device. Computer logic reviews subsequent stored image data and makes comparisons of picture contrast differences. By enhancing inspection areas with certain lighting and optical conditions, a VIDOMET system can be capable of inspecting many parts per minute at 100% reliability.

We will take a close look at what an optical inspection system is and how these tools can be configured to solve high speed production line inspection and gaging problems.

The first part of any optical system is the inspector's eyes or vision. Image cameras are categorized into three (3) distinct groups. They are: line scan, solid state matrix, and analog. Each type offers unique advantages and disadvantages (See Figure 1) when applied to specific inspection problems. There currently is no one type which is best suited for each different application, however, the analog camera can offer more potential to commonality if certain improvements could be made to its weaknesses.

ITI has made significant improvements to the analog camera's deficiencies which complement its advantages. Therefore, we will use the analog camera as the major image sensor throughout this discussion.

Figure 2 is an image of a ball taken from a normal TV monitor. All TV monitors in your home have pictures made from analog signals. In your TV, a series of horizontal scan lines from top to bottom make up the picture. Each point along the line is a continuous change in light to dark intensity depending on the image being shown. Figure 3 is such a line taken directly through the ball which shows a uniform or continual darkening of the analog video signal as the image changes from white to black.

Since computers can only work with digital information, the Figure 3 video signal will have to be transformed into a digital format.

Image Sensor Choices

	Advantages	Disadvantages
Line Scan	Resolution Scan Speed	Requires Pixel Normalization Intense Lighting Constant Part Speed Cost
Solid State Matrix	Excellent Linearity Cannot Image "Burn"	Requires Pixel Normalization Unusable Pixels Resolution Cost
Analog	Resolution Low Lighting Contrast Sensitivity Capability To Preprocess Cost	Poor Linearity Image Drift Image Burn

Figure 1

Analog Image

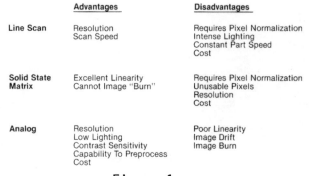

Signal Sample

Figure 2

Analog Video Signal

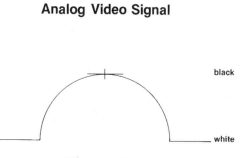

black

white

Figure 3

Digital Video Signal

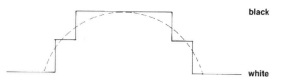

black

white

Figure 4

If we were to sample equal segments along the video line and assign average contrast intensities to each signal, this finite unitary segment can be transferred into a digital value (See Figure 4).

Figure 5 shows the same picture of the ball as in Figure 2, but the picture has been disected by rows and columns into finite segments, each having a known size and contrast value. Current hardware available to transfer analog video signals to digital matrix arrays are available in 128 x 128, 256 x 256, 512 x 512, and other formats. Additionally, this same hardware can distinguish between many shades of gray. Available hardware shades are 16, 64, and 256 contrast differences from white to black. As can be seen in Figure 5, increased part definition improves as the resolution increases and available inspection problems can be handled by a 256 x 256 x 6 piece of hardware.

A block diagram showing components required for optical inspections is shown in Figure 6. The image is acquired, digitized, and stored in memory. With the video in digital, the processor can access each segment, or pixel, by row and column address. Any one (1) pixel can be reviewed by the computer for contrast data.

A common production line problem would be to check for component presence or absence in an automatic assembly line. Presence/absence inspection is one of the easiest checks for an optical system. Figure 7 depicts a typical presence/absence inspection of a large part requiring several camera inputs for sufficient resolution. Figure 8 is the resultant digital image from one of the cameras. The inspection requirement is to check for bracket and nut presence.

Digital Image Matrix

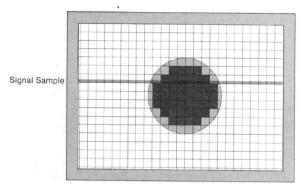

Signal Sample

Figure 5

Analog Frame Grab Block Diagram

Figure 6

Check for presence/absence

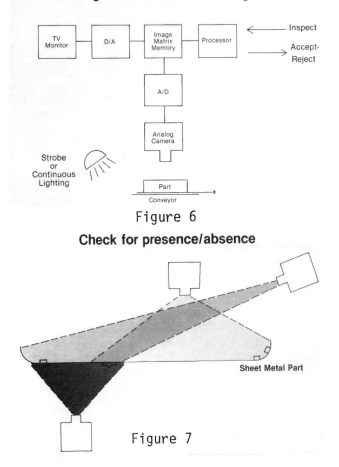

Sheet Metal Part

Figure 7

411

Figure 8

Most earlier video analysis techniques would have made a comparison or subtraction, pixel by pixel, of the part being inspected to an image of a good part in memory. If the images did not identically correspond, then the part being inspected was interpreted as a bad part. This technique, called pattern reconnection, has had limited success for industrial inspections since most factory parts are not made identical, but are still considered acceptable.

A new technique called gray scale analysis has been developed which copes with large good part variations, but still finds defects with 100% reliability. This technique is shown in Figure 9. Discrete areas, or windows, are formed around only the portions of the image to be inspected. For determining if brackets are present, high intensity lighting is positioned so that a bracket, when present, will cast a dark shadow. When the bracket is missing, no shadow will be cast. As shown in Figure 10, when the bracket is present, a large number of darker pixels can be observed in the window due to the cast shadow than when a bracket is missing. An easy analysis of bracket presence or absence would be to set a contrast threshold between the dark and light pixel value area and have the computer count up the number of pixels which must meet this threshold requirement. Figure 11 shows such a threshold value. Any discrete area which meets a minimum number of pixels exceeding this set threshold will be interpreted as bracket presence. Discrete area analysis is a powerful tool and can most generally be used for inspections of presence/absence, correct part assembly,

orientation, part integrity, etc. Figure 12 shows such an application.

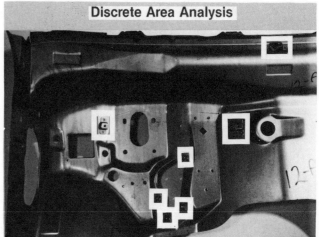

Discrete Area Analysis

Figure 9

Pixel Frequency Distribution

Figure 10

Number of Pixels

Grey Shades

1   64

without bracket

with bracket

1. Set a Contrast threshhold
2. Set a minimum number of pixels to meet #1 condition

Figure 11

25   36   47

Test would verify bracket presence if 30 pixels were a contrast value of 36 or less

Figure 12

One of the problems with area analysis is that the parts must be presented to the inspection system in a repetitive fixtured position. If the part cannot be precisely fixtured, subsequent part movement would most certainly cause an inaccurate window analysis interpretation (See Figure 13). Precise part fixturing is usually expensive, but has always been necessary for hard gaging techniques. With optical inspection systems, we have the capability to reorient a non-fixtured part to a precise fixtured position within the digital image memory. Figure 14 shows the same part as in Figure 13. The arrow is one line of video pixel elements. If we were to take a magnified look at that line of digital video memory, Figure 15 would occur.

Part Placement Inaccuracies

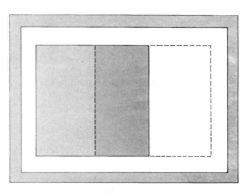

Figure 13

Part Placement Inaccuracies

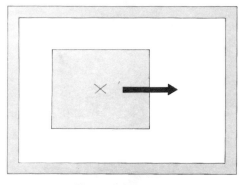

Figure 14

All precisely placed or fixtured parts would have the part edge transition repetitively in position at one pixel location. Parts not fixtured or inaccurately placed, would vary in part edge transition. For the computer to reorient

the part in memory, a simple set of pre-inspection instructions can be executed. First, have the computer access one line of digital video encompassing the worst case part placement positions. Then, let the computer know what the pixel contrast values are with a part present and without at the edge transition point. The computer will now remember the edge transition pixel position and subtract subsequent part edge transition values for the first. Figure 16 shows how each discrete window can now be adjusted with this new delta difference before making this interpretation.

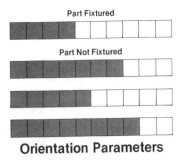

Part Fixtured

Part Not Fixtured

Orientation Parameters

1. Set start and stopping points
2. At start and stop location determine contrast values
3. Set contrast threshhold value

Figure 15

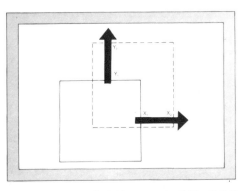

All windows would be adjusted in memory by (X₂-X₁) and (Y₂-Y₁) values before image analysis determination.

Figure 16

This reorientation technique can be used for circular transition parameters as seen in Figure 17. Differences occur only in transition analysis. Search rows and columns would expand out from design part placement center looking for both end transition points. When they are found, the total number of pixels in the row and column individually are halved to find the center of each search row and column.

When this is found, a new center position is determined and offset values are used to adjust each discrete area analysis before gray scale evaluation.

Circular Orientation

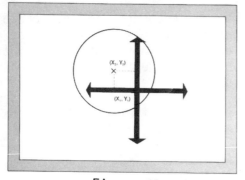

Figure 17

Applications of this technique can be found for many industrial parts. One such part is the inspection of engine cylinder heads. Figure 18 shows two (2) valves from a cylinder head. When assembled automatically, the half moon retaining keys must be inspected for proper seating position. It was observed (See Figure 19) that when the keys were properly seated, the total gap area between the two half moon parts remained constant. Mispositioned keys, or not properly seated keys would be found to have a higher total gap area than properly seated keys.

Figure 20 places a circular window around the two half moon keys only, then discrete area analysis is used for gap area determination. Gaps exceeding a minimum value would have an excess of darker pixels and would be cause for failed part interpretation. Since this inspection is particularly critical on part positioning, because of the surrounding block areas. Circular orientation parameters are required (See Figure 21). Figure 22 shows this inspection system incorporated into the actual production line.

Another powerful application of optical gaging is high speed, non-contact measurement of parts. Figure 23 shows how the cameras field of views can be disected by a relatively large value matrix.

A 2 inch field of view (Figure 24) would be divided by 512 to determine pixel resolution. Each pixel is of equal size

Figure 18

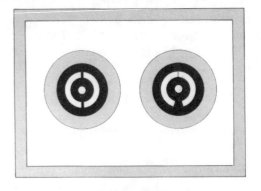

Gap Spacing Varies
Total gap area does not vary

Figure 19

Circular Window

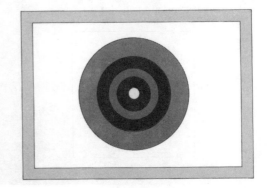

Figure 20

Circular Orientation added to Circular Window

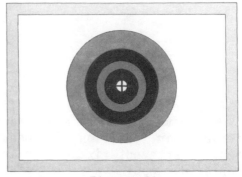

Figure 21

and can be easily determined (See Figure 25).

Figure 22

Non-Contact Measurement

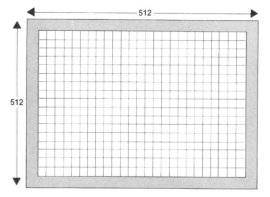

512

512

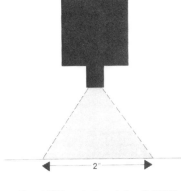

2"

Then 2"/512 = pixel resolution ≅ 0.004"

Figure 24

Non-Contact Measurement

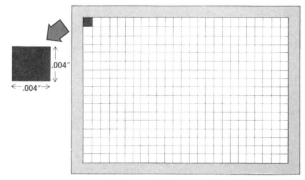

.004"

.004"

Figure 25

Non-Contact Measurement

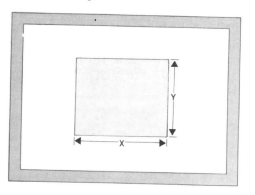

Y

X

If we want to know X & Y dimensions.......

Figure 26

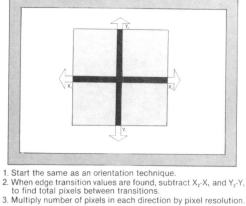

Y_2

X_1

X_2

Y_1

1. Start the same as an orientation technique.
2. When edge transition values are found, subtract X_2-X_1 and Y_2-Y_1 to find total pixels between transitions.
3. Multiply number of pixels in each direction by pixel resolution.
   $(X_2$-$X_1) \times$ Pixel Resolution = X
   $(Y_2$-$Y_1) \times$ Pixel Resolution = Y

Figure 27

Applying non-contact measurement for determining "X" and "Y" dimensions (Figure 26) is a further refinement to the circular orientation technique. Figure 27 shows both a row and column searching technique looking for edge transitions. When found, the total number of pixels in each row or column can be multiplied by the pre-determined pixel resolution value.

Taking a careful and magnified look at each edge transition would reveal that if the edge broke a portion of a pixel, there would be a question as to whether the pixel should be counted or left out. Therefore, any resultant measurement value must contain an uncertainty of at least ± 1 pixel resolution value and must be factored into overall system accuracies (See Figure 28).

Applications of non-contact measurements are numerous. One such application is the measurement of the inside diameter and wall thickness of ceramic tubes. Figure 29 shows the live video picture and Figure 30 is the resultant computer analyzed (digitized) picture. To achieve Figure 30 conditions, single black/white threshold parameters were applied to the whole image before searching in both the "X" and "Y" dimensions for contrast transition values.

This overall turnkey system also required outside diameter measurement at points in from the tube ends and overall length dimensions. Figure 31 shows how a turnkey application can be configured and demonstrates how the various image cameras must be used for specific application problems.

foot distance are required to be referenced back to a bench mark point (See Figure 32)

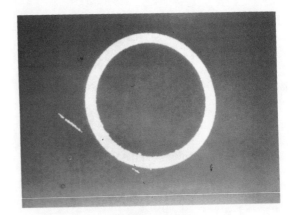

Figure 30

Frame Grab — I.D. and Wall Thickness
Line Scan — O.D. @ point in from ends
Photo Diode — Length

Figure 31

To increase resolution Multiplex Cameras

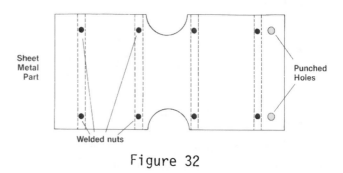

Figure 32

However, we must take into account inaccuracies

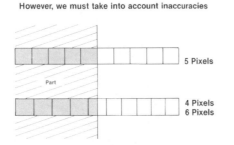

Each transition has at least ±1 pixel accuracy

Figure 28

Figure 29

Lastly, optical measurements are not necessarily constrained to resolution obtained from one camera's field of view only. Several cameras can be multiplexed, each having a portion of the total measurement problem. An example of this technique is a large part where several individual measurements over a seven (7)

The part in this example is a sheet metal truck bed panel which has many tack welded nut locations. The inspection parameters were the absence/presence of the nuts and the dimensional measurement of the nut position to be over a punched hole location. Because one camera's field of view did not have nearly enough resolution for this precise measurement, a camera was positioned under each sill and nut location (See Figure 33). Additionally, solid state camera resolution was not sufficient and analog techniques were required.

As you probably know, analog camera images have a tendency to drift with time and temperature. A technique had to be found which would negate or compensate drift if we were to meet the measurement accuracies.

Patented techniques for drift compensation

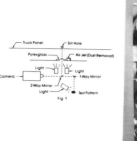

Figure 34

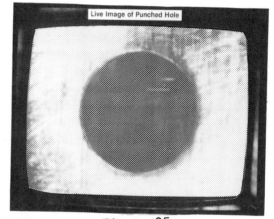

Figure 35

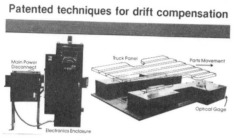

Figure 33

Figure 34 shows ITI's solution to camera drift. Before each measurement, the lights which focus camera image up to the panel are turned off and other internal lights are turned on. With these lights on, the image passes through a half silvered mirror and sees a test pattern. The test pattern image is used as a master for drift position analysis by doing a reorientation on the test pattern. Once drift compensation has been computed by orientation techniques, the lights are switched and the camera can view its correspondent nut or hole location. (Figures 30, 35, 36, and 37 show the steps taken during image analysis for nut presence and hole dimensional analysis.) Drift compensation values are added to part orientation values to correct for part placement inaccuracies before analyzing nut presence and hole positional dimensions. Each hole position required its true dimension referenced to a base hole location.

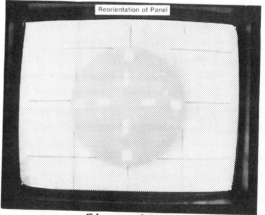

Figure 36

OPTICAL INSPECTION α GAGING IS A POWERFUL TOOL

Advantages Over Hard Gaging

- Inspection Speed
- Tolerate Little or No Part Fixturing
- Reprogramming for Various Parts

Past Disadvantages Have Been Improved For Industrial Applications

Inspection Technology, Incorporated, designs, develops, and manufactures industrial non-destructive inspection equipment. ITI has the resources and capabilities to design complete turnkey NDT inspection devices using the latest state-of-the-art techniques and equipment to insure cost effective systems.

One of ITI's product lines is based on video image analysis for on-line inspection of parts. A video image analysis system generally consists of a TV camera, video digitizing hardware, solid state storage capabilities, and computer processing devices. In many instances, ITI's equipment can be adaptable over existing production lines. In these applications, a reject mechanism is supplied. In other cases where the customer wishes to presort or batch load parts, ITI will design complete mechanical handling systems for specific tasks.

ITI's unique ability to provide custom-designed turnkey systems reduces the customer's responsibility and concern that his inspection problem may not be quite ideal for an off-the-shelf application. ITI takes the customer's part and specific pass/fail criteria and insures working turnkey performance.

Figure 37

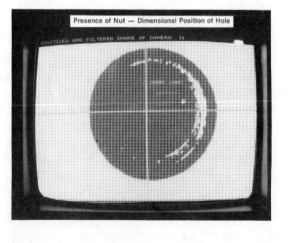

Figure 38

APPENDIX

Military Standard:
Gage Inspection

(Abbreviated for publication in *Gaging: Practical Design and Application*

I

PURPOSE AND SCOPE

1.1 GENERAL

1.1.1 Scope. This standard is to provide correlated technical information applicable to the inspection of gages, special tools, and measuring devices. The principal subjects covered are nomenclature, tolerances and fits, measuring tools and equipment, gages, and methods of measurement and inspection. Details of gage design are not included.

1.1.2 Purpose. The purpose of gaging is to determine compliance of the component parts of a mechanical product with the dimensional requirements of contracts, drawings, and specifications, whether such product is made by a single manufacturer or whether parts to be assembled are made by several manufacturers variously located. Gaging thereby assures the proper functioning and interchangeability of parts, that is, one part will fit in the same place as any similar part and perform the same function, whether the part is for the original assembly or replacement in service.

1.1.3 Classification of gages and measuring equipment. Gages and measuring equipment are classified as follows:

1.1.3.1 Length standards. Standards of length and angle from which all measurements of gages are derived consist of precision gage blocks, end measuring rods, line-graduated standards, master disks, calibrated wires and rolls, precision squares, graduated circles, and similar items.

1.1.3.2 Master gages. Master gages are made to their basic dimensions as accurately as possible and are used for reference, such as for checking or setting inspection or manufacturer's gages.

1.1.3.3 Inspection gages. Inspection gages are used by the representative of the purchasing agency to inspect products for *acceptance*. These gages are made in accordance with established design requirements. Inasmuch as inspection gages are subjected to continuous use, a gage maker's tolerance will always be applied and a wear allowance, where applicable, may be included in the design of these gages. Tolerances of inspection gages are prescribed by specified drawing limits.

1.1.3.4 Manufacturer's gages. Manufacturer's gages are used by the manufacturer or contractor for inspection of parts during *production*. In order that the product will be within the limits of the inspection gages, manufacturer's or working gages should have dimensional limits, resulting from gage tolerances and wear allowances, slightly farther from the specified limits of the parts inspected.

1.1.3.5 Nonprecision measuring equipment. Nonprecision measuring equipment are simple tools such as rules and plain protractors used to measure by means of line graduations such as those on a scale.

1.1.3.6 Precision measuring equipment. Precision measuring equipment are tools used to measure in thousandths of an inch or finer and usually employ a mechanical, electrical or optical means of magnification to facilitate reading.

1.1.3.7 Comparators. Comparators are precision measuring equipment used for comparative measurements between the work and a contact standard, such as a gage or gage blocks.

1.1.3.8 Optical comparators and gages. Optical comparators and gages are those which apply optical methods of magnification exclusively. Examples are optical flats, tool maker's microscopes, and projection comparators. Projection comparators usually provide for absolute measurements as well as for comparative measurements.

DIMENSIONS AND TOLERANCES

2.1 GENERAL. This section presents definitions relating to dimensions and tolerances and standard practice relative to gage tolerances and wear allowances. This subject is dealt with further in MIL–STD–8.

2.2 DEFINITIONS

2.2.1 Dimension. A dimension is a specified size of a geometrical characteristic such as a diameter, length, angle, circumference, or center distance.

2.2.2 Nominal size. The nominal size is the designation which is used for the purpose of general identification.

2.2.3 Basic size. The basic size of a dimension is the theoretical size from which the limits of size for that dimension are derived by the application of the allowance and tolerances.

2.2.4 Design size. The design size of a dimension is the size in relation to which the limits of tolerance for that dimension are assigned.

2.2.5 Actual size. The actual size of a dimension is the measured size of that dimension on an individual part.

2.2.6 Limits of size. These limits are the maximum and minimum sizes permissible for a specific dimension.

2.2.7 Fit. The fit between two mating parts is the relationship existing between them with respect to the amount of clearance or interference which is present when they are assembled.

2.2.8 Tolerance. The tolerance on a dimension is the total permissible variation in its size. The tolerance is the difference between the limits of size.

2.2.9 Allowance. An allowance is an intentional difference in correlated dimensions of mating parts. It is the minimum clearance (positive allowance) or maximum interference (negative allowance) between such parts.

2.3 TOLERANCES. Variation in size can be restricted but not avoided. However, a certain amount of variation can be tolerated with-

out impairing the functioning of a part. The purpose of tolerances is to confine such variation within definite limits.

2.3.1 Unilateral tolerances. Unilateral tolerances are those which are applied to the basic or design size in one direction only. The unilateral system of tolerances will generally be used on all drawings. The application of unilateral tolerances is as shown on figure 1.

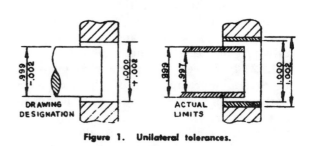

Figure 1. Unilateral tolerances.

2.3.2 Bilateral tolerances. Bilateral tolerances are those which are applied to the basic or design size in both directions. The basic or design size may or may not be the mean size, as the plus and minus tolerances are not necessarily equal. The application of bilateral tolerances is as shown on figures 2 and 3.

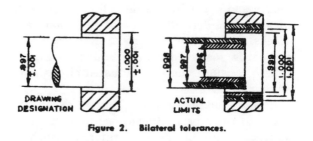

Figure 2. Bilateral tolerances.

2.3.3 Gage tolerances. Gage tolerances are applied to the gaging dimensions of gages in

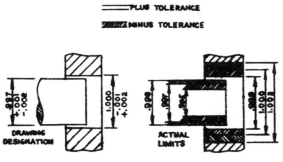

PLUS TOLERANCE
MINUS TOLERANCE

DRAWING DESIGNATION — ACTUAL LIMITS

Figure 3. Unequal bilateral tolerances.

order to limit variations in size during their manufacture. In general practice, gage tolerances should not exceed 10 percent of the tolerance of the part to be gaged.

2.3.3.1 Tolerances on plain gages. Tolerances for plain cylindrical plug and ring gages have been standardized in four classes according to nominal size and the degree of accuracy necessary to cover most gaging operations, as shown in table I.

TABLE I. TOLERANCES FOR PLAIN CYLINDRICAL PLUG AND RING GAGES

Size range	Total tolerance	Wear allowance	Tolerance Go	Tolerance Not go
Up to 0.825	0.0005	0.00000	0.00004	0.00004
0.825 to 1.510		.00000	.00006	*No not go gages, snug fit on go*
1.510 to 2.510		.00000	.00008	
2.510 to 4.510		.00000	.00010	
4.510 to 6.510		.00000	.00013	
6.510 to 8.510		.00000	.00015	
8.510 to 10.510		.00000	.00017	
10.510 up		.00000	.00020	
Up to 0.825	.0010	.00010	.00005	.00005
0.825 to 1.510		.00010	.00006	.00006
1.510 to 2.510		.00000	.00006	.00008
2.510 to 4.510		.00000	.00010	.00010
4.510 to 6.510		.00000	.00013	*No not go gages, snug fit on go*
6.510 to 8.510		.00000	.00015	
8.510 to 10.510		.00000	.00017	
10.510 up		.00000	.00020	
Up to 0.825	.0020	.00010	.00010	.00010
0.825 to 1.510		.00010	.00010	.00010
1.510 to 2.510		.00010	.00010	.00010
2.510 to 4.510		.00010	.00010	.00010
4.510 to 6.510		.00000	.00020	.00013
6.510 to 8.510		.00000	.00020	.00016
8.510 to 10.510		.00000	.00020	.00020
10.510 up		.00000	.00020	.00020
Up to 0.825	.0030	.00010	.00010	.00010
0.825 to 1.510		.00010	.00020	.00010
1.510 to 2.510		.00010	.00020	.00010
2.510 to 4.510		.00010	.00020	.00010
4.510 to 6.510		.00010	.00020	.00020
6.510 to 8.510		.00010	.00020	.00020
8.510 to 10.510		.00000	.00030	.00020
10.510 up		.00000	.00030	.00030
Up to 0.825	.0040	.00020	.00020	.00010
0.825 to 1.510		.00020	.00020	.00010
1.510 to 2.510		.00020	.00020	.00010
2.510 to 4.510		.00020	.00020	.00020
4.510 to 6.510		.00020	.00030	.00020
6.510 to 8.510		.00020	.00030	.00020
8.510 to 10.510		.00010	.00040	.00020
10.510 to 12.510		.00010	.00040	.00020
12.510 up		.00010	.00050	.00020
Up to 0.825	.005	.00030	.00020	.00010
0.825 to 1.510		.00030	.00020	.00010
1.510 to 2.510		.00030	.00020	.00010
2.510 to 4.510		.00020	.00020	.00020
4.510 to 6.510		.00020	.00030	.00020
6.510 to 8.510		.00020	.00030	.00020
8.510 to 10.510		.00020	.00040	.00020
10.510 to 12.510		.00010	.00050	.00030
12.510 up		.00010	.00050	.00030
Up to 0.825	.006	.00030	.00020	.00010
0.825 to 1.510		.00030	.00020	.00010
1.510 to 2.510		.00030	.00020	.00010
2.510 to 4.510		.00030	.00020	.00020
4.510 to 6.510		.00020	.00030	.00020
6.510 to 8.510		.00020	.00030	.00020
8.510 to 10.510		.00020	.00040	.00030
10.510 to 12.510		.00020	.00050	.00040
12.510 up		.00010	.00060	.00040

Size range	Total tolerance	Wear allowance	Tolerance Go	Tolerance Not go
Up to 0.825	0.007	0.00040	0.00020	0.00010
0.825 to 1.510		.00040	.00020	.00010
1.510 to 2.510		.00030	.00020	.00020
2.510 to 4.510		.00030	.00030	.00020
4.510 to 6.510		.00030	.00030	.00030
6.510 to 8.510		.00030	.00040	.00030
8.510 to 10.510		.00020	.00050	.00030
10.510 to 12.510		.00020	.00060	.00040
12.510 to 14.510		.00020	.00070	.00040
14.510 up		.00010	.00080	.00050
Up to 0.825	.008	.00040	.00020	.00010
0.825 to 1.510		.00040	.00020	.00010
1.510 to 2.510		.00040	.00020	.00020
2.510 to 4.510		.00030	.00030	.00030
4.510 to 6.510		.00030	.00040	.00030
6.510 to 8.510		.00030	.00040	.00040
8.510 to 10.510		.00030	.00050	.00040
10.510 to 12.510		.00020	.00060	.00050
12.510 to 14.510		.00020	.00070	.00050
14.510 up		.00020	.00080	.00060
Up to 0.825	.009	.00040	.00020	.00010
0.825 to 1.510		.00040	.00030	.00010
1.510 to 2.510		.00040	.00030	.00020
2.510 to 4.510		.00040	.00040	.00030
4.510 to 6.510		.00040	.00040	.00030
6.510 to 8.510		.00040	.00050	.00040
8.510 to 10.510		.00030	.00060	.00040
10.510 to 12.510		.00030	.00070	.00050
12.510 to 14.510		.00030	.00080	.00050
14.510 up		.00030	.00090	.00060
Up to 0.825	.010	.00040	.00030	.00010
0.825 to 1.510		.00040	.00030	.00020
1.510 to 2.510		.00040	.00040	.00030
2.510 to 4.510		.00040	.00040	.00030
4.510 to 6.510		.00040	.00050	.00040
6.510 to 8.510		.00040	.00060	.00040
8.510 to 10.510		.00040	.00070	.00050
10.510 to 12.510		.00040	.00080	.00050
12.510 to 14.510		.00030	.00090	.00060
14.510 up		.00030	.00100	.00060
Up to 1.510	.012	.00040	.00060	.00020
1.510 to 4.510		.00040	.00070	.00030
4.510 to 8.510		.00040	.00080	.00040
8.510 to 12.510		.00040	.00090	.00050
12.510 up		.00030	.00100	.00060
Up to 1.510	.014	.00040	.00060	.00020
1.510 to 4.510		.00040	.00080	.00030
4.510 to 8.510		.00040	.00100	.00040
8.510 to 12.510		.00040	.00120	.00050
12.510 up		.00040	.00140	.00060
Up to 2.510	.016	.00040	.00080	.00020
2.510 to 6.510		.00040	.00100	.00040
6.510 to 12.510		.00040	.00120	.00060
12.510 up		.00040	.00140	.00080
Up to 4.510	.020	.00040	.00100	.00040
4.510 to 6.510		.00040	.00100	.00060
6.510 to 12.510		.00040	.00150	.00080
12.510 up		.00040	.00200	.00100
Up to 4.510	.025 up	.00040	.00100	.00080
4.51 to 6.51		.00040	.00100	.00100
6.51 to 12.51		.00040	.00150	.00150
12.51 up		.00040	.00200	.00200

INDEX

D

H

I

J

L

M

N

Predicted drift, 21
Preloaded design, 132
Pressure, 89, 116
Pressure/clearance curve, 89, 92
Pressure/clearance circuits, 90
Pretax profit, 104
Prices, 38
Primary datum, 244, 246, 247
Principles, 88-94
Probe types, 372
Process capability studies, 79
Process control, 218
Process equipment, 50
Product description, 38, 39, 40
Product design, 43, 45, 60, 61, 77, 106-113
Product development, 37
Product liability, 388
Product performance, 50
Product teams, 238
Productivity, 77, 97, 103, 105, 106, 113, 235, 237, 350, 352
Profile filters, 159
Profile tolerances, 206
Profiling, 168
Program schedule, 40
Programming, 234, 236, 353, 372
PROMS/programmable read-only memory units, 234
Protective guards, 385
Prototype testing, 42
Prototypes, 38, 41

Q

Qualification, 44
Qualification program, 44
Quality assurance, 37, 40, 44, 194-212, 376
Quality control, 56, 57, 350, 352, 362, 387
Quality control charts, 54, 56
Quality control engineer, 151
Quality
    acceptable limits, 76
    air plug gages, 92
    charts, 54, 56
    demand for, 39
    dimensional, 218
    engineers, 151
    functional unit, 37
    gears, 52
    independent metrology labs, 87
    product development, 40
    problems, 38
    surface, 60
    tests, 45

R

Radiation, 16
Random environment, 18
Random sampling, 54
Range, 55
Reaction errors, 189
Receiver gages, 239
Recess pad diameter, 137
Recessed-pad hydrostatic gas bearings, 115
Record keeping, 80
Red and green limits, 109
Reject rates, 42

Reject station, 378
Rejected parts, 26-36, 50
Relative humidity, 82, 86
Remote control, 363
Repair, 39
Repeatability, 329
Reproducibility, 317
Research and development, 77
Retirement, 77, 106
Return on investment, 355, 357, 358, 359
Rework, 42, 152
Rework determination, 152
RFS datum specifications, 154
RFS feature control frame, 242
Rheostat knob, 54, 71
Rigid body kinematics, 314
Rigid body model, 178
Ring gage tolerances, 423
Ring gages, 422, 423
Risk, 105
Risk capital, 103
Robotics, See: Robots
Robots, 371, 373, 395
Roller, 134
Rotary table, 321, 324, 331, 334
Rotating devices, 156
Roughness, 167, 169
Round-shaft-design transducer, 116
Roundness, 156-160, 214, 216, 217, 351
Roundness equipment, 216
Roundness measurement systems, 160
Roundness of pitch cylinder, 215, 217

S

Safety, 43, 385, 387, 393
Sample lots, 55
Scale factor, 150
Scaling dimensions, 147-148
Scaling tolerances, 147
Scanning, 372
Scanning laser beam, 172
Scanning systems, 173
Schematic diagrams, 314
Scrap, 42, 353
Screw machines, 56
Screw thread specification, 249
Screw threads, 213-219, 226, 287
Screws, 12
Secondary datum, 244, 246, 247
Secondary springs, 180
Segregation of good parts, 378
Self-aligning pads, 140
Sensitivity, 90
Service objectives, 38
Servo motors, 144
Setup time, 350, 369
Shaft alignment, 185-193
Shaft diameter, 128
Shaft/piston intersection, 124
Shaft shear, 186
Shaft sticking, 114
Shaft strain curve, 186
Shaft transducer, 125
Shaft travel, 123
Shafting arrangement requiring a free section, 187
Shape tolerance zone, 296
Shear error(s), 189

T

Tool design, 107
Tool designers, 108
Tool life, 369
Toolmaker's microscopes, 421
Toolholders, 369
Tooling, 43, 110, 218
Tooling capabilities, 296
Tooling check, 152
Torque loadings, 109
Total thermal error, 20
Training, 51, 77, 86
Transducer compliance, 122
Transducer design, 114
Transducer response, 122, 123
Transducer testing, 121
Transducer time response, 129
Transducers, 114-129
Transfer machines, 368
Transient environment, 18
Truck bevel-gear pinions, 369
Tube wall, 7
Tungsten carbide, 93, 127
Tungsten carbide inserts, 93
Turning, 232, 367, 368, 369, 388
Turnkey systems, 416, 418
Turnkey tool design, 111
TV, 410
TV camera, 418
Twin jet air plug gauge, 91, 92
Two-gage configurations, 186

U

Ultrasonic inspection, 104
Ultrasonics, 100
Uncertainty of nominal expansion, 6, 19
Uncertainty of nominal differential expansion, 6, 14, 19
Uncertainty of nominal coefficient of expansion, 19
Undulations, 157
UNF-UNC, 85
Unit assembly, 118
Upper tolerance zone overlay gage, 269
User-friendly prompting, 376

V

V blocks, 157, 164, 197, 199, 205
V-block clamp, 390
Vacuum, 116, 125, 126, 323

Vacuum chucks, 323
Variations of thermal environments, 18
Venture capital, 101, 102
Vibration, 82, 86, 133, 325, 329
Video picture, 416
Video pixel elements, 413
Virtual condition, 249
Virtual diameters, 224
Virtual size check, 346
Vision, 410-418
Vision systems, 395, 402, 404
Volumetric length measuring accuracy, 374
Vulcanization extrusion lines, 101

W

Wall thickness, 370
Water, 332
Waviness, 167
Wear, 133. 156, 216, 228, 229, 367, 370, 371, 385
Welding, 56
Wheel hub, 367
Wheels, 156
Work problems, 274
Working standard conditions, 86

X

X-ray intensifier fluoroscopic devices, 410
X-ray microscope, 229, 231
X-ray microscope design, 230

Y

Y-14.5, 321
Y-Z Cutting Machine, 127
Y-Z measuring machine, 321
Yaw errors, 135
Yield curves, 132

Z

Zero drawing, 106
Zero position tolerance concept, 261
Zinc, 64